Hydrogen-Transfer Reactions

Edited by
James T. Hynes,
Judith P. Klinman,
Hans-Heinrich Limbach,
Richard L. Schowen

1807–2007 Knowledge for Generations

Each generation has its unique needs and aspirations. When Charles Wiley first opened his small printing shop in lower Manhattan in 1807, it was a generation of boundless potential searching for an identity. And we were there, helping to define a new American literary tradition. Over half a century later, in the midst of the Second Industrial Revolution, it was a generation focused on building the future. Once again, we were there, supplying the critical scientific, technical, and engineering knowledge that helped frame the world. Throughout the 20th Century, and into the new millennium, nations began to reach out beyond their own borders and a new international community was born. Wiley was there, expanding its operations around the world to enable a global exchange of ideas, opinions, and know-how.

For 200 years, Wiley has been an integral part of each generation's journey, enabling the flow of information and understanding necessary to meet their needs and fulfill their aspirations. Today, bold new technologies are changing the way we live and learn. Wiley will be there, providing you the must-have knowledge you need to imagine new worlds, new possibilities, and new opportunities.

Generations come and go, but you can always count on Wiley to provide you the knowledge you need, when and where you need it!

William J. Pesce
President and Chief Executive Officer

Peter Booth Wiley
Chairman of the Board

Hydrogen-Transfer Reactions

Volume 3

Edited by

James T. Hynes, Judith P. Klinman,
Hans-Heinrich Limbach, Richard L. Schowen

WILEY-VCH Verlag GmbH & Co. KGaA

The Editors

Prof. James T. Hynes
Department of Chemistry and Biochemistry
University of Colorado
Boulder, CO 80309-0215
USA

Prof. Judith P. Klinman
Departments of Chemistry and
Molecular and Cell Biology
University of California
Berkeley, CA 94720-1460
USA

Département de Chimie
Ecole Normale Supérieure
24 rue Lhomond
75231 Paris
France

Prof. Hans-Heinrich Limbach
Institut für Chemie und Biochemie
Freie Universität Berlin
Takustrasse 3
14195 Berlin
Germany

Prof. Richard L. Schowen
Departments of Chemistry, Molecular
Biosciences, and Pharmaceutical Chemistry
University of Kansas
Lawrence, KS 66047
USA

Cover
The cover picture is derived artistically from the potential-energy profile for the dynamic equilibrium of water molecules in the hydration layer of a protein (see A. Douhal's chapter in volume 1) and the three-dimensional vibrational wavefunctions for reactants, transition state, and products in a hydride-transfer reaction (see the chapter by S.J. Benkovic and S. Hammes-Schiffer in volume 4).

Library of Congress Card No.: applied for

British Library Cataloguing-in-Publication Data
A catalogue record for this book is available from the British Library.

Bibliographic information published by Die Deutsche Bibliothek
Die Deutsche Nationalbibliothek lists this publication in the Deutsche Nationalbibliografie; detailed bibliographic data are available in the Internet at <http://dnb.d-nb.de>.

Typesetting Kühn & Weyh, Freiburg; Asco Typesetters, Hongkong
Printing betz-druck GmbH, Darmstadt
Bookbinding Litges & Dopf GmbH, Heppenheim
Cover Design Adam-Design, Weinheim

Printed in the Federal Republic of Germany.
Printed on acid-free paper.

ISBN: 978-3-527-30777-7

Foreword
The Remarkable Phenomena of Hydrogen Transfer

Ahmed H. Zewail*
California Institute of Technology
Pasadena, CA 91125, USA

Life would not exist without the making and breaking of chemical bonds - chemical reactions. Among the most elementary and significant of all reactions is the transfer of a hydrogen atom or a hydrogen ion (proton). Besides being a fundamental process involving the smallest of all atoms, such reactions form the basis of general phenomena in physical, chemical, and biological changes. Thus, there is a wide-ranging scope of studies of hydrogen transfer reactions and their role in determining properties and behaviors across different areas of molecular sciences.

Remarkably, this transfer of a small particle appears deceptively simple, but is in fact complex in its nature. For the most part, the dynamics cannot be described by a classical picture and the process involves more than one nuclear motion. For example, the transfer may occur by tunneling through a reaction barrier and a quantum description is necessary; the hydrogen is not isolated as it is part of a chemical bond and in many cases the nature of the bond, "covalent" and/or "ionic" in Pauling's valence bond description, is difficult to characterize; and the description of atom movement, although involving the local hydrogen bond, must take into account the coupling to other coordinates. In the modern age of quantum chemistry, much has been done to characterize the rate of transfer in different systems and media, and the strength of the bond and degree of charge localization. The intermediate bonding strength, directionality, and specificity are unique features of this bond.

* The author is currently the Linus Pauling Chair Professor of chemistry and physics and the Director of the Physical Biology Center for Ultrafast Science & Technology and the National Science Foundation Laboratory for Molecular Sciences at Caltech in Pasadena, California, USA. He was awarded the 1999 Nobel Prize in Chemistry.
Email: zewail@caltech.edu
Fax: 626.792.8456

Hydrogen-Transfer Reactions. Edited by J. T. Hynes, J. P. Klinman, H. H. Limbach, and R. L. Schowen

ISBN: 978-3-527-30777-7

The supreme example for the unique role in specificity and rates comes from life's genetic information, where the hydrogen bond determines the complementarities of G with C and A with T and the rate of hydrogen transfer controls genetic mutations. Moreover, the not-too-weak, not-too-strong strength of the bond allows for special "mobility" and for the potent hydrophobic/hydrophilic interactions. Life's matrix, liquid water, is one such example. The making and breaking of the hydrogen bond occurs on the picosecond time scale and the process is essential to keeping functional the native structures of DNA and proteins, and their recognition of other molecules, such as drugs. At interfaces, water can form ordered structures and with its amphiphilic character, utilizing either hydrogen or oxygen for bonding, determines many properties at the nanometer scale.

Hydrogen transfer can also be part of biological catalysis. In enzyme reactions, a huge complex structure is involved in bringing this small particle of hydrogen into the right place at the right time so that the reaction can be catalytically enhanced, with rates orders of magnitude larger than those in solution. The molecular theatre for these reactions is that of a very complex energy landscape, but with guided bias for specificity and selectivity in function. Control of reactivity at the active site has now reached the frontier of research in "catalytic antibody", and one of the most significant achievements in chemical synthesis, using heterogeneous catalysis, has been the design of site-selective reaction control.

Both experiments and theory join in the studies of hydrogen transfer reactions. In general, the approach is of two categories. The first involves the study of prototypical but well-defined molecular systems, either under isolated (microscopic) conditions or in complexes or clusters (mesoscopic) with the solvent, in the gas phase or molecular beams. Such studies over the past three decades have provided unprecedented resolution of the elementary processes involved in isolated molecules and en route to the condensed phase. Examples include the discovery of a "magic solvent number" for acid-base reactions, the elucidation of motions involved in double proton transfer, and the dynamics of acid dissociation in finite-sized clusters. For these systems, theory is nearly quantitative, especially as more accurate electronic structure and molecular dynamics computations become available.

The other category of study focuses on the nature of the transfer in the condensed phase and in biological systems. Here, it is not perhaps beneficial to consider every atom of a many-body complex system. Instead, the objective is hopefully to project the key electronic and nuclear forces which are responsible for behavior. With this perspective, approximate, but predictive, theories have a much more valuable outreach in applications than those simulating or computing bonding and motion of all atoms. Computer simulations are important, but for such systems they should be a tool of guidance to formulate a predictive theory. Similarly for experiments, the most significant ones are those that dissect complexity and provide lucid pictures of the key and relevant processes.

Progress has been made in these areas of study, but challenges remain. For example, the problem of vibrational energy redistribution in large molecules, although critical to the description of rates, statistical or not, and to the separation of intra and intermolecular pathways, has not been solved analytically, even in an approximate but predictive formulation. Another problem of significance concerns the issue of the energy landscape of complex reactions, and the question is: what determines specificity and selectivity?

This series edited by prominent players in the field is a testimony to the advances and achievements made over the past several decades. The diversity of topics covered is impressive: from isolated molecular systems, to clusters and confined geometries, and to condensed media; from organics to inorganics; from zeolites to surfaces; and, for biological systems, from proteins (including enzymes) to assemblies exhibiting conduction and other phenomena. The fundamentals are addressed by the most advanced theories of transition state, tunneling, Kramers' friction, Marcus' electron transfer, Grote-Hynes reaction dynamics, and free energy landscapes. Equally covered are state-of-the-art techniques and tools introduced for studies in this field and including ultrafast methods of femtochemistry and femtobiology, Raman and infrared, isotope probes, magnetic resonance, and electronic structure and MD simulations.

These volumes are a valuable addition to a field that continues to impact diverse areas of molecular sciences. The field is rigorous and vigorous as it still challenges the minds of many with the fascination of how the physics of the smallest of all atoms plays in diverse applications, not only in chemistry, but also in life sciences. Our gratitude is to the Editors and Authors for this compilation of articles with new knowledge in a field still pregnant with challenges and opportunities.

Pasadena, California
August, 2006

Ahmed Zewail

Contents

Hydrogen-Transfer Reactions. Edited by J. T. Hynes, J. P. Klinman, H. H. Limbach, and R. L. Schowen

ISBN: 978-3-527-30777-7

Preface

As one of the simplest of chemical reactions, pervasive on this highly aqueous planet populated by highly aqueous organisms, yet still imperfectly understood, the transfer of hydrogen as a subject of scientific attention seems hardly to require defense. This claim is supported by the readiness with which the editors of this series of four volumes on *Hydrogen-transfer Reactions* accepted the suggestion that they organize a group of their most active and talented colleagues to survey the subject from viewpoints beginning in physics and extending into biology. Furthermore, forty-nine authors and groups of authors acceded, with alacrity and grace, to the request to contribute and have then supplied the articles that make up these volumes.

Our scheme of organization involved an initial division into physical and chemical aspects on the one hand, and biological aspects on the other hand (and one might well have said biochemical and biological aspects). In current science, such a division may provide an element of convenience but no-one would seriously claim the segregation to be either easy or entirely meaningful. We have accordingly felt quite entitled to place a number of articles rather arbitrarily in one or the other category. It is nevertheless our hope that readers may find the division adequate to help in the use of the volumes. It will be apparent that the division of space between the two categories is unequal, the physical and chemical aspects occupying considerably more pages than the biological aspects, but our judgment is that this distribution of space is proper to the subjects treated. For example, many of the treatments of fundamental principles and broadly applicable techniques were classified under physical and chemical aspects. But they have powerful implications for the understanding and use of the matters treated under biological aspects.

Within each of these two broad disciplinary categories, we have organized the subject by beginning with the simple and proceeding toward the complex. Thus the physical and chemical aspects appear as two volumes, volume1 on simple systems and volume 2 on complex systems. Similarly, the biological aspects appear as volume 3 on simple systems and volume 4 on complex systems.

Volume 1 then begins with isolated molecules, complexes, and clusters, then treats condensed-phase molecules, complexes, and crystals, and finally reaches

Hydrogen-Transfer Reactions. Edited by J. T. Hynes, J. P. Klinman, H. H. Limbach, and R. L. Schowen

ISBN: 978-3-527-30777-7

treatments of molecules in polar environments and in electronic excited states. Volume 2 reaches higher levels of complexity in protic systems with bimolecular reactions in solution, coupling of proton transfer to low-frequency motions and proton-coupled electron transfer, then organic and organometallic reactions, and hydrogen-transfer reactions in solids and on surfaces. Thereafter articles on quantum tunneling and appropriate theories of hydrogen transfer complete the treatment of physical and chemical aspects.

Volume 3 begins with simple model (i.e., non-enzymic) reactions for proton-transfer, both to and from carbon and among electronegative atoms, hydrogen-atom transfer, and hydride transfer, as well as the extension to small, synthetic peptides. It is completed by treatments of how enzymes activate C-H bonds, multiple hydrogen transfer reactions in enzymes, and theoretical models. Volume 4 moves then into enzymic reactions and a thorough consideration of quantum tunneling and protein dynamics, one of the most vigorous areas of study in biological hydrogen transfer, then considers several specific enzyme systems of high interest, and is completed by the treatment of proton conduction in biological systems.

While we do not claim any sort of comprehensive coverage of this large subject, we believe the reader will find a representative treatment, written by accomplished and respected experts, of most of the matters currently considered important for an understanding of hydrogen-transfer reactions. I am enormously grateful to James T. (Casey) Hynes and Hans-Heinrich Limbach, who saw to the high quality of the volumes on the physical and chemical aspects, and to Judith Klinman, who gave me a nearly free pass as her co-editor of the volumes on biological aspects. We are all grateful indeed to the authors who contributed their wisdom and eloquence to these volumes. It has been a very great pleasure to be assisted, encouraged, and supported at every turn by the outstanding staff of VCH-Wiley in Weinheim, particularly (in alphabetical order) Ms. Nele Denzau, Dr. Renate Dötzer, Dr. Tim Kersebohm, Dr. Elke Maase, Ms. Claudia Zschernitz, and – of course – Dr. Peter Gölitz.

Lawrence, Kansas, USA, September 2006 *Richard L. Schowen*

Preface to Volumes 3 and 4

These volumes together address the rather enormous subject of hydrogen transfer in biological systems, volume 3 presenting the role of relatively simple systems in the understanding of hydrogen transfer while volume 4 considers complex systems, for the most part enzymes.

Volume 3 contains two parts that treat basic concepts and systems not limited to a single enzyme or class of enzymes in their significance. Part I consists of five chapters on the chemistry of the transfer of hydrogen in biological model systems: as a proton to and from carbon (Amyes and Richard, Ch. 1); as a proton in acid-base catalysis; i.e., largely among electronegative atoms (Kirby, Ch. 2); as a hydrogen atom (Schöneich, Ch. 3); as a hydride ion (Schowen, Ch..4); as a proton in acid-base catalysis in designed peptides (Baltzer, Ch. 5). Part II is composed of three chapters on generally significant features of biological hydrogen-transfer reactions: in enzyme-catalyzed proton transfer from carbon (Gerlt, Ch. 6); in multiple proton transfers in enzymic systems (Spies and Toney, Ch. 7); and in computer simulations of enzymic hydrogen transfer (Braun-Sand, Olsson, Mavri, and Warshel, Ch. 8).

Volume 4, consisting of three parts, then proceeds to studies in enzyme and protein systems that for the most part serve well as paradigms for broader groups in which hydrogen transfer is important. Part III brings together seven chapters on the subject of quantum tunneling in enzymic hydrogen-transfer and its relationship to protein motions. A relative new theoretical approach is described by Schwartz (Ch. 9), leading into a general consideration of the existing evidence and its significance for the tunneling/dynamics nexus (Knapp, Meyer, and Klinman, Ch. 10), and articles by Huskey (Ch. 11) on the importance of multiple-isotope labeling for characterization of tunneling phenomena, by Kohen on kinetic isotope effects (Ch.12) and by Basran, Hothi, Masgrau, Sutcliffe, and Scrutton on the opportunities afforded by flavoprotein systems (Ch. 13). This part is closed by articles on two important experimental approaches, isotope exchange with solvent as a probe of protein motion (Lee, Croy, Resing, and Ahn, Ch. 14) and resonance Raman spectroscopy as a probe of active-site dynamical properties (Callender and Deng, Ch. 15). Part IV brings into focus several central examples of important enzyme classes: thiamin-dependent enzymes (Ch. 16 by Hübner, Golbik, and

Hydrogen-Transfer Reactions. Edited by J. T. Hynes, J. P. Klinman, H. H. Limbach, and R. L. Schowen

ISBN: 978-3-527-30777-7

Tittmann), dihydrofolate reductase (Ch. 17 by Benkovic and Hammes-Schiffer), hydrolases (Ch. 18 by Stein), and vitamin B_{12} enzymes (Ch. 19 by Banerjee, Truhlar, Dybala-Defratyka, and Paneth). The volume is the closed by a one-chapter Part V on proton conduction in biology, in which Gutman and Nachliel (Ch. 20) treat the subject of proton conductance at protein surfaces and interfacial regions.

JPK acknowledges the support of grant MCB 0446395 from the US National Science Foundation and of grant GM 025765 from the US National Institutes of Health.

Berkeley, California, USA, September 2006 — *Judith P. Klinman*
Lawrence, Kansas, USA, September 2006 — *Richard L. Schowen*

List of Contributors to Volumes 3 and 4

Natalie G. Ahn
Department of Chemistry and Biochemistry
Howard Hughes Medical Institute
University of Colorado
Boulder, CO 80309-0215
USA

Tina L. Amyes
Department of Chemistry
University at Buffalo
SUNY
Buffalo, NY 14260-3000
USA

Lars Baltzer
Department of Chemistry
Uppsala University
Box 599
75124 Uppsala
Sweden

Ruma Banerjee
Biochemistry Department
University of Nebraska
Lincoln, NE 68588-0664
USA

Jaswir Basran
Department of Biochemistry
University of Leicester
University Road
Leicester LE1 7RH
UK

Stephen J. Benkovic
Department of Chemistry
104 Chemistry Building,
Pennsylvania State University
University Park, PA 16802
USA

Sonja Braun-Sand
University of Southern California
Department of Chemistry
3620 McClintock Avenue, SGM 418
Los Angeles, CA 90089-1062
USA

Robert Callender
Department of Biochemistry
Albert Einstein College of Medicine
1300 Morris Park Avenue
Bronx, NY 10461
USA

Carrie H. Croy
Department of Chemistry and Biochemistry
Howard Hughes Medical Institute
University of Colorado
Boulder, CO 80309-0215
USA

Hua Deng
Department of Biochemistry
Albert Einstein College of Medicine
1300 Morris Park Avenue
Bronx, NY 10461
USA

Hydrogen-Transfer Reactions. Edited by J. T. Hynes, J. P. Klinman, H.-H. Limbach, and R. L. Schowen

ISBN: 978-3-527-30777-7

Agnieszka Dybala-Defratyka
Faculty of Chemistry
Technical University of Lodz
90-924 Lodz
Poland

John A. Gerlt
University of Illinois,
Urbana-Champaign
Departments of Biochemistry and
Chemistry
600 South Mathews Avenue
Urbana, IL 61801
USA

Ralph Golbik
Institute of Biochemistry
Martin Luther University
Halle-Wittenberg
Kurt-Mothes-Strasse 3
06120 Halle/Saale
Germany

Menachem Gutman
Laser Laboratory for Fast Reactions
in Biology
Department of Biochemistry
George S. Wise Faculty of Life Sciences
Tel Aviv University
Tel Aviv 69978
Israel

Sharon Hammes-Schiffer
Department of Chemistry
104 Chemistry Building
Pennsylvania State University
University Park, PA 16802
USA

Parvinder Hothi
Faculty of Life Sciences and
Manchester Interdisciplinary Biocentre
University of Manchester
131 Princess Street
Manchester M1 7ND
UK

Gerhard Hübner
Institute of Biochemistry
Martin Luther University
Halle-Wittenberg
Kurt-Mothes-Strasse 3
06120 Halle/Saale
Germany

W. Phillip Huskey
Department of Chemistry
Rutgers University – Newark
73 Warren Street
Newark, NJ 07102
USA

Anthony J. Kirby
University Chemical Laboratory
Cambridge CB2 1EW
UK

Judith P. Klinman
Departments of Chemistry and
Molecular and Cell Biology
University of California
Berkeley, CA 94720-1460
USA

Michael J. Knapp
Department of Chemistry
710 N. Pleasant Street
University of Massachusetts
Amherst, MA 01003-9336
USA

Amnon Kohen
Department of Chemistry
University of Iowa
Iowa City, IA 52242
USA

Thomas Lee
Department of Chemistry and
Biochemistry
Howard Hughes Medical Institute
University of Colorado
Boulder, CO 80309-0215
USA

Laura Masgrau
School of Chemical Engineering and Analytical Science Manchester Interdisciplinary Biocentre
University of Manchester
131 Princess Street
Manchester M1 7ND
UK

Janez Mavri
National Institute of Chemistry
P.O.B. 660
Hajarihova 19
SI-1001 Ljubljana
Slovenia

Matthew Meyer
Merced School of Natural Sciences
University of California
P.O. Box 2039
Merced, CA 95344
USA

Esther Nachliel
Laser Laboratory for Fast Reactions in Biology
Department of Biochemistry
George S. Wise Faculty of Life Sciences
Tel Aviv University
Tel Aviv 69978
Israel

Mats H. M. Olsson
University of Southern California
Department of Chemistry
3620 McClintock Avenue, SGM 418
Los Angeles, CA 90089-1062
USA

Piotr Paneth
Faculty of Chemistry
Technical University of Lodz
90-924 Lodz
Poland

Katheryn A. Resing
Department of Chemistry and Biochemistry
Howard Hughes Medical Institute
University of Colorado
Boulder, CO 80309-0215
USA

John P. Richard
Department of Chemistry
University at Buffalo
SUNY
Buffalo, NY 14260-3000
USA

Christian Schöneich
Department of Pharmaceutical Chemistry
University of Kansas
2095 Constant Avenue
Lawrence, KS 66047
USA

Richard L. Schowen
Departments of Chemistry, Molecular Biosciences, and Pharmaceutical Chemistry
University of Kansas
Lawrence, KS 66047
USA

Steven D. Schwartz
Departments of Biophysics and Biochemistry
Seaver Center for Bioinformatics
Albert Einstein College of Medicine
Bronx, New York
USA

Nigel S. Scrutton
Faculty of Life Sciences and Manchester Interdisciplinary Biocentre
University of Manchester
131 Princess Street
Manchester M1 7ND
UK

Michael Ashley Spies
Department of Biochemistry
University of Illinois
600 South Mathews Avenue
Urbana, IL 61801
USA

Ross L. Stein
Laboratory for Drug Discovery in Neurodegeneration
Harvard Center for Neurodegeneration and Repair
Department of Neurology
Harvard Medical School
65 Landsdowne Street, Fourth Floor
Cambridge, MA 02129
USA

Michael J. Sutcliffe
School of Chemical Engineering and Analytical Science Manchester
Interdisciplinary Biocentre
University of Manchester
131 Princess Street
Manchester M1 7ND
UK

Kai Tittmann
Martin Luther University
Halle-Wittenberg
Institute of Biochemistry
Kurt-Mothes-Strasse 3
06120 Halle/Saale
Germany

Michael D. Toney
Department of Chemistry
University of California, Davis
1-Shields Avenue
Davis, CA 96616
USA

Donald G. Truhlar
Chemistry Department
University of Minnesota
Minneapolis, MN 55455-0431
USA

Arieh Warshel
University of Southern California
Department of Chemistry
3620 McClintock Avenue, SGM 418
Los Angeles, CA 90089-1062
USA

II Biological Aspects

Part I
Models for Biological Hydrogen Transfer

This section contains, in five chapters, treatments in model systems of the distinguishable classes of biological hydrogen-transfer reactions: proton transfer to and from carbon (Ch. 1 by Amyes and Richard), proton transfer among electronegative atoms as is typical in acid-base catalysis (Ch. 2 by Kirby), hydrogen-atom transfer (Ch. 3 by Schöneich), and hydride transfer (Ch. 4 by Schowen). Baltzer (Ch.5) then extends the important subject of acid-base catalysis from simple models toward the complexity of proteins by describing studies in designed peptides.

Amyes and Richard's treatment in Ch. 1 suggests that the correlation of C–H fission rates either in solution or in enzyme active sites with the thermodynamic acidity of the proton donor is not a simple matter. The Marcus-theory separation into intrinsic and thermodynamic barriers is rendered more complicated by the perhaps surprising observation that the intrinsic barrier rises as the acidity *increases*. This is consistent with the view that intrinsic barriers are small when the liberated electron pair is localized (as with electronegative atoms) and larger as the work of reorganization and delocalization becomes greater. Kirby's presentation and analysis in Ch. 2 of the phenomenology of acid-base catalysis as a potential contributor to enzyme catalysis notes first the entropic cost of producing a potentially high-efficiency catalytic array from aqueous solution and the consequent utility of intramolecular reactions in exploring the preorganization strategy of proteins. The information to date is then found to leave a considerable gap between the efficiency of models and the efficiency of enzymes. Other points of note are the still incompletely understood superiority of intramolecular nucleophilic over intramolecular acid-base catalysis, and the fact that strong hydrogen bonds, if they play a catalytic role, must necessarily do so in the transition state and not in stable states. Schöneich's Ch. 3 on hydrogen-atom transfers takes the reader through the comparative phenomenology of transfer of hydrogen atoms to the likely acceptors in the biological context (radicals centered on O, N, S, or C). A cautionary note is sounded on the dangers of uncritical extrapolation of model studies to the biological context. In Ch. 4, Schowen reviews on hydride-transfer models, principally

Hydrogen-Transfer Reactions. Edited by J. T. Hynes, J. P. Klinman, H.-H. Limbach, and R. L. Schowen

ISBN: 978-3-527-30777-7

those related to nicotinamide and flavin cofactors, noting that these have indicated a major role for hydrogen-tunneling in the non-enzymic reactions. This thus indicates that the corresponding enzymes have not evolutionarily "invented" tunneling as a mechanism but rather have accelerated an existing tunneling pathway or diverted the system to a different, more rapid tunneling pathway. From these studies of relatively simple molecular species, Baltzer in Ch. 5 takes the subject of acid-base catalysis into the realm of designed peptides. The field is reviewed generally but proper emphasis is given peptides with the helix-loop-helix motif that dimerize into a four-helix bundle. These species are extraordinary in their susceptibility to imaginative introduction of catalytic functional groups.

1
Proton Transfer to and from Carbon in Model Reactions

Tina L. Amyes and John P. Richard

1.1
Introduction

Much of what is known about the mechanism for proton transfer to and from carbon in aqueous solution has come through experimental studies of model reactions. This work is, for several reasons, invaluable to biochemists interested in understanding the mechanism for proton transfer reactions at carbon in biological systems, virtually all of which are enzyme-catalyzed. First, model studies may be used to define the activation barrier for nonenzymatic proton transfer which must be lowered by the enzyme to obtain a catalytic rate acceleration. Second, the results of these studies help elucidate strategies which enzyme catalysts might follow to lower this barrier. Third, these results help to define the roles for various amino acid side-chains at an enzyme active site in the catalysis of proton transfer at carbon. This chapter will highlight recent model studies of proton transfer to and from carbon that we consider to be helpful in either defining the problems faced by enzyme catalysts of these reactions, or suggesting solutions to these problems.

1.2
Rate and Equilibrium Constants for Carbon Deprotonation in Water

The most fundamental experimental determinations in model studies of proton transfer at weakly basic carbon are of the rate and equilibrium constants for carbon deprotonation to form an unstable carbanion (Eq. (1.1)). These parameters define the kinetic and thermodynamic barriers to proton transfer (Eq. (1.2) for Fig. 1.1). They are of interest to enzymologists because they specify the difficulty of the problem that must be solved in the evolution of proteins which catalyze proton transfer with second-order rate constants k_{cat}/K_m of 10^6–10^8 M^{-1} s^{-1} that are typically observed for enzymatic reactions [1, 2]. The barrier to thermodynamically unfavorable deprotonation of carbon acids ($\Delta G_f^{\dagger}$, Fig. 1.1) in water is equal to the sum of the thermodynamic barriers to proton transfer (ΔG°) and the barrier to downhill protonation of the carbanion in the reverse direction ($\Delta G_r^{\dagger}$, Eq. (1.2)). The thermo-

Hydrogen-Transfer Reactions. Edited by J. T. Hynes, J. P. Klinman, H.-H. Limbach, and R. L. Schowen

ISBN: 978-3-527-30777-7

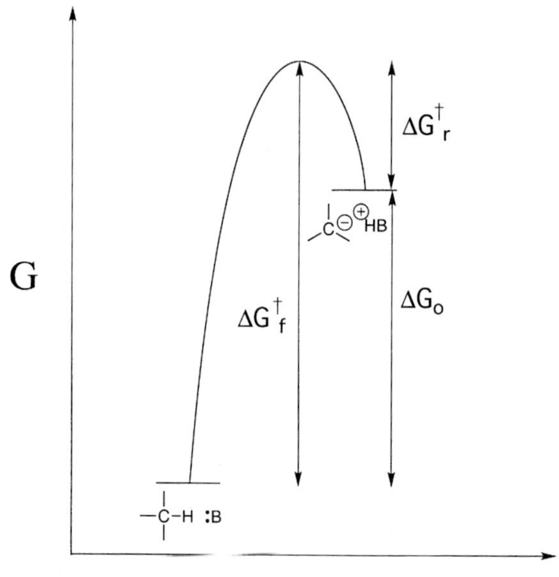

Figure 1.1. Free energy profile for deprotonation of a weak carbon acid (Eq. (1.1)) which shows that the barrier to thermodynamically unfavorable proton transfer ($\Delta G_f^{\dagger}$) is equal to the sum of the thermodynamic barriers to proton transfer (ΔG°) and the barrier to downhill protonation of the carbanion in the reverse direction ($\Delta G_r^{\dagger}$).

dynamic barrier can be calculated directly from the equilibrium constant for ionization of the carbon acid in water. This barrier is the dominant term in Eq. (1.2) ($\Delta G^{\circ} \gg \Delta G_r^{\dagger}$) for strongly unfavorable ionization of weak carbon acids to form highly unstable carbanions.

$$\text{—C—H} \;\; \text{B} \; \underset{}{\overset{K_{eq} = k_f/k_r}{\rightleftharpoons}} \; \text{C}^{\ominus} \;\; \text{HB}^{\oplus} \tag{1.1}$$

$$\Delta G_f^{\dagger} = \Delta G_{eq}{}^{\circ} + G_r^{\dagger} \tag{1.2}$$

Acidity constants for ionization of weak carbon acids in water cannot be determined by direct measurement when the strongly basic carbanion is too unstable to exist in detectable concentrations in this acidic solvent. Substituting dimethylsulfoxide (DMSO) for water causes a large decrease in the solvent acidity because, in contrast with water, the aprotic cosolvent DMSO does not provide hydrogen-bonding stabilization of hydroxide ion, the conjugate base of water. This allows the determination of the pK_as of a wide range of weak carbon acids in mixed DMSO/water solvents by direct measurement of the relative concentrations of the carbon acid and the carbanion at chemical equilibrium [3, 4]. The pK_as determined for weak carbon acids in this mixed solvent can be used to estimate pK_as in water, subject to the uncertainty of the effect of the DMSO cosolvent on the carbon acid pK_a [5].

The equilibrium constant for deprotonation of carbon acids is equal to the ratio of the rate constants for formation and reaction of the product carbanion (Scheme 1.1A–C). In recent years, kinetic methods have been used to provide solid values of the pK_as for ionization of a wide range of weak carbon acids. These experiments are, in principle, straightforward and require *only* the determination or estimate of two rate constants – one for the slow and thermodynamically unfavorable generation of the carbanion, and a second for fast downhill carbanion protonation. The observed first-order rate constant for carbanion formation may be controlled through the choice of the basic proton acceptor. Relatively strong carbon acids undergo detectable deprotonation by the weak base water in a pseudo-first-order reaction (Scheme 1.1A), but stronger general bases (Scheme 1.1B) or hydroxide ion (Scheme 1.1C) are required to give detectable deprotonation of weaker carbon acids in bimolecular reactions.

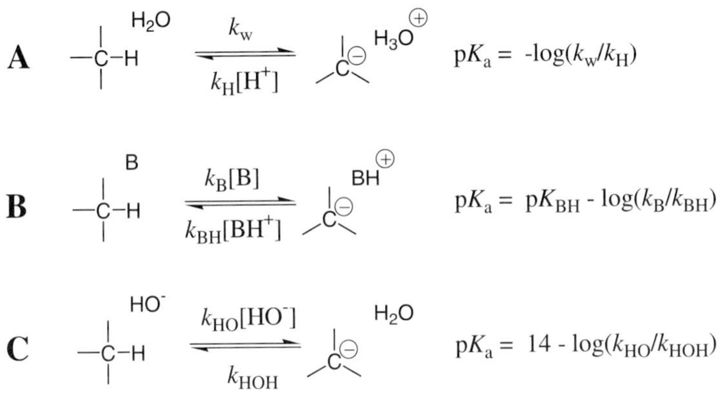

A $\quad$ C–H + H_2O $\underset{k_H[H^+]}{\overset{k_w}{\rightleftharpoons}}$ C^- + H_3O^+ $\qquad pK_a = -\log(k_w/k_H)$

B $\quad$ C–H + B $\underset{k_{BH}[BH^+]}{\overset{k_B[B]}{\rightleftharpoons}}$ C^- + BH^+ $\qquad pK_a = pK_{BH} - \log(k_B/k_{BH})$

C $\quad$ C–H + HO^- $\underset{k_{HOH}}{\overset{k_{HO}[HO^-]}{\rightleftharpoons}}$ C^- + H_2O $\qquad pK_a = 14 - \log(k_{HO}/k_{HOH})$

Scheme 1.1

1.2.1
Rate Constants for Carbanion Formation

Rate constants for deprotonation of carbon acids are determined under conditions where the carbanion is generated effectively irreversibly and then undergoes a fast reaction to form a detectable product. The most *general* fast reaction of a carbanion is "trapping" by a deuterium or tritium derived from solvent to give isotopically labeled product (Eq. (1.3)) [6, 7]. Tritium has the advantage of a high sensitivity for detection in hydron exchange. However, experiments to monitor tritium exchange reactions require quantitative separation of the tritium labeled solvent from the labeled carbon acid. This is difficult for volatile simple carbon acids such as ethyl acetate and acetonitrile when the solvent is water. In recent years high resolution proton NMR has been shown to be a simple and effective method for directly monitoring the incorporation of deuterium into weak carbon acids [8–20]. This analytical method has been used to determine rate constants for deprotonation of carbon acids with pK_as as high as 33.5 for proton transfer reactions at room temperature [17]. The pK_as of weaker carbon acids, which do not undergo detectable deprotonation at room temperature, may be determined by monitoring hydron transfer at a higher reaction temperature and making the appropriate temperature correction [21].

Studies of multistep chemical reactions that proceed through carbanion intermediates such as those shown by Eq. (1.4)–(1.6) have provided a rich and informative body of rate data for deprotonation of biologically important carbon acids in water. Carbon deprotonation to form a carbanion intermediate is effectively irreversible and rate determining for each of the reactions shown in Eq. (1.4)–(1.6). Eq. (1.4) is an example of an alkene-forming elimination reaction [22–24], where formation of the carbanion is effectively irreversible and is followed by rapid expulsion of a phosphate dianion or trianion leaving group [24]. In Eq. (1.5) and (1.6) the carbanion is trapped by an electrophilic carbonyl group in either an intramolecular aldol (Eq. (1.5)) [10, 25] or a bimolecular Claisen-type (Eq. (1.6)) [26] condensation reaction. Not shown is the classic diffusion-controlled trapping of a carbanion by a halogen, which has been used in the determination of the pK_a of acetone [27, 28].

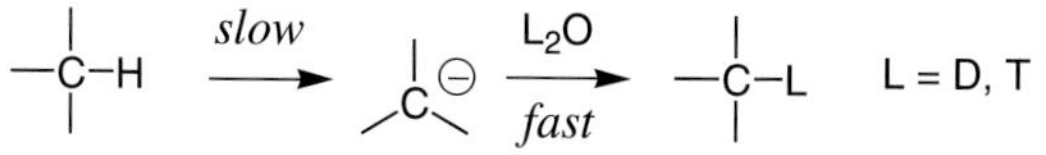

$$\text{C–H} \xrightarrow{slow} \text{C}^{\ominus} \xrightarrow[fast]{L_2O} \text{C–L} \quad L = D, T \tag{1.3}$$

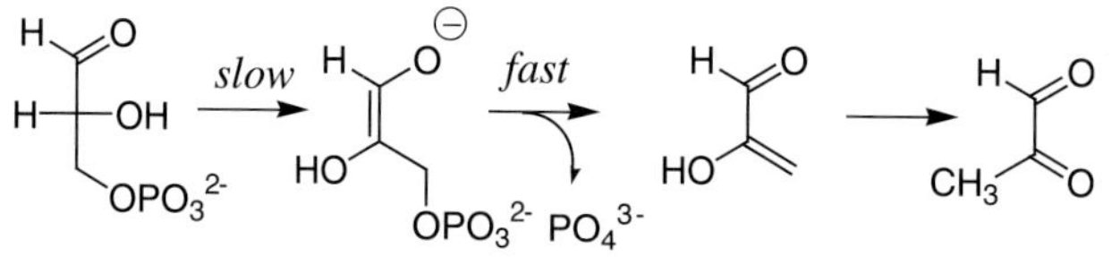

(1.4)
slow
fast
OPO_3^{2-}
PO_4^{3-}

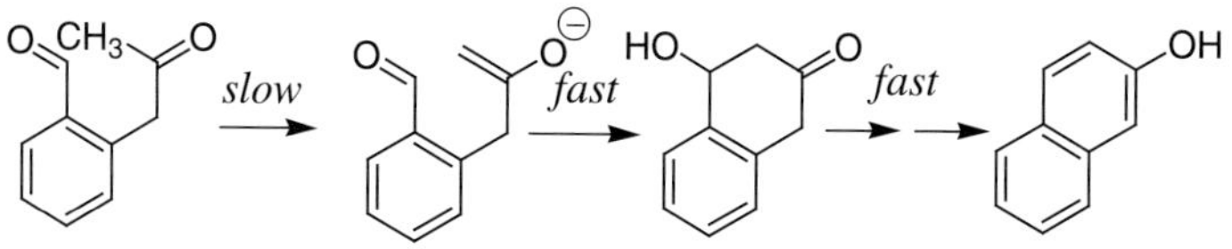

(1.5)
slow
fast
fast

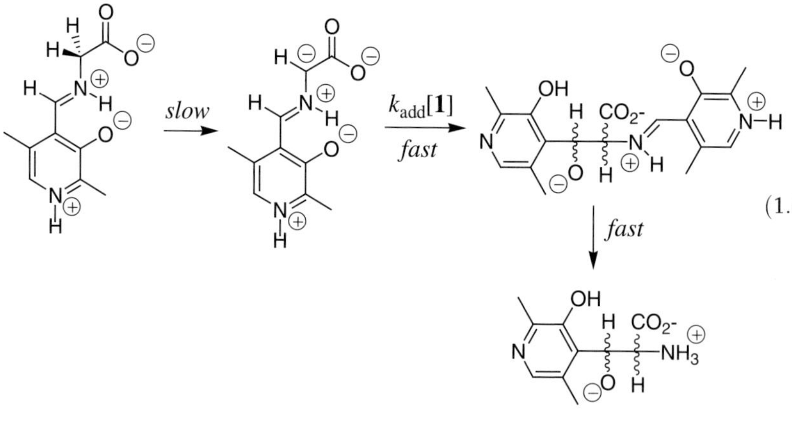

(1.6)
slow
$k_{add}[\mathbf{1}]$
fast
fast

1.2.2
Rate Constants for Carbanion Protonation

Rate constants for thermodynamically favorable protonation of unstable carbanions are typically very large. These may be determined by direct methods. A description of these direct methods, the most important of which use laser flash photolysis in carbanion generation [29, 30], is outside the scope of this chapter. The indirect methods used to estimate rate constants for carbanion protonation will be described in greater detail, because they provide insight into the nature of the rate determining step for carbanion protonation in water.

Carbanion protonation in water is a two-step reaction: (i) movement of a Brønsted acid into a reactive position, and (ii) proton transfer to carbon. The overall rate constant for carbanion protonation may be limited by either the rate constant for formation of the reactive complex, in which case the overall rate constant for proton transfer can be estimated by using a representative rate constant for the rate-determining transport step, or by the rate constant for proton transfer to carbon.

The limiting rate constants for thermodynamically favorable protonation of carbanions, and the observations from experiments that provide evidence for these limiting reactions are different, depending upon the type of acid that protonates the carbanion.

1.2.2.1 Protonation by Hydronium Ion

The microscopic reverse of deprotonation of a carbon acid by water is protonation of the product carbanion by hydronium ion (Scheme 1.1A), with a limiting rate constant of $(k_d)_H \approx 10^{10}$ M^{-1} s^{-1} for diffusional encounter of the carbanion and hydronium ion (Scheme 1.2A). A value of $k_H = 7 \times 10^9$ M^{-1} s^{-1} has been determined by direct measurement for protonation of the enolate of acetone by hydronium ion, which is downhill by ca. 30 kcal mol^{-1} (Table 13 of Ref. [30]). This

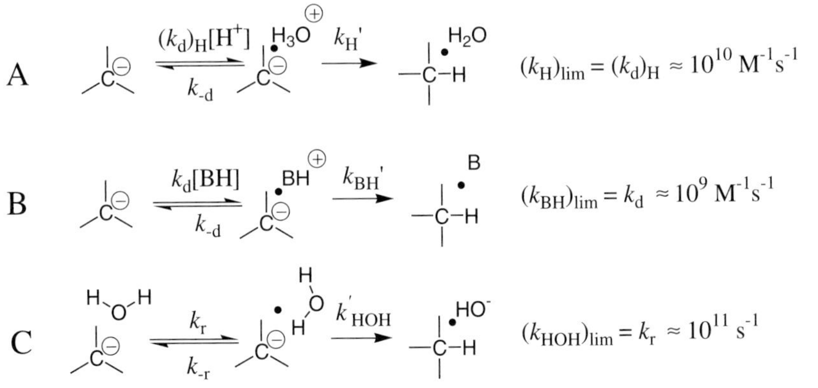

A $\quad (k_d)_H[H^+]$, k_{-d}, H_3O^+, k_H', H_2O $\qquad (k_H)_{lim} = (k_d)_H \approx 10^{10}\ M^{-1}s^{-1}$

B $\quad k_d[BH]$, k_{-d}, BH^+, k_{BH}', B $\qquad (k_{BH})_{lim} = k_d \approx 10^{9}\ M^{-1}s^{-1}$

C $\quad k_r$, k_{-r}, k'_{HOH}, HO^- $\qquad (k_{HOH})_{lim} = k_r \approx 10^{11}\ s^{-1}$

Scheme 1.2

provides good justification for the use of a similar limiting rate constant for protonation of enolates of like thermodynamic stability. For example, a pK_a of 18.0 (Scheme 1.1A) for the α-hydrogen of *N,N,N* trimethylammonium glycine methyl ester has been determined from the ratio of the experimentally determined rate constant $k_w = 5 \times 10^{-9}$ s^{-1} for water-catalyzed deprotonation of the carbon acid and an estimated limiting rate constant of $k_H = 5 \times 10^9$ M^{-1} s^{-1} for protonation of the enolate by hydronium ion [14].

1.2.2.2 Protonation by Buffer Acids

A Brønsted coefficient of $\beta = 1.1$ has been determined for deprotonation of ethyl acetate by 3-substituted quinuclidines to form the free enolate (Scheme 1.3) [11]. The microscopic reverse of deprotonation of a carbon acid by a buffer general base is protonation of the product carbanion by the conjugate acid of the Brønsted base (Scheme 1.1B). The limiting rate constant for exothermic proton transfer is $k_{enc} \approx 10^9$ M^{-1} s^{-1} when encounter of the Brønsted acid and base is rate determining (Scheme 1.3) [31]. Now, the observed Brønsted coefficient of $\beta > 1.1$ for deprotonation of ethyl acetate shows that the base catalyst bears a net positive charge at the transition state for carbon deprotonation which is greater than the unit positive charge at the conjugate acid [11, 32]. This is consistent with reversible deprotonation of the carbon followed by rate determining separation of the cation–anion pair intermediate (k'_{-d}, Scheme 1.3). Protonation of the enolate of ethyl acetate by the Brønsted acid is rate determining for reaction in the microscopic reverse direction ($k_{enc} \approx 10^9$ M^{-1} s^{-1}, Scheme 1.3), with $\alpha = -0.1$ for formation of the encounter complex between the enolate ion and buffer acid ($\alpha + \beta = 1.0$).

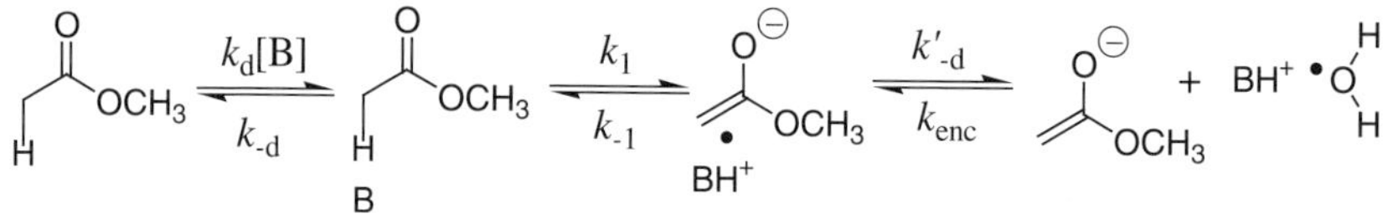

Scheme 1.3

The Brønsted parameters of $\beta > 1.0$ and $\alpha < 0$ proton transfer at ethyl acetate (Scheme 1.3) show that the barrier to formation of encounter complexes between the enolate of ethyl acetate (k_{enc}, Scheme 1.3) and the quinuclidinone cation catalyst increases with the acidity of the tertiary ammonium ion. This has been proposed to reflect the increasing strength of the hydrogen bond to water that is cleaved upon formation of the encounter complex [11, 32]. The small uncertainty in the barrier to desolvation of the Brønsted acid introduces a corresponding uncertainty into the value of the limiting rate constant for the encounter-limited reaction. The limits of $k_{BH} = 2\text{–}5 \times 10^9$ M^{-1} s^{-1} for the encounter-limited reaction of the simple oxygen ester enolate with protonated quinuclidine (p$K_{BH} = 11.5$) were combined with $k_B = 2.4 \times 10^{-5}$ M^{-1} s^{-1} for deprotonation of ethyl acetate by quinuclidine (Scheme 1.1B), to give $pK_a = 25.6 \pm 0.5$ for ionization of ethyl acetate as a carbon acid in aqueous solution [11].

1.2.2.3 **Protonation by Water**

The microscopic reverse of deprotonation of a carbon acid by hydroxide ion is protonation of the product carbanion by water (Scheme 1.1C). The limiting rate constant for strongly exothermic carbanion protonation is $k_r \approx 10^{11}$ s^{-1} (Scheme 1.4) for a reaction in which rotation of water into a reactive position is the rate determining step [33–35]. The failure to observe a normal primary kinetic isotope effect on lyoxide-catalyzed hydron exchange between solvent and a carbon acid provides evidence that the rate determining step for exchange is solvent reorganization. For example, most of the 3-fold difference in the rate constants for hydroxide ion catalyzed exchange of H for D at CD_3CN (Scheme 1.4A) compared with deuteroxide-catalyzed exchange of D for H at CH_3CN (Scheme 1.4B) is due to the 2.4-fold greater basicity of HO^- compared with that of DO^-. There is only a small primary kinetic isotope effect on the hydron exchange reaction [13]. This provides strong evidence that hydron transfer to lyoxide ion is reversible and that reorganization of solvent ($k_r \approx 10^{11}$) is largely rate determining for the lyoxide ion-catalyzed exchange reaction, so that $k_{-p} > k_r \approx 10^{11}$ for protonation of the α-cyanomethyl carbanion (Scheme 1.4). A pK_a of 29 for deprotonation of acetonitrile (CH_3CN) was calculated from the ratio of $k_{HO} = 1.1 \times 10^{-4}$ M^{-1} s^{-1} and $k_r = k_{HOH} \approx 10^{11}$ (Scheme 1.1C) [13].

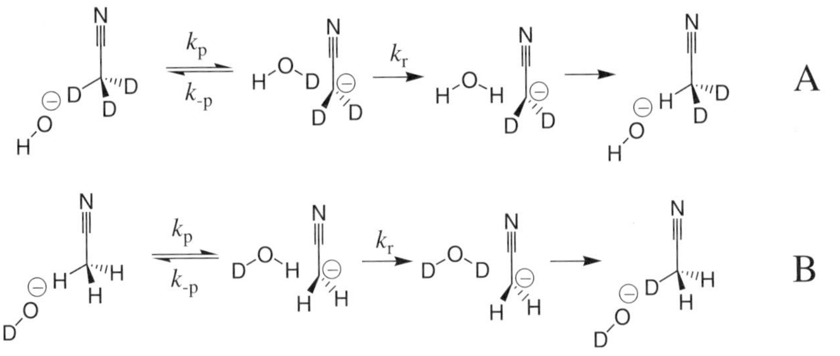

Scheme 1.4

1.2.3
The Burden Borne by Enzyme Catalysts

The pK_as for ionization of several biologically important carbon acids are summarized in Scheme 1.5. The pK_as of 17 for pyruvate **2** [36] and 18 for dihydroxyacetone phosphate **3** [24] are close to the pK_a of 19 for the parent ketone acetone **4** [37]. The α-protons of carboxylate anions are much less acidic than those of the

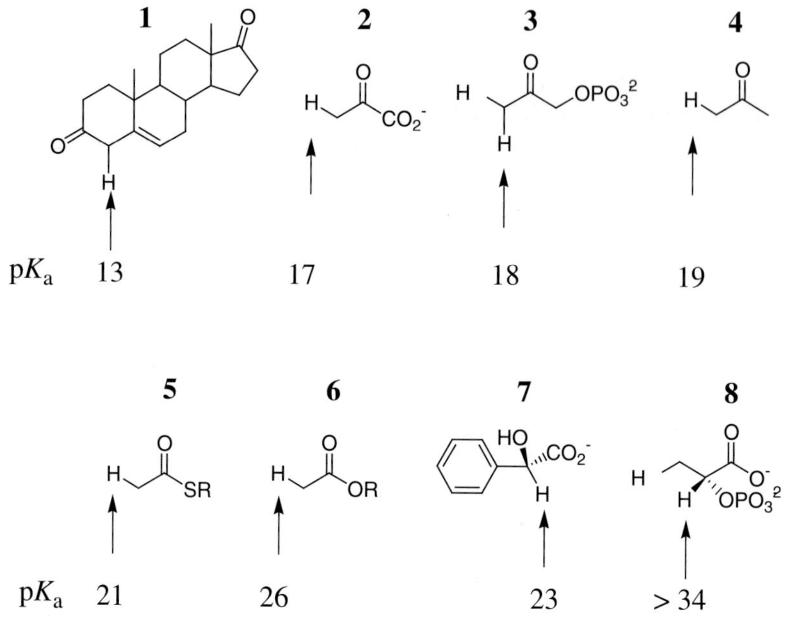

	1	2	3	4
pK_a	13	17	18	19

	5	6	7	8
pK_a	21	26	23	> 34

Scheme 1.5

corresponding aldehyde. For example, a pK_a of 23 has been determined for the benzylic α-proton of mandelic acid **7** [38], whose carbon deprotonation is catalyzed by mandelate racemase; and the pK_a of the α-proton of 2-phosphoglycerate **8** must be at least as large as the pK_a of 33.5 estimated for acetate anion [17]. The >21 unit difference in $pK_a \approx 13$ for the β,γ-unsaturated steroid **1** [39] and $pK_a \approx 34$ for the α-proton of 2-phosphoglycerate **8** corresponds to a greater than 29 kcal mol^{-1} difference in the thermodynamic barriers to deprotonation of these substrates that must be surmounted by the enzymes ketosteroid isomerase and enolase, respectively.

By contrast, k_{cat}/K_m for enzymatic catalysis of deprotonation at carbon is *not* strongly dependent on intrinsic carbon acidity. For example, k_{cat}/K_m is close to the diffusion-controlled limit for both the ketosteroid-isomerase-catalyzed deprotonation of the ketone **1** ($pK_a \approx 13$) [39] and the triosephosphate-isomerase-catalyzed deprotonation of the ketone **3** ($pK_a \approx 18$) [40]. An extreme example is the small difference in the values of $k_{cat}/K_m = 3 \times 10^8$ and 1.4×10^6 M^{-1} s^{-1} for enzyme-catalyzed isomerization of **1** [39] and the elimination reaction of **8** [41], respectively, both of which proceed by C–H bond cleavage. This corresponds to a ca. 3 kcal mol^{-1} difference in the activation barriers for the enzyme-catalyzed reactions, but the corresponding difference in the activation barriers for nonenzymatic proton transfer in water will be similar to the >29 kcal mol^{-1} difference in the thermodynamic barriers to these proton transfer reactions.

Efficient catalysis of deprotonation of strongly acidic carbon that undergoes relatively rapid deprotonation in water should be *easier* to achieve than catalysis of deprotonation of weakly acidic carbon acids. However it does not appear any *easier* to understand the mechanism for enzymatic catalysis of deprotonation of strong compared with weak carbon acids, perhaps because such explanations are not fully formulated. A simple test for quantitative explanations for enzyme catalysis of proton transfer is whether they provide a simple rationalization for the differences in the catalytic power of enzymes that catalyze deprotonation of carbon acids of widely different pK_a with similar second-order rate constants k_{cat}/K_m.

1.3 Substituent Effects on Equilibrium Constants for Deprotonation of Carbon

The pK_as for simple alkanes have been estimated to be ca. 50 in water [42], and their deprotonation in this solvent has not been observed experimentally. The majority of enzyme-catalyzed proton transfer reactions are at α-carbonyl carbon and give as product enolates, which are strongly stabilized by delocalization of negative charge from carbon to the more electron-withdrawing oxygen (Eq. (1.7)). The α-carbonyl substrates for enzyme-catalyzed proton transfer reactions span a wide range of acidity (Scheme 1.5). However, even the α-carbon of acetate anion ($pK_a = 33.5$) undergoes slow, but detectable, deuteroxide-ion catalyzed proton transfer with a half time of ca. 60 years for reaction at 25 °C in the presence of 1.0 M KOD [17].

B H H H (C)–C(=O)X ⇌ BH⁺ H H ⊖C–C(=O)X ⟷ H H C=C(X)O⊖ (1.7)

Organic chemists and biochemists are comfortable referring to the product of deprotonation of α-carbonyl carbon as a carbanion, because most important organic reactions of this delocalized anion with electrophiles occur at carbon. However, the preponderance of negative charge at these *alkenyl oxide anions* lies on the more electronegative oxygen [43]. There is good evidence that the large activation barriers observed for thermodynamically favorable protonation of enolate anions and other resonance stabilized carbanions ($\Delta G_r^{\dagger}$, Fig. 1.1) are caused in some way by the requirement that movement of an electron pair from the enolate oxygen to carbon be coupled to C–H bond formation at this carbon (Section 1.4.3.2).

The pK_as of simple carbon acids are also influenced by polar substituents. These substituent effects are significant, but are generally smaller than for the resonance effect of the carbonyl group which is mostly responsible for the 33 unit difference in the pK_as of ethane ($pK_a \approx 50$) [42] and acetaldehyde ($pK_a = 16.7$) [44]. For example, the pK_a for the α-carbonyl hydrogen of the amino acid glycine **9** decreases by 13 units upon protonation of the α-amino group **10** and methylation of the α-carboxylate group **11** [14]. A notable exception is the large stabilizing polar interaction between localized positive and negative charge at adjacent carbon. For exam-

	9	10	11
pK_a	≈34	29	21

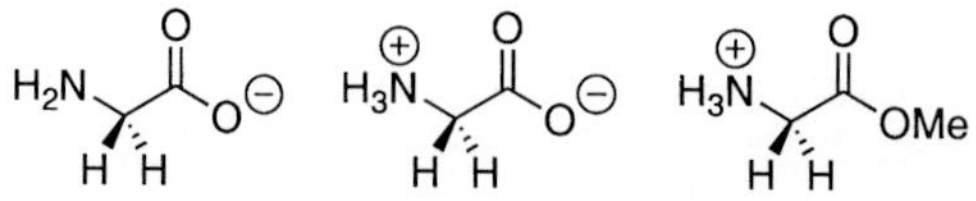

ple the acidic hydrogen of the thiazolium group of thiamine (Eq. (1.8)), has a pK_a of 18 [45] which is similar to that for the α-carbonyl hydrogen of a simple ketone.

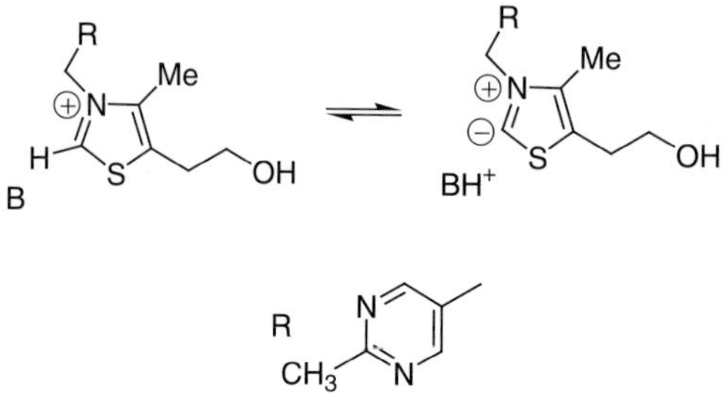

(1.8)

1.4
Substituent Effects on Rate Constants for Proton Transfer at Carbon

1.4.1
The Marcus Equation

The barrier to thermodynamically unfavorable deprotonation of carbon acids ($\Delta G_f{}^{\dagger}$, Fig. 1.1) in water is equal to the sum of the thermodynamic barrier to proton transfer (ΔG°) and the barrier to downhill protonation of the carbanion in the reverse direction ($\Delta G_r{}^{\dagger}$, Eq. (1.2)). The observation of significant activation barriers $\Delta G_r{}^{\dagger}$ for strongly thermodynamically favorable protonation or resonance stabilized carbanions shows that there is some intrinsic *difficulty* to proton transfer. The Marcus equation defines this *difficulty* with greater rigor as the intrinsic barrier Λ, which is the activation barrier for a related but often hypothetical thermoneutral proton transfer reaction (Fig. 1.2B) [46].

$$\Delta G^{\mp} = \Lambda(1 + \Delta G^{\circ}/4\Lambda)^2 \tag{1.9}$$

The Marcus equation was first formulated to model the dependence of rate constants for electron transfer on the reaction driving force [47–49]. Marcus assumed in his treatment that the energy of the transition state for electron transfer can be calculated from the position of the intersection of parabolas that describe the reactant and product states (Fig. 1.2A). This equation may be generalized to proton transfer (Fig. 1.2A) [46, 50, 51], carbocation-nucleophile addition [52], bimolecular nucleophilic substitution [53, 54] and other reactions [55–57] by assuming that their reaction coordinate profiles may also be constructed from the intersection of

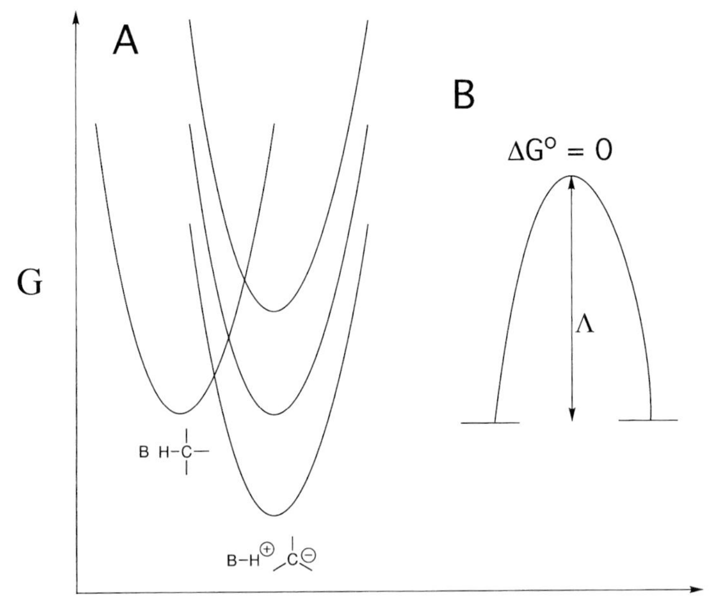

Figure 1.2. A, Reaction coordinate profiles for proton transfer at carbon constructed from the intersection of parabolas for the reactant and product states. B, The reaction coordinate profile for a reaction where $\Delta G^{\circ} = 0$ and $\Delta G^{\dagger}$ is equal to the Marcus intrinsic barrier Λ.

parabolas that describe the reactant and product states. This assumption is likely to be only approximately correct. Also, the profound differences between reaction coordinate profiles for the transfer of a *light* electron between two metal cations and for the transfer of the nearly 2000-times heavier proton between two heavy atoms are ignored by this simple model. More rigorous treatments of these differences would serve to emphasize the superficial nature of the similarities between electron and proton transfer reactions that allow for their common treatment by the Marcus equation.

Many laboratories, including our own, have used the Marcus equation empirically as a relatively simple and convenient framework for describing the differences in the intrinsic *difficulty* for related reactions, after correction for differences in the reaction thermodynamic driving force. This has led to the determination of the Marcus intrinsic barriers for a variety of proton transfer reactions by experiment and through calculations [58–65]. This compilation of intrinsic reaction barriers represents an attempt to compress an essential feature of these kinetic barriers to a single experimental parameter. An examination of the substituent effects on these intrinsic barriers has provided useful insight into the transition state for organic reactions [66].

1.4.2
Marcus Intrinsic Barriers for Proton Transfer at Carbon

There is only a small barrier for thermoneutral proton transfer between electronegative oxygen or nitrogen acids and bases [31]. These reactions proceed by encounter-controlled formation of a hydrogen-bonded complex between the acid and base (k_d, Scheme 1.6), proton transfer across this complex (k_p, Scheme 1.6), followed by diffusional separation to products (k_{-d}, Scheme 1.6) [31]. Much larger Marcus intrinsic barriers are observed for proton transfer to and from carbon [67]. There are at least two causes for this difference in intrinsic barriers for proton transfer between electronegative atoms and proton transfer at carbon.

$$\text{A–H} + \text{:B} \underset{k_{-d}}{\overset{k_d}{\rightleftharpoons}} \text{A–H :B} \underset{k_{-p}}{\overset{k_p}{\rightleftharpoons}} \text{A:}^{\ominus}\ \text{H–B}^{\oplus} \underset{k_d}{\overset{k_{-d}}{\rightleftharpoons}} \text{A:}^{\ominus} + \text{H–B}^{\oplus} \qquad \text{A}$$

$$\text{N≡C–H} + \text{:B} \underset{k\text{-}d}{\overset{kd}{\rightleftharpoons}} \text{N≡C–H :B} \underset{k\text{-}p}{\overset{kp}{\rightleftharpoons}} \text{N≡C}^{\ominus}\ \text{H–B}^{\oplus} \underset{kd}{\overset{k\text{-}d}{\rightleftharpoons}} \text{N≡C}^{\ominus} + \text{H–B}^{\oplus} \qquad \text{B}$$

$B = RCOO^{\ominus}, RO^{\ominus}$ log k_p = 9.1 for thermoneutral proton transfer.

$B = RS^{\ominus}$ log k_p < 6.8 for thermoneutral proton transfer.

$B = CN^{\ominus}$ log kp < 6.1 for thermoneutral proton transfer.

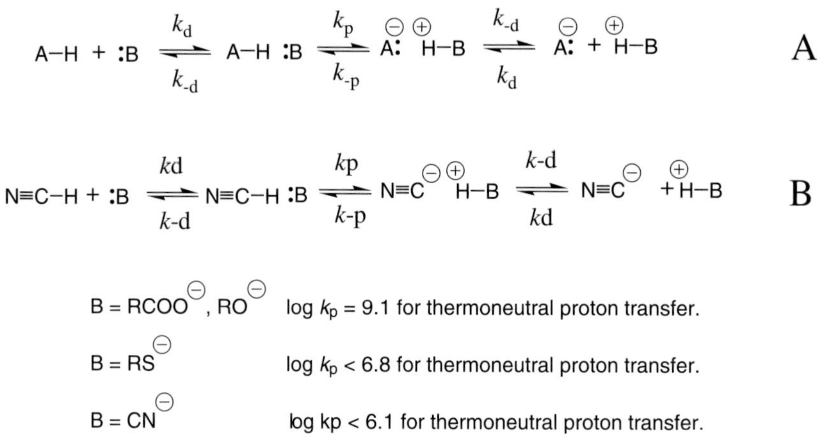

Scheme 1.6

1.4.2.1 Hydrogen Bonding

The first step in proton transfer between electronegative atoms is the formation of a hydrogen-bonded encounter complex between the proton donor and acceptor (Scheme 1.6A). The three-centered hydrogen bond is maintained during the proton transfer reaction, which proceeds through a symmetrical transition state in which there is approximately equal partial bonding of the proton to the donor and acceptor atom. Hydrogen cyanide is a simple and moderately strong carbon acid ($pK_a = 9.4$) for which there is little or no electron delocalization or changes in bond angles or bond lengths on ionization (Section 1.4.2.2.). The observation that the rate constant for thermoneutral deprotonation of HCN by oxygen anions is close to the diffusion-controlled limit ($k_p \approx 10^8$ M^{-1} s^{-1}) shows that there is only a small intrinsic barrier to these reactions [68]. By contrast, there is no detectable deprotonation of HCN in water by the thermoneutral reaction of a thiol anion or by CN^-. This corresponds to >200- and >1000-fold smaller rate constants, respectively, for thermoneutral deprotonation of HCN by carbon and sulfur bases (Scheme 1.6B) [68].

Proton transfer between electronegative atoms may be thought of as the movement of a hydrogen across the potential energy surface for a hydrogen bond, where the relative energies of the symmetrical transition state for proton transfer in water and the asymmetric H-bond with hydrogen localized at a single atom is strongly dependent upon the medium [69–71]. The symmetrical hydrogen bond is almost always a local maximum for the transition state for proton transfer between electronegative atoms in water [72, 73]; and the symmetrical species changes from a local maximum to a local minimum for a single potential minimum hydrogen bond as the medium is changed to a vacuum for proton transfer in the gas phase [69–71]. The order of decreasing reactivity for thermoneutral deprotonation of HCN, O > S > C, parallels the decreasing hydrogen bonding ability of these atoms [74]. This trend suggests that the three-centered symmetrical transition state for proton transfer is strongest relative to the asymmetric hydrogen bond when the donor and acceptor are electronegative atoms such as O and N, and that the symmetric species becomes relatively more unstable with the change to less electronegative atoms such as C and S.

1.4.2.2 Resonance Effects

Figure 1.3 shows three distinct correlations on a plot of rate constants k_{HO} (M^{-1} s^{-1}) for carbon acid deprotonation by hydroxide anion against carbon acid acidity for deprotonation of a variety of carbon acids [20]. The upper correlation

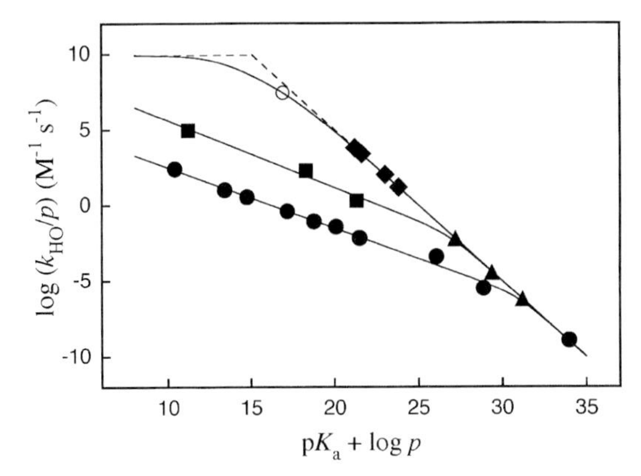

Figure 1.3. Rate-equilibrium correlations of k_{HO} (M^{-1} s^{-1}) for deprotonation of carbon acids by hydroxide ion with the pK_a of the carbon acid in water at 25 °C. The values of k_{HO} and pK_a were statistically corrected for the number of acidic protons p at the carbon acid. (●) Correlation for neutral monocarbonyl carbon acids. (■) Correlation for cationic monocarbonyl carbon acids. (▲) Data for cyanoalkanes which define a slope of −1.0. (◆) Data for simple imidazolium cations which define a slope of −1.0. (○) Data for the 3-cyanomethyl-4-methylthiazolium cation **1b**. The Eigen/Marcus curve through the data for the imidazolium and 3-cyanomethyl-4-methylthiazolium cations was constructed using an estimated Marcus intrinsic barrier of 5.0 kcal mol^{-1}, as described in Ref. 20. Reprinted with permission from *J. Am. Chem. Soc.* **2004**, *126*, 4366–4374. Copyright (2004) American Chemical Society.

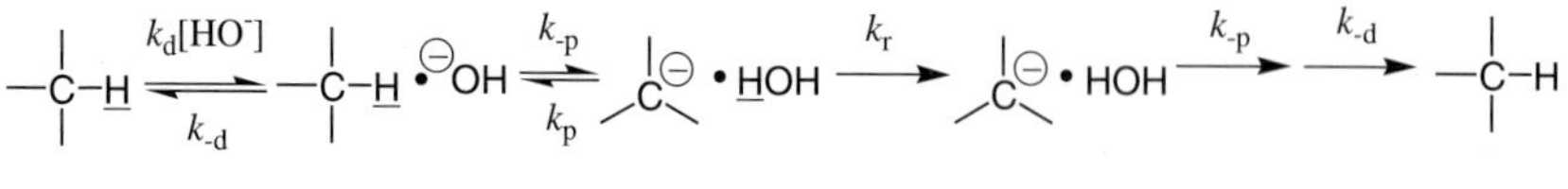

Scheme 1.7

with a Brønsted slope of 1.0 is for deprotonation of cyanoalkanes (**12–14**), imidazolium cations **15–18**, and the 3-cyanomethyl-4-methylthiazolium cations **19**. These data have been fit to an Eigen-type mechanism (Scheme 1.7) for hydroxide ion-catalyzed (B = HO^-, Scheme 1.7) exchange of hydrons between the carbon acid and solvent water (Scheme 1.7) [31], where k_r (s^{-1}) (Scheme 1.4) is the rate constant for the dielectric relaxation of water [33–35]. This value for k_r is assumed to be equal to the rate constant for reorganization of the surrounding aqueous solvation shell which leads to exchange of water labeled with the hydron derived from the carbon acid for the bulk solvent, as shown in Scheme 1.4 for the hydron exchange reactions of acetonitrile. The solid curve shows the calculated fit of the experimental data to the mechanism in Scheme 1.7 that is obtained using $k_d = k_{HO} - 10^{9.9}$ M^{-1} s^{-1} for the thermodynamically favorable diffusion-limited proton transfer between HCN ($pK_a = 9.0$) and hydroxide ion [68], $k_r = 10^{11}$ s^{-1}, and a Marcus intrinsic barrier for the actual proton transfer step ($k_{-p} = 10^9$ s^{-1}) of 5.0 kcal mol^{-1} [20].

12 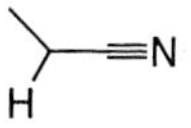 **13** 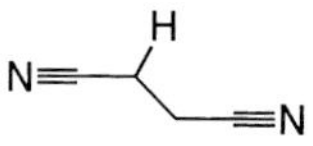 **14**

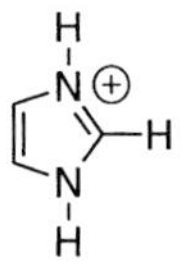

15 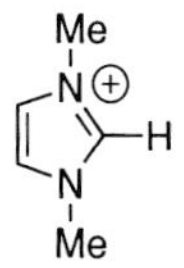 **16** 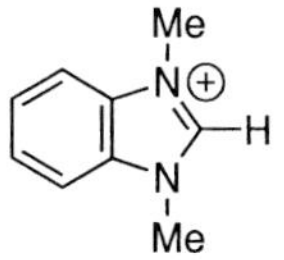 **17** 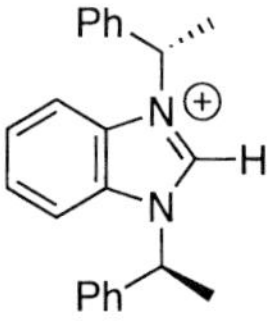 **18** 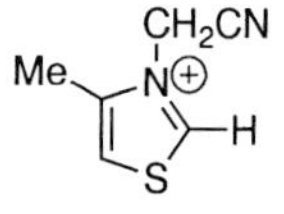 **19**

Eigen-type curvature is observed in Fig. 1.3 for reactions that undergo a change from a rate determining chemical step (k_p, Scheme 1.7) to a rate determining transport step (k_r). Surprisingly, there is no evidence for curvature in plots of data for wholly chemically-limited reactions that is predicted by Marcus theory (see below). The figure shows that progressively smaller values of log k_{HO} (M^{-1} s^{-1}) for thermoneutral deprotonation of carbon acids of $pK_a = 15.7$ by hydroxide ion, and a systematic shift towards the right-hand side of the graph in the position of the downward break to slope of -1.0 are observed on moving from the top correlation line of Fig. 1.3 to the middle correlation line for hydroxide ion deprotonation of cationic esters (e.g. **11**) and ketones and then to the bottom correlation line for deprotonation of neutral carbonyl compounds. A variety of effects are manifested by the decreasing intrinsic reactivity and increasing Marcus intrinsic barrier Λ to proton transfer for these series of carbon acids [20].

1. The small Marcus intrinsic barrier to proton transfer from C(2) of imidazolium and thiazolium cations is consistent with a high degree of localization of the lone pair at the in-plane sp^2-orbital of the carbene/ylide conjugate base, similar to the localization of charge at electronegative atoms. The intrinsic barriers for these proton transfer reactions presumably are larger than for proton transfer at electronegative atoms, because of the relatively weak stabilization of the transition state by hydrogen bonding to carbon (see above).
2. A related, but more involved explanation has been offered to account for the small intrinsic barrier for deprotonation of α-cyano carbon compared with α-carbonyl carbon [13].
3. The Marcus intrinsic barriers for deprotonation of carbon acids to form enolates that are stabilized by resonance delocalization of negative charge from carbon to oxygen are larger than for deprotonation of carbon acids to form carbanions where the charge is localized mainly at carbon.
4. The difference in the Marcus intrinsic barriers for deprotonation of cationic (middle correlation) and neutral (bottom correlation) α-carbonyl carbon is consistent with a greater localization of negative charge at the α-carbon (right-hand resonance structure for a simple ester enolate, Scheme 1.8) of the formally neutral enolate zwitterions of cationic monocarbonyl carbon acids compared with the anionic enolates of monocarbonyl carbon acids. A simple explanation for this difference in resonance delocalization of charge is that it is due to the enhancement of polar interactions between opposing charges that occurs as negative charge is shifted from oxygen to the cation-bearing carbon of the enolate zwitterion [12, 14].

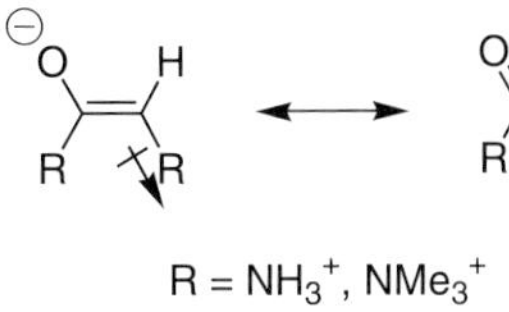

R = NH_3^+, NMe_3^+

Scheme 1.8

5. The lower correlation on Fig. 1.3 for deprotonation of neutral α-carbonyl carbon is linear, with a slope of 0.4 for carbon acids of $pK_a < 30$. By comparison, the simple Marcus equation (Eq. (1.9)) requires curvature for such rate equilibrium correlations and tangential slopes of >0.5 for thermodynamically unfavorable proton transfer [46, 50]. The absence of curvature and the reduced slope for this lower correlation are consistent with an increasing intrinsic carbon acid

Table 1.1. Rate constants, equilibrium constants, and Marcus intrinsic reaction barriers for deprotonation of α-carbonyl carbon by hydroxide ion in water[a].

Carbon acid	Carbanion	log K_{eq}	log k_{OH} (M^{-1} s^{-1})	Λ (kcal mol^{-1})
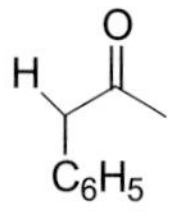 H, O, C_6H_5	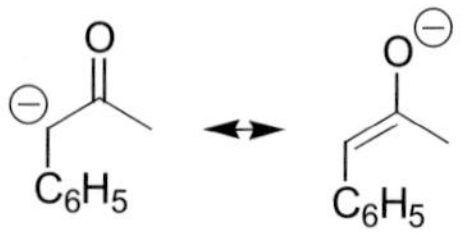 O, ⊖, C_6H_5 ↔ $O^{\ominus}$, C_6H_5	2.3	1.0	17.6
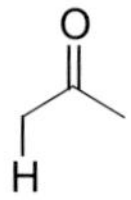 O, H	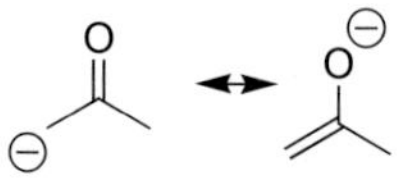 O, ⊖ ↔ $O^{\ominus}$	−4.3	−1.4	16.3
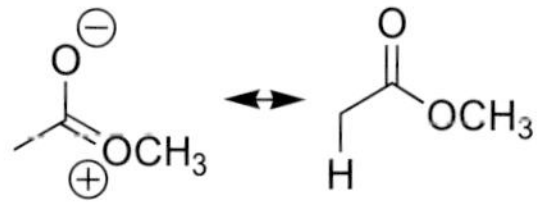 $O^{\ominus}$, OCH_3, ⊕ ↔ O, OCH_3, H	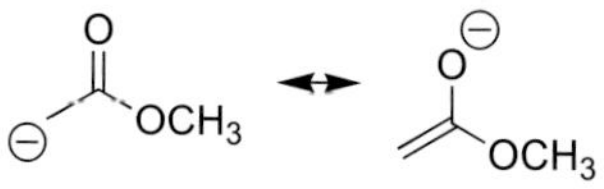 O, OCH_3, ⊖ ↔ $O^{\ominus}$, OCH_3	−10.3	−3.4	14.2

[a] The intrinsic barriers Λ were calculated using the rate and equilibrium constants summarized in Ref. [11] and Eq. (1.10).

reactivity with decreasing carbon acidity, due to a decreasing Marcus intrinsic barrier for proton transfer [11]. The magnitude of this decrease is shown in Table 1.1 which reports intrinsic barriers Λ calculated using Eq. (1.10) [75] and representative individual rate and equilibrium data for the lower correlation from Fig. 1.3.

$$\log k_p = \frac{1}{1.36}\left\{17.44 - \Lambda\left(1 - \frac{1.36\log(K_W/K_a)}{4\Lambda}\right)\right\} \tag{1.10}$$

Table 1.1 shows that there is a significant *decrease* in the Marcus intrinsic barrier Λ with *decreasing* acidity of α-carbonyl hydrogen that correlates well with the *decreasing* resonance stabilization of the product carbanion [11]. Many such correlations between intrinsic reaction barrier and resonance delocalization of charge at the product carbanion have been observed for proton transfer reactions at carbon [67, 76, 77]. The increase in the Marcus intrinsic barrier for deprotonation of carbon with increasing resonance stabilization of the carbanion product is observed because the *fractional* expression of the carbanion-stabilizing resonance substituent effect at the reaction transition state is *smaller* than predicted by the simple Marcus equation, which assumes that the intrinsic reaction barrier is independent of driving force. This has been described by Jencks as an imbalance between the relatively *small* expression of resonance substituent effects at the reaction transition state (these effects cause the intrinsic reaction barrier to change), compared with the *larger* expression of polar substituent effects (these effects do not greatly affect the intrinsic reaction barrier) [78]. Bernasconi refers to the same phenomenon as non-perfect synchronization of polar and resonance substituent effects at the transition state [67, 76, 77].

Kresge has proposed that imbalances between the expression of polar and resonance substituent effects are observed at the transition state for deprotonation of carbon because [79]: (i) The fraction of the effect of polar electron-withdrawing substituents X on the equilibrium constant for proton transfer that is expressed at the reaction transition state is roughly *proportional* to the fractional buildup of negative charge at the reacting carbon (α), which in turn depends upon the fractional bonding between hydrogen and carbon at this transition state $(1 - \alpha)$ (**20**). (ii) The fraction of the overall effect of resonance electron-withdrawing substituents Y on the equilibrium constant for proton transfer that is expressed at the transition state is less than expected for a transition state with fractional charge α, because the resonance interaction depends not only upon this fractional transition state charge, but is further reduced because delocalization of charge at the partly sp^3-hybridized carbon of the transition state will be less effective than delocalization at the planar sp^2-hybrized carbon for the product enolate.

α = Fractional negative charge at transition state
$1 - \alpha$ = C-H bond order at transition state

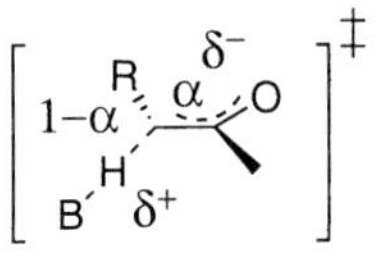

R = X, Polar electron-withdrawing group.

R = Y, Resonance electron-withdrawing group.

20

It has been proposed that part or all of the intrinsic barrier for deprotonation of α-carbonyl carbon is associated with the requirement for solvation of the negatively charged oxygen of the enolate anion [80]. However, the observation of small intrinsic barriers for deprotonation of oxygen acids by electronegative bases to form solvated anions [31] suggests that the requirement for a similar solvation of enolate anions should not make a large contribution to the intrinsic barrier for deprotonation of α-carbonyl carbon.

1.5 Small Molecule Catalysis of Proton Transfer at Carbon

Deprotonation of α-carbonyl carbon is catalyzed by small Brønsted bases, which react directly to abstract a proton from carbon (Scheme 1.9A); by small Brønsted

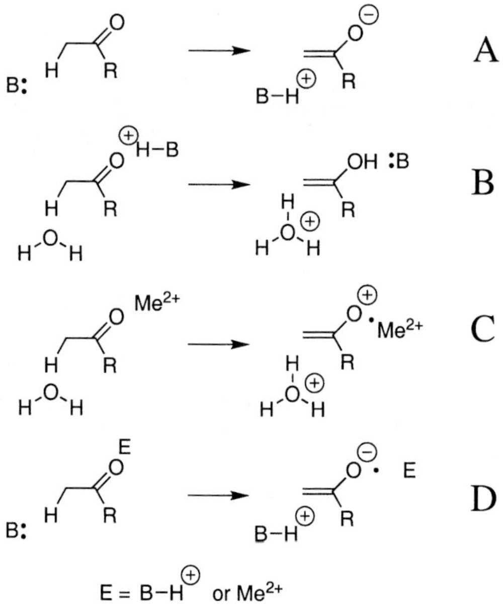

E = B–H$^{\oplus}$ or Me^{2+}

Scheme 1.9

acids, which stabilize negative charge at the enolate oxygen by proton transfer (Scheme 1.9B); and by metal cations, which provide electrostatic stabilization of charge at the enolate oxygen (Scheme 1.9C). Finally, catalysis by the concerted reaction of a Brønsted base to abstract a proton from carbon and a Brønsted acid (Scheme 1.9D) or metal cation (Scheme 1.9E) electrophile to stabilize negative charge at oxygen is sometimes observed.

1.5.1
General Base Catalysis

Scheme 1.10 shows the relative importance of general base catalysis of deprotonation of several carbon acids, where the catalytic effectiveness is defined as the rela-

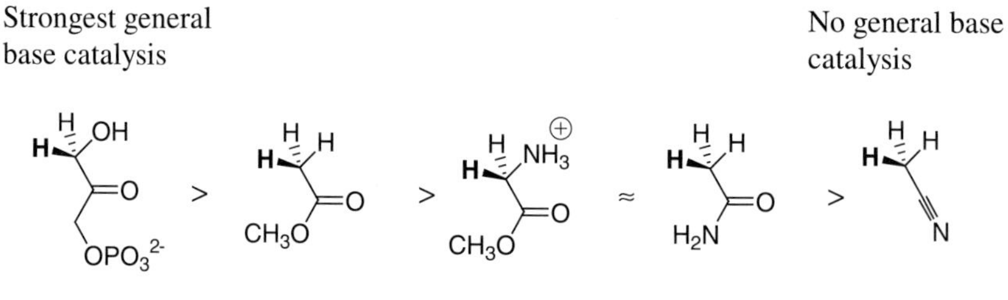

Strongest general base catalysis

No general base catalysis

Scheme 1.10

tive acceleration of the solvent reaction observed at 1 M buffer catalyst. Deprotonation of dihydroxyacetone phosphate (DHAP) by 1.0 M quinuclidinone buffer ($pK_a = 7.5$) at pH 7.0 and 25 °C is 10^5-times faster than deprotonation by hydroxide ion under the same conditions [24]. The greater reactivity of quinuclidinone compared with hydroxide ion toward deprotonation of DHAP is due to [81]: (i) hydroxide ion being an intrinsically unreactive base for its pK_a and, (ii) the relatively small value of $\beta = 0.5$ [24] for this proton transfer reaction, which causes the importance of general base catalysis to increase with decreasing pH and pK_a of the buffer catalyst.

The low intrinsic reactivity of hydroxide ion compared with other buffer bases toward deprotonation of a variety of carbon acids is known as the lyoxide ion anomaly [50, 82]. Good buffer catalysis is observed for deprotonation of ethyl acetate by substituted quinuclidinone due to the anomalously low reactivity of hydroxide ion. However, this catalysis is much weaker than for deprotonation of DHAP (Scheme 1.10) due to the larger value of $\beta = 1.09$ for proton transfer [11]. There is only weak catalysis of deprotonation of the cationic amino acid ester **18** by quinuclidinol ($pK_a = 10.0$) [14]. Here the value of $\beta = 0.92$ for proton transfer is large, and the intrinsic reactivity of hydroxide ion compared to tertiary amines toward deprotonation of cationic carbon acid is much greater than for deprotonation of neutral α-carbonyl carbon. The difference in the magnitude of the lyoxide ion anomaly for deprotonation of cationic (small anomaly) and neutral carbon acids (large anomaly) shows that this anomaly is partly electrostatic in origin [14].

There is no detectable buffer catalysis of exchange of deuterium for the α-methyl hydrogen of acetonitrile in D_2O (Scheme 1.10) [13]. This is because the rate-determining step for the competing DO^--catalyzed reaction is rotation of the D_2O into a reactive conformation with a rate constant $k_r \approx 10^{11}$ s^{-1} (Scheme 1.4) that is independent of the concentration of buffer bases [13]. The observation that buffer catalysis of exchange of deuterium for α-methyl hydrogen of the neutral carbon acid acetamide (Scheme 1.10) is just barely detectable provides evidence that this reaction also proceeds through a very reactive amide enolate, that is protonated by water with a rate constant that is approaching the value for a rotation limited reaction [17].

1.5.2 Electrophilic Catalysis

Brønsted general acid catalysis of the deprotonation of acetone by water (Scheme 1.9B) can only be detected for strong buffer acids such as acetic acid ($pK_a = 4.8$) [83], that provide a strong thermodynamic driving force for protonation of the relatively weakly basic enolate ion ($pK_a = 10.9$ for enol acetone [37]) [84]. Again, general acid catalysis is weak, because of the high reactivity of hydronium ion in the competing solvent reaction.

There is no obvious pattern in the metal ion requirements of enzymes that catalyze proton transfer at carbon. For example, mannose 6-phosphate isomerase [85] is a metalloenzyme while triosephosphate isomerase [86] and glucose 6-phosphate isomerase [87] are not. The observation that enzyme catalysts may use either metal cations or Brønsted acids to stabilize negative charge that develops at the enolate oxygen shows that both types of catalysis are viable, and raises questions about the imperatives for the observation of catalysis by one mechanism rather than another.

In fact, there are only small differences between the second-order rate constant for nonenzymatic deprotonation of acetone by acetate anion and the third-order rate constants for catalysis of this reaction by acetic acid and Zn^{2+}, so that the stabilities of the transition states for the acetate-ion-promoted reactions assisted by Zn^{2+}, acetic acid and solvent water (55 M) are similar (Scheme 1.11) [19]. Scheme 1.11 shows that acetic acid and Zn^{2+}, stabilize the transition state for proton transfer from acetone to acetate anion by 1.9 and 3.3 kcal mol^{-1}, respectively, relative to a common standard state of 1 M water and 1 M electrophile.

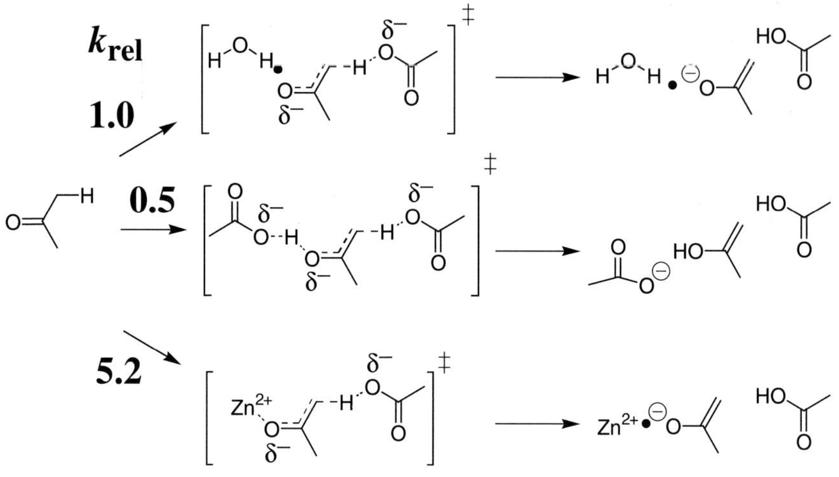

k_{rel} 1.0, 0.5, 5.2

Scheme 1.11

There is an increase in the importance of electrophilic catalysis by zinc cation relative to acetic acid for deprotonation of the α-carbonyl carbons of hydroxyacetone, a substrate which provides a second stabilizing chelate interaction between the hydroxy group at the substrate and the metal dication that is expressed at transition state for proton transfer [19]. For example, the third-order rate constants k_T for the Zn^{2+}-assisted acetate-ion-promoted deprotonation of the α-CH_3 and α-CH_2OH groups of hydroxyacetone are 32-fold and 770-fold larger, respectively, than the corresponding second-order rate constants k_{AcO} for proton transfer to acetate anion "assisted" by solvent water that is present at 55 M (Scheme 1.12). This shows that Zn^{2+} stabilizes the transition state for proton transfer from the α-CH_3

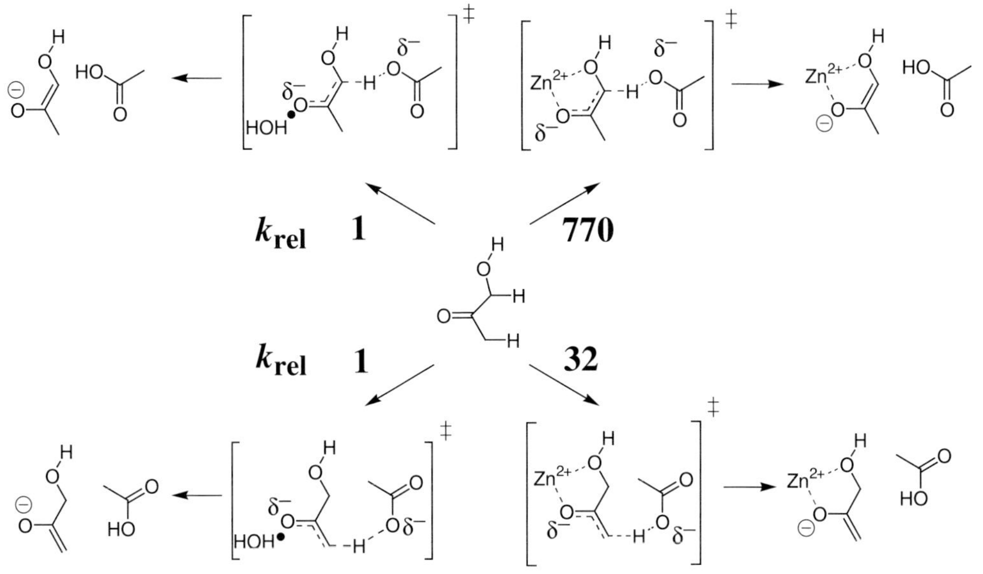

Scheme 1.12

and α-CH_2OH groups of hydroxyacetone by 4.4 and 6.3 kcal mol^{-1}, respectively, relative to a common standard state of 1 M water and 1 M Zn^{2+}.

A similar chelation of metal to enzyme-bound substrate may also contribute to enzyme catalysis of proton transfer at carbon. For example, X-ray crystallographic analysis of complexes between 3-keto-L-gulonate 6-phosphate decarboxylase and analogs of the 1,2-enediolate reaction intermediate provide evidence that the essential magnesium dication is stabilized by coordination to both the C-2 oxygen and the nonreacting C-3 hydroxy of the reaction intermediate [88].

In summary, catalysis of proton transfer at carbon in water by the small molecule reactions shown in Scheme 1.9 is generally weak. Small Brønsted acid and base buffer catalysts do not act to reduce the large thermodynamic barrier to endothermic proton transfer reactions (ΔG°, Fig. 1.1), which constitutes most of the observed activation barrier ($\Delta G_f^\dagger$). Buffer catalysis is the result of the lower Marcus intrinsic barrier for the buffer compared to the competing solvent-catalyzed reaction and catalysis is weak because the effect of these buffers on the intrinsic reaction barrier is small. The formation of a stable chelate between a metal cation and the product enolate anion may reduce the thermodynamic driving force for deprotonation at the α-carbonyl carbon compared with the solvent reaction. This results in effective metal cation catalysis when there is a second group to chelate the metal cation, such as the hydroxy group of hydroxyacetone [19]. The relatively weak catalysis by Zn^{2+} observed in the absence of a second chelating group shows that this cation does not cause a large reduction in the intrinsic barrier for the competing solvent-catalyzed proton transfer reaction.

1.6
Comments on Enzymatic Catalysis of Proton Transfer

Studies on proton transfer to and from carbon in model reactions have shown that the activation barrier to most enzyme-catalyzed reactions is composed mainly of the thermodynamic barrier to proton transfer (Fig. 1.1), so that in most cases this barrier for proton transfer at the enzyme active site will need to be reduced in order to observe efficient catalysis. A smaller part of the activation barrier to deprotonation of α-carbonyl carbon is due to the intrinsic difficulty of this reaction to form a resonance stabilized enolate. There is evidence that part of the intrinsic barrier to proton transfer at α-carbonyl carbon reflects the intrinsic instability of negative charge at the transition state of mixed sp^2–sp^3-hybridization at carbon [79]. Small buffer and metal ion catalysts do not cause a large reduction in this intrinsic reaction barrier.

There is extensive evidence from site-directed mutagenesis and other studies of enzymes that catalyze proton transfer that acidic and basic amino side chains and, in some cases, metal cations, are required for the observation of efficient catalysis. However, catalysis of the deprotonation of α-carbonyl by small molecule analogs of these side chains, and by metal cations is generally weak. Relatively little attention has been directed towards understanding the mechanism for the "enhancement" of Brønsted acid/base and electrophilic catalysis for enzyme-catalyzed reactions [89].

An apparent enhancement of Brønsted acid base catalysis will result if there is a greater driving force for proton transfer to the catalytic base at the enzyme active site compared with solution. One mechanism to increase the thermodynamic driving force for deprotonation of α-carbonyl carbon at an enzyme-bound substrate compared to proton transfer in solution is to use an enzyme active site of low overall dielectric constant where there are several precisely oriented polar groups of opposite charge or dipole moment from the enolate anion to provide electrostatic stabilization of this anion. In addition, catalysis of deprotonation of cationic carbon acids will be strongly favored at a nonpolar enzyme active site by the strong stabilizing intramolecular electrostatic interaction at the product zwitterionic enolate anion [14, 18, 90].

There may also be a reduction in the intrinsic barrier for proton transfer at the enzyme active site compared to solution [80]. This possibility is intriguing; however, we are unable to offer a convincing mechanism for such a reduction of intrinsic reaction barrier.

Acknowledgment

We acknowledge the National Institutes of Health Grant GM 39754 for its generous support of the work from our laboratory described in this review.

References

1 R. Wolfenden, M. J. Snider *Acc. Chem. Res.* **2001**, *34*, 938–945.
2 S. L. Bearne, R. Wolfenden *J. Am. Chem. Soc.* **1995**, *117*, 9588–9589.
3 W. S. Matthews, J. E. Bares, J. E. Bartmess, F. G. Bordwell, F. J. Cornforth, G. E. Drucker, Z. Margolin, R. J. McCallum, G. J. McCollum, N. R. Vanier *J. Am. Chem. Soc.* **1975**, *97*, 7006–7014.
4 R. W. Taft, F. G. Bordwell *Acc. Chem. Res.* **1988**, *21*, 463–469.
5 M. M. Kreevoy, E. H. Baughman *J. Am. Chem. Soc.* **1973**, *95*, 8178–8179.
6 T. Aroella, C. H. Arrowsmith, M. Hojatti, A. J. Kresge, M. F. Powell, Y. S. Tang, W. H. Wang *J. Am. Chem. Soc.* **1987**, *109*, 7198–7199.
7 A. J. Kresge, M. F. Powell *Int. J. Chem. Kinet.* **1982**, *14*, 19–34.
8 R. L. D'Ordine, B. J. Bahnson, P. J. Tonge, V. E. Anderson *Biochemistry* **1994**, *33*, 14734–14742.
9 T. L. Amyes, J. P. Richard *J. Am. Chem. Soc.* **1992**, *114*, 10297–10302.
10 R. W. Nagorski, T. Mizerski, J. P. Richard *J. Am. Chem. Soc.* **1995**, *117*, 4718–4719.
11 T. L. Amyes, J. P. Richard *J. Am. Chem. Soc.* **1996**, *118*, 3129–3141.
12 A. Rios, J. P. Richard *J. Am. Chem. Soc.* **1997**, *119*, 8375–8376.
13 J. P. Richard, G. Williams, J. Gao *J. Am. Chem. Soc.* **1999**, *121*, 715–726.
14 A. Rios, T. L. Amyes, J. P. Richard *J. Am. Chem. Soc.* **2000**, *122*, 9373–9385.
15 A. Rios, J. Crugeiras, T. L. Amyes, J. P. Richard *J. Am. Chem. Soc.* **2001**, *123*, 7949–7950.
16 A. Rios, J. P. Richard, T. L. Amyes *J. Am. Chem. Soc.* **2002**, *124*, 8251–8259.
17 J. P. Richard, G. Williams, A. C. O'Donoghue, T. L. Amyes *J. Am. Chem. Soc.* **2002**, *124*, 2957–2968.
18 G. Williams, E. P. Maziarz, T. L. Amyes, T. D. Wood, J. P. Richard *Biochemistry* **2003**, *42*, 8354–8361.
19 J. Crugeiras, J. P. Richard *J. Am. Chem. Soc.* **2004**, *126*, 5164–5173.
20 T. L. Amyes, S. T. Diver, J. P. Richard, F. M. Rivas, K. Toth *J. Am. Chem. Soc.* **2004**, *126*, 4366–4374.
21 A. Sievers, R. Wolfenden *J. Am. Chem. Soc.* **2002**, *124*, 13986–13987.
22 R. C. Cavestri, L. R. Fedor *J. Am. Chem. Soc.* **1970**, *92*, 4610–4613.
23 L. R. Fedor, W. R. Glave *J. Am. Chem. Soc.* **1971**, *93*, 985–989.
24 J. P. Richard *J. Am. Chem. Soc.* **1984**, *106*, 4926–4936.
25 J. P. Richard, R. W. Nagorski *J. Am. Chem. Soc.* **1999**, *121*, 4763–4770.
26 K. Toth, T. L. Amyes, J. P. Richard, J. P. G. Malthouse, M. i. E. Ni'Beilliu *J. Am. Chem. Soc.* **2004**, *126*, 10538–10539.
27 E. Tapuhi, W. P. Jencks *J. Am. Chem. Soc.* **1982**, *104*, 5758–5765.
28 J. P. Guthrie, J. Cossar, A. Klym *J. Am. Chem. Soc.* **1984**, *106*, 1351–1360.
29 Y. Chiang, A. J. Kresge *Science* **1991**, *253*, 395–400.
30 J. R. Keeffe, A. J. Kresge, in *The Chemistry of Enols*, Rappoport, Z. (Ed.), John Wiley and Sons, Chichester, 1990, pp. 399–480.
31 M. Eigen *Angew. Chem., Int. Ed. Engl.* **1964**, *3*, 1–72.
32 W. P. Jencks, M. T. Haber, D. Herschlag, K. L. Nazaretian *J. Am. Chem. Soc.* **1986**, *108*, 479–483.
33 U. Kaatze *J. Chem. Eng. Data* **1989**, *34*, 371–374.
34 U. Kaatze, R. Pottel, A. Schumacher *J. Phys. Chem.* **1992**, *96*, 6017–6020.
35 K. Giese, U. Kaatze, R. Pottel *J. Phys. Chem.* **1970**, *74*, 3718–3725.
36 Y. Chiang, A. J. Kresge, P. Pruszynski *J. Am. Chem. Soc.* **1992**, *114*, 3103–3107.
37 Y. Chiang, A. J. Kresge, Y. S. Tang, J. Wirz *J. Am. Chem. Soc.* **1984**, *106*, 460–462.
38 Y. Chiang, A. J. Kresge, V. V. Popik,

N. P. Schepp *J. Am. Chem. Soc.* **1997**, *119*, 10203–10212.

39 D. C. Hawkinson, T. C. M. Eames, R. M. Pollack *Biochemistry* **1991**, *30*, 10849–10858.

40 W. J. Albery, J. R. Knowles *Biochemistry* **1976**, *15*, 5631–5640.

41 R. K. Poyner, W. W. Cleland, G. H. Reed *Biochemistry* **2001**, *40*, 8009–8017.

42 W. L. Jorgensen, J. M. Briggs, J. Gao *J. Am. Chem. Soc.* **1987**, *109*, 6857–6858.

43 K. B. Wiberg, H. Castejon *J. Org. Chem.* **1995**, *60*, 6327–6334.

44 Y. Chiang, M. Hojatti, J. R. Keeffe, A. J. Kresge, N. P. Schepp, J. Wirz *J. Am. Chem. Soc.* **1987**, *109*, 4000–4009.

45 M. W. Washabaugh, W. P. Jencks *Biochemistry* **1988**, *27*, 5044–5053.

46 R. A. Marcus *J. Phys. Chem.* **1968**, *72*, 891–899.

47 R. A. Marcus *J. Chem. Phys.* **1956**, *24*, 966–978.

48 R. A. Marcus *J. Chem. Phys.* **1957**, *26*, 872–877.

49 R. A. Marcus *J. Chem. Phys.* **1957**, *26*, 867–871.

50 A. J. Kresge *Chem. Soc. Rev.* **1974**, *2*, 475–503.

51 R. A. Marcus *J. Am. Chem. Soc.* **1969**, *91*, 7224–7225.

52 J. P. Richard *Tetrahedron* **1995**, *51*, 1535–1573.

53 W. J. Albery, M. M. Kreevoy *Adv. Phys. Org. Chem.* **1978**, *16*, 87–157.

54 E. S. Lewis, D. D. Hu *J. Am. Chem. Soc.* **1984**, *106*, 3292–3296.

55 J. P. Guthrie *ChemPhysChem* **2003**, *4*, 809–816.

56 J. P. Guthrie *J. Am. Chem. Soc.* **1997**, *119*, 1151–1152.

57 E. Grunwald *J. Am. Chem. Soc.* **1985**, *107*, 125–133.

58 C. F. Bernasconi, P. J. Wenzel *J. Am. Chem. Soc.* **1994**, *116*, 5405–5413.

59 C. F. Bernasconi, P. J. Wenzel *J. Am. Chem. Soc.* **1996**, *118*, 10494–10504.

60 C. F. Bernasconi, P. J. Wenzel, J. R. Keeffe, S. Gronert *J. Am. Chem. Soc.* **1997**, *119*, 4008–4020.

61 C. F. Bernasconi, P. J. Wenzel *J. Am. Chem. Soc.* **2001**, *123*, 7146–7153.

62 C. F. Bernasconi, P. J. Wenzel *J. Am. Chem. Soc.* **2001**, *123*, 2430–2431.

63 C. F. Bernasconi, P. J. Wenzel *J. Org. Chem.* **2001**, *66*, 968–979.

64 J. E. Van Verth, W. H. Saunders, Jr. *J. Org. Chem.* **1997**, *62*, 5743–5747.

65 J. W. Bunting, J. P. Kanter *J. Am. Chem. Soc.* **1993**, *115*, 11705–11715.

66 J. P. Richard, T. L. Amyes, M. M. Toteva *Acc. Chem. Res.* **2001**, *34*, 981–988.

67 C. F. Bernasconi *Tetrahedron* **1985**, *41*, 3219–3234.

68 R. A. Bednar, W. P. Jencks *J. Am. Chem. Soc.* **1985**, *107*, 7117–7126.

69 Y. Pan, M. A. McAllister *J. Am. Chem. Soc.* **1998**, *120*, 166–169.

70 Y. Pan, M. A. McAllister *J. Org. Chem.* **1997**, *62*, 8171–8176.

71 S. Scheiner, T. Kar *J. Am. Chem. Soc.* **1995**, *114*, 6970–6975.

72 C. L. Perrin *Science* **1994**, *266*, 1665–1668.

73 C. L. Perrin, J. B. Nielson *J. Am. Chem. Soc.* **1997**, *119*, 12734–12741.

74 S. N. Vinogradov, R. H. Linnell *Hydrogen Bonding*, Van Nostrand-Reinhold, New York, 1971, pp. 120–124.

75 J. P. Guthrie *J. Am. Chem. Soc.* **1991**, *113*, 7249–7255.

76 C. F. Bernasconi *Acc. Chem. Res.* **1987**, *20*, 301–308.

77 C. F. Bernasconi *Adv. Phys. Org. Chem.* **1992**, *27*, 119–238.

78 D. A. Jencks, W. P. Jencks *J. Am. Chem. Soc.* **1977**, *99*, 7948–7960.

79 A. J. Kresge *Can. J. Chem.* **1974**, *52*, 1897–1903.

80 J. A. Gerlt, P. G. Gassman *J. Am. Chem. Soc.* **1993**, *115*, 11552–11568.

81 W. P. Jencks *Acc. Chem. Res.* **1976**, *9*, 425–432.

82 M. W. Washabaugh, W. P. Jencks *J. Am. Chem. Soc.* **1989**, *111*, 683–692.

83 A. F. Hegarty, J. P. Dowling, S. J. Eustace, M. McGarraghy *J. Am. Chem. Soc.* **1998**, *120*, 2290–2296.

84 W. P. Jencks *J. Am. Chem. Soc.* **1972**, *94*, 4731–4732.

85 R. W. Gracy, E. A. Noltmann *J. Biol. Chem.* **1968**, *243*, 4109–4116.

86 K. A. Komives, L. C. Chang, E. Lolis, R. F. Tilton, G. A. Petsko, J. R. Knowles *Biochemistry* **1991**, *30*, 3011–3019.

87 C. J. Jeffrey, R. Hardré, L. Salmon *Biochemistry* **2001**, *40*, 1560–1564.

88 E. L. Wise, W. S. Yew, J. A. Gerlt, I. Rayment *Biochemistry* **2003**, *42*, 12133–12142.

89 J. P. Richard *Biochemistry* **1998**, *37*, 4305–4309.

90 J. P. Richard, T. L. Amyes *Bioorg. Chem.* **2004**, *32*, 354–366.

2
General Acid–Base Catalysis in Model Systems

Anthony J. Kirby

2.1
Introduction

Proton transfer is the most common reaction in living systems, in which reactions have to be strictly controlled, and most are catalyzed by enzymes. The great majority of enzyme catalyzed reactions are ionic, involving heterolytic bond making and breaking, and thus the creation or neutralization of charge. Under conditions of constant pH this requires the transfer of protons (Eq. (2.1)).

$$X{-}Y \longrightarrow \left[\overset{+}{X} \quad Y^{-} \quad \underset{pH \sim 7}{\overset{H_2O}{\rightleftharpoons}} \right] XOH,\ HY \qquad (2.1)$$

General acid and general base catalysis are terms commonly used to describe two different characteristics of reactions, the (observable) form of the rate law or a (hypothetical) reaction mechanism proposed to account for it. It is important to be aware of (and for authors to make clear) which is meant in a particular case.

General acid–base catalysis provides mechanisms for bringing about the necessary proton transfers without involving hydrogen or hydroxide ions, which are present in water at concentrations of only about 10^{-7} M under physiological conditions. At pHs near neutrality relatively weak acids and bases can compete with lyonium or lyate species because they can be present in much higher concentrations.

2.1.1
Kinetics

The basics of general acid and general base catalysis are described clearly and in detail in Chapter 8 of Maskill [1]. Acid–base catalysis is termed *specific* if the rate of the reaction concerned depends only on the acidity (pH, etc.) of the medium. This is the case if the reaction involves the conjugate acid or base of the reactant preformed in a rapid equilibrium process – normal behavior if the reactant is weakly basic or acidic. The conjugate acid or base is then, by definition, a strong

Hydrogen-Transfer Reactions. Edited by J. T. Hynes, J. P. Klinman, H.-H. Limbach, and R. L. Schowen

ISBN: 978-3-527-30777-7

acid or base, and the reverse proton transfer to solvent is thus rapid, probably diffusion-controlled – and certainly faster than a competing forward reaction involving the making or breaking of covalent bonds. This forward reaction of the conjugate acid or base of the reactant is therefore rate determining, and the rate expression – for example for the hydrolysis of an unreactive ester (Scheme 2.1) – contains only a single term in (lyonium) acid concentration:

$$-\mathrm{d}[\mathbf{1.1}]/\mathrm{d}t = k_{\mathrm{H}}[\mathbf{1.1}][\mathrm{H_3O^+}]$$

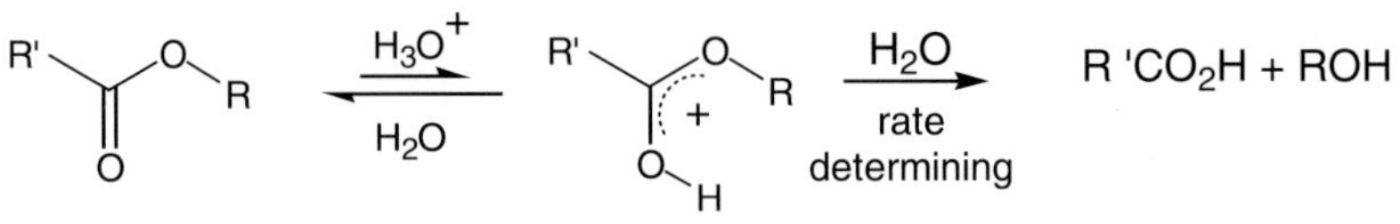

R' – C(=O) – O – R (**1.1**) ⇌ (H_3O^+ / H_2O) **1.1.H⁺** → (H_2O, rate determining) $R'CO_2H + ROH$

Scheme 2.1

General acid–base catalysis is defined experimentally by the appearance in the rate law of acids and/or bases other than lyonium or lyate ions. For example, the hydrolysis of enol ethers **1.2** (Scheme 2.2) is general acid-catalyzed. In strong acid the rate expression will be the same as in Scheme 2.1, but near neutral pH the rate is found to depend also on the concentration of the buffer ($HA + A^-$) used to maintain the pH. Measurements at different buffer ratios show that the catalytic species is the acid HA. (If more than one acid is present there will be an additional term $k_{\mathrm{HA}i}[\mathrm{HA}_i][\mathbf{1.2}]$ for each.)

$$-\mathrm{d}[\mathbf{1.2}]/\mathrm{d}t = k_{\mathrm{H}}[\mathbf{1.2}][\mathrm{H_3O^+}] + k_{\mathrm{HA}}[\mathbf{1.2}][\mathrm{HA}]$$

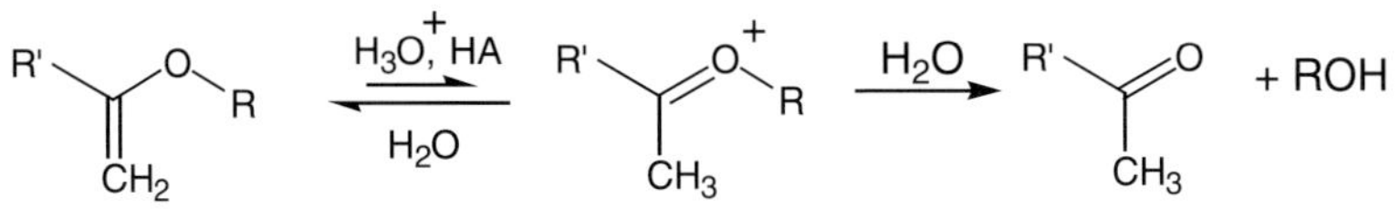

R' – C(=CH₂) – O – R (**1.2**) ⇌ (H_3O^+, HA / H_2O) **1.2.H⁺** → (H_2O) R' – C(=O) – CH_3 + ROH

Scheme 2.2

If in these experiments the measurements at different buffer ratios showed that the catalytic species was the conjugate base A^- the reaction would be kinetically general base catalyzed. In which case HA and A^- would probably subsequently be referred to as BH^+ and B. Thus the enolisation of ketones is general base catalysed (Scheme 2.3).

$$-\mathrm{d}[\mathbf{1.3}]/\mathrm{d}t = k_{\mathrm{H}}[\mathbf{1.3}][\mathrm{HO^-}] + k_{\mathrm{B}i}[\mathbf{1.3}][\mathrm{B}_i]$$

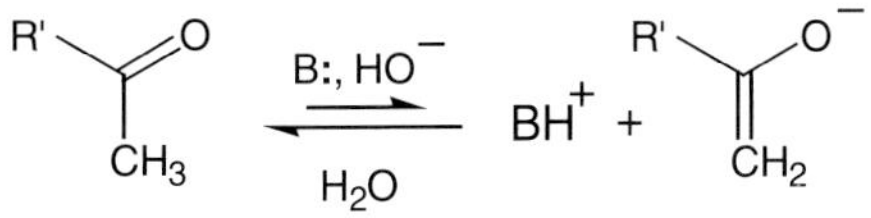

Scheme 2.3

The rate constants k_{HA} and k_B depend on the strength of the acid or base, and for a given reaction are correlated by the Brönsted equation: conventionally written for general acid and general base catalyzed reactions, respectively:

$$\log k_{HA} = \alpha \log K_{HA} + \text{constant} = -\alpha pK_{HA} + \text{constant}$$

$$\log k_{B} = \beta \log K_{B} + \text{constant} = \beta\, pK_{B} + \text{constant}$$

The pK_as used are those of the conjugate acids, HA and BH^+.

2.1.2 Mechanism

Enzymes have evolved highly efficient mechanisms for catalysis under physiological conditions. Such mechanisms must avoid high energy intermediates, with their associated high energy barriers. So potential high energy species – such as the ions X^+ and Y^- in Eq. (2.1), above, need to be neutralised as part of the reaction. This is accomplished in water by the very general mechanism outlined in Scheme 2.4 (the bond that breaks may be either a σ- or π-bond).

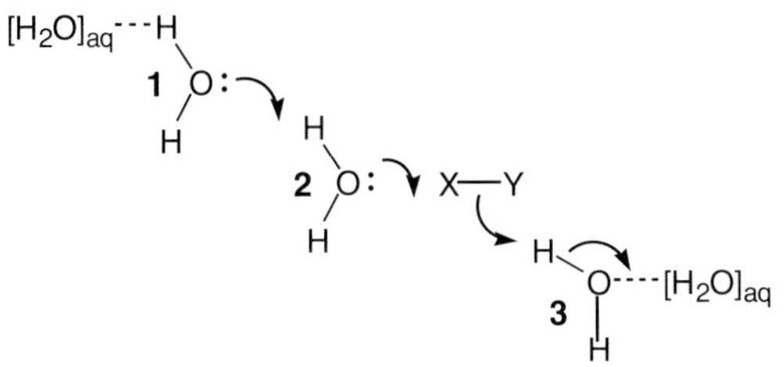

Scheme 2.4

Here a water molecule **2** acts as a nucleophile, generating the potentially strongly acidic H_2O^+–X; but in a suitable buffered solution this can be neutralised, as part of the reaction, by a series of rapid proton transfers. Variants of this general mechanism account for almost all solvolyses in protic solvents giving, for sufficiently reactive systems, reactions which can be observed and studied in mechanistic detail.

The mechanism of Scheme 2.4 is generalised further in Scheme 2.5. Water molecule **1** (Scheme 2.4), which removes the proton from the incipient H_2O^+–X, acts formally as a *general base*. Water molecule **3** (Scheme 2.4) acts as a *general acid*, transferring a proton to the potential strong base Y^-. ($Y = C$ is a special case because the proton is transferred directly to an X–Y bonding orbital rather than to a lone pair on Y. See Section 2.4, below.)

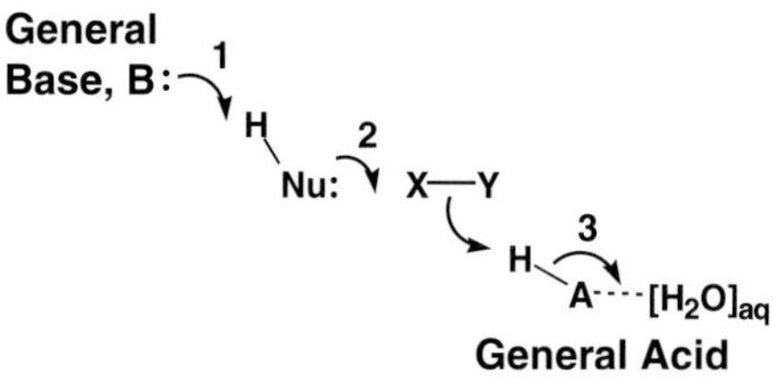

General Base, B: 1 H Nu: 2 X—Y H 3 A····$[H_2O]_{aq}$ General Acid

Scheme 2.5

The mechanism of Scheme 2.5 is a general solution to the problem of avoiding high energy intermediates, and offers the prospect of seriously low activation energies: because a wide range of weak acids and bases are nevertheless much stronger acids, bases and nucleophiles than water. It is not however observed in solution under normal conditions because it requires the (entropically) prohibitively unfavorable encounter of four separate molecules – general acid, general base, nucleophile and substrate. Most observed reactions involve successful bimolecular encounters, and there is an entropic price to pay for the specific involvement even of solvent molecules. However, the mechanism of Scheme 2.5 might have been designed for an enzyme reaction (rather than, presumably, vice versa), since in the active site all the components bar the substrate (the molecule X–Y) come preassembled. Scheme 2.5 does indeed describe the mechanisms of many enzyme reactions, of which the serine proteases are perhaps the most familiar [2].

(Concerted) steps 1 and 2 of Scheme 2.5 define the "classical" general base catalysis mechanism, and step 3 the "classical" general acid catalysis mechanism. When step 3 is rate determining the "*general acid*" HA is present in the rate determining transition state, and thus appears in the observed rate law. The same applies to the "*general base*" B, when (concerted) steps 1 and 2 are rate determining.

Thus the defining element of general acid–base catalysis is a rate determining step involving proton transfer. Proton transfers between electronegative centers, especially O and N, are known to be so fast in the thermodynamically favorable direction that they are diffusion-controlled, so are likely to be rate determining only if they involve species – particularly high energy intermediates – that are present in only very low concentration. For example, the very fast hydrolysis of dialkyl maleamic acids 1 (half-life < 1 s at 15 °C) is general acid catalysed Scheme 2.6) [3]. The rate determining step was identified as the proton transfer that converts the tetrahedral intermediate $\mathbf{T^0}$ to the zwitterion $\mathbf{T^\pm}$ (and thus the amine to a viable leaving

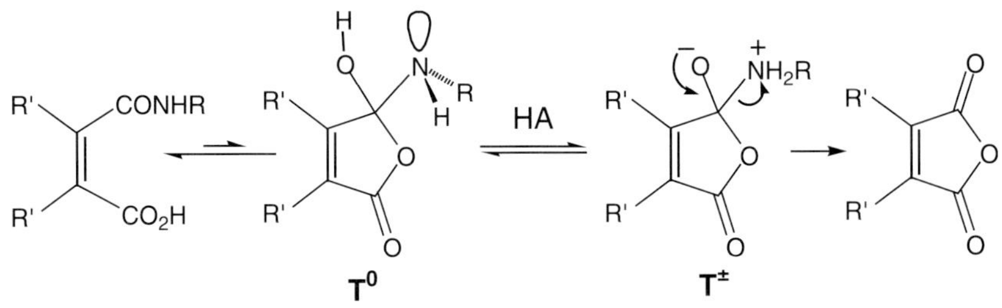

Scheme 2.6

group), and the reaction behaves, as expected, as a diffusion-controlled reaction. A similar, more recent, example is the general acid catalyzed cyclization of 1-amino-8-trifluoroacetylaminonaphthalene [4].

In "classical" general-acid–base catalysis (Scheme 2.5) thc proton transfer step is slow because it is concerted with the formation or cleavage of a bond between heavy (non-hydrogen) atoms. This broad generalisation includes the familiar general base catalyzed enolisation and related processes involving proton transfer to and from carbon. Such reactions are often considered to be "intrinsically" slow, but this is not fundamentally because of the involvement of a C–H bond as such, but because a carbanion is generally formed only in situations where the negative charge can be delocalised on to a more electronegative center (see, for example, Scheme 2.3): as before, the proton transfer step is relatively slow because it is concerted with the formation of a bond between heavy (non-hydrogen) atoms, which requires geometrical changes. Where no such geometrical changes are involved – for example in the ionization of H–CN or the C(2)–H bond of the thiazolium system, the proton transfers are (more or less) normal diffusion-controlled processes [5, 6].

Detailed mechanisms for proton transfers from carbon do of course show significant differences from those between two electronegative centers. These include the shape and height of the energy barrier to the reaction, and the absence of significant hydrogen bonding between C–H and solvent or general base in protic solvents. For these reasons they are discussed separately, in Section 2.4 below.

2.1.3 Kinetic Equivalence

The simple examples quoted so far might suggest that the observation of general acid or general base kinetics is prima facie evidence for the mechanisms of the same name. This is not the case, for the usual reasons of (i) kinetic equivalence (the proton is a uniquely mobile species), and (ii) the absence of direct evidence from the rate law of the involvement of the solvent (for example, water molecule **2** in Scheme 2.5). Thus (i) the kinetically observed general acid catalysis of ketone enolisation is explained not by the general acid catalysis mechanism **a** (Scheme 2.7) but by the kinetically equivalent specific acid–general base catalysis mecha-

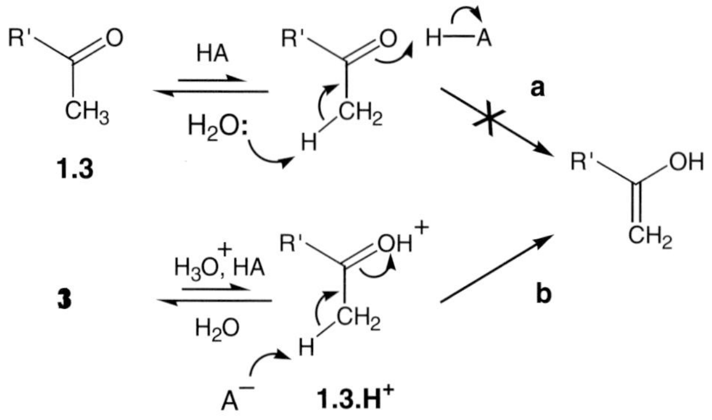

Scheme 2.7

nism **b** (which requires only bimolecular encounters). Similarly, general base catalysis of the breakdown of acetaldehyde hemiacetals is accounted for by general acid catalysis of the reaction of the hemiacetal anion [7].

Finally (ii), a common mechanistic problem is to distinguish between nucleophilic and general base catalysis in cases where the products are the same. A strong base is generally a good nucleophile (depending on the electrophilic center concerned), and the rate expressions will be identical for the two mechanisms. A classical example is catalysis of the hydrolysis of substituted acetate esters by acetate anion (Scheme 2.8): acetate acts as a nucleophile for esters with very good leaving groups like 2,4-dinitrophenolate, but as a general base for poor leaving groups like phenolate. For leaving groups of intermediate basicity, such as *p*-nitrophenolate, both mechanisms are observed.

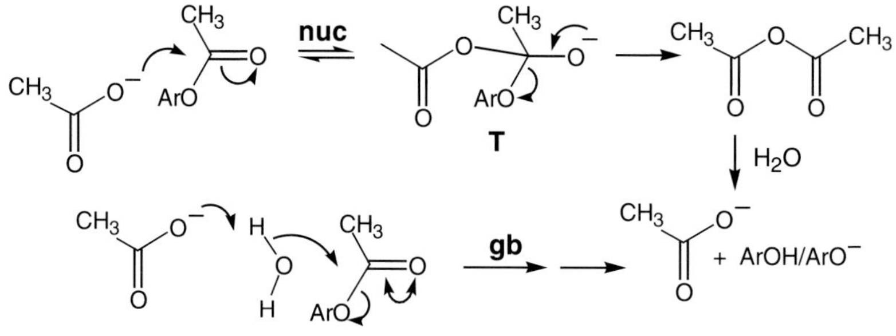

Scheme 2.8

The partitioning of the tetrahedral intermediate **T** of the nucleophilic mechanism is the key: acetate is eliminated preferentially, to regenerate the starting ester, if the leaving group is poor; but the elimination of better aryloxide leaving groups, to generate acetic anhydride (as a second intermediate, which can be trapped) becomes increasingly competitive. Acetate acting as a nucleophile can displace a phenolate of pK_a some 3 units higher: the general base catalysis mechanism delivers hydroxide, effectively irreversibly, but is at an entropic disadvantage. Other things being equal, nucleophilic catalysis wins, and careful experimental design may be necessary to isolate general base catalysis. (A simple example is the work of Butler and Gold [8] on the hydrolysis of acetic anhydride: catalysis by acetate anion can *only* be due to general base catalysis because the nucleophilic mechanism simply regenerates acetic anhydride.)

2.2 Structural Requirements and Mechanism

The central reaction in the general mechanism of Scheme 2.5 involves two proton transfers (1 and 3 in Scheme 2.9), supporting the transfer of the group X to the nucleophile Nu. (Note that in the reverse reaction the original general base becomes the general acid, and vice versa: general acid catalysis is the microscopic reverse of general base catalysis, and establishing a mechanistic pathway for one identifies it also for the other.) In the transition complex **TC** (Scheme 2.9) both protons are involved in hydrogen bonds: typical proton transfers between electronegative centers take place within hydrogen bonds.

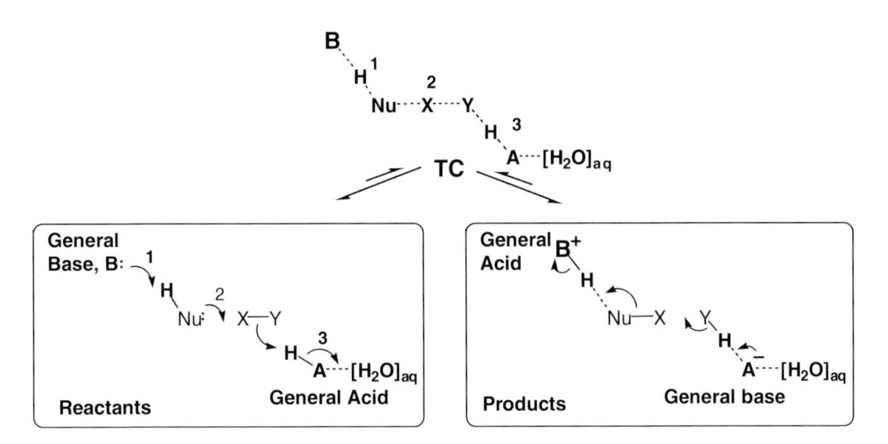

Scheme 2.9

Jencks' "libido rule" [9, 10] attempts to identify situations in which proton transfer can be expected to be concerted with the making or breaking of covalent bonds:

"Concerted general acid–base catalysis of complex reactions in aqueous solution can occur only (a) at sites that undergo a large change in pK_a in the course of the reaction, and (b) when this change in pK converts an unfavorable to a favorable proton transfer with respect to the catalyst; i.e., the pK of the catalyst is intermediate between the initial and final pK_a values of the substrate site." [9].

2.2.1 General Acid Catalysis

These conditions are, in broad terms, necessary but not sufficient. Thus general acids with pK_as of 7 ± 4, of potential interest in biological systems, are well qualified to assist in the cleavage of bonds to oxygen: since the pK_as of ester oxygens are negative but those of their alcohol cleavage products typically >14. For example, general acid catalysis is readily observed in the hydrolysis of orthoesters, and indeed of most systems with three (or more) O, N or S atoms attached to a central carbon atom: including the tetrahedral intermediates involved in the acyl transfer reactions of esters and amides. However, the hydrolysis of acetals is typically specific acid catalyzed, and general acid catalysis is observed only for special cases. The two mechanisms are always in competition, but specific acid catalysis involves at least one, and most often two, intermediates of relatively high energy, the conjugate acid and the oxocarbocation (Scheme 2.10).

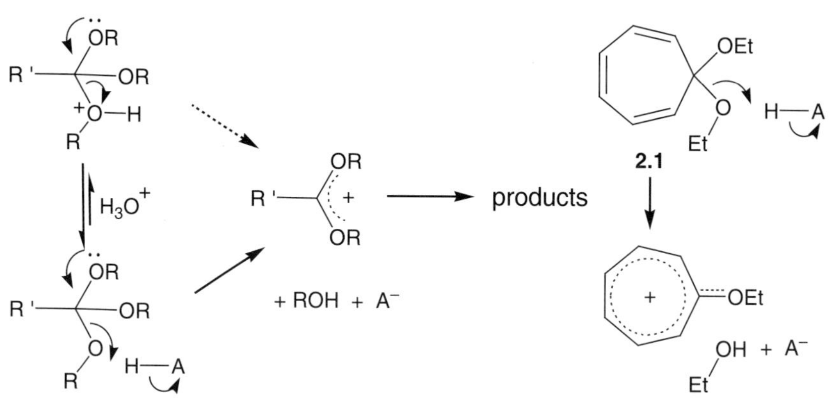

Scheme 2.10

If either intermediate is too unstable general acid catalysis is observed. In the case of orthoesters (Scheme 2.10) the electronic effect of the third OR group stabilizes the oxocarbocation *and* makes the oxygen centers less basic: so that C–O cleavage occurs before proton transfer is complete (Scheme 2.10, HA = H_3O^+ or a general acid) [11, 12].

Aryl but not alkyl tetrahydropyranyl acetals show general acid catalysis, for the same reason [13]; but aryl methyl acetals do not, because the methoxymethyl carbenium ion is not sufficiently stable. (This situation can lead to "enforced" general acid catalysis, when the specific acid catalyzed reaction *requires* nucleophilic assistance: if the nucleophile is the conjugate base of the general acid this will be observed as general acid catalysis.) At the other extreme, sufficient stabilization of the carbenium ion can have the same effect, as shown by the observation of general acid catalysis of tropolone diethyl acetal **2.1** (Scheme 2.10) [14]. And even steric effects can shift the balance, as revealed by the appearance of general acid catalysis in the hydrolysis of benzaldehyde acetals derived from tertiary alcohols [15, 16].

The reverse of reactions of this sort, the general base catalyzed neutralization of carbenium ions by alcohols and water has been studied in some detail by Jencks and Richard [17]. Catalysis is seen only with the more stable substituted 1-phenylethyl carbocations, is most important for the reactions with weakly basic alcohols and disappears when diffusion processes begin to compete. Thus the reaction of CF_3CH_2OH with the 4-dimethylamino compound has a Brönsted β of 0.33, but the low β of 0.08 for the trifluorethanolysis of the 4-methoxy compound is consistent with a transition state with no more than hydrogen-bonding between the general base and the nucleophilic alcohol, which is itself only weakly involved in bonding in the very early transition state for this reaction. This behavior marks the borderline with specific acid catalysis of the hydrolysis reaction of the trifluorethyl ether: and probably also that of the corresponding fluoride [18]. HF has a pK_a in the region of 3, depending on the solvent, so proton transfer to fluoride is not favorable from a general acid with $pK_a > 3$, and none is observed by cyanoacetic acid ($pK_a = 2.2$) [18]. General acid catalysis of the hydrolysis of α-glucosyl fluoride by phosphate and phosphonate monoanions is characterized by a low Brönsted β-value of 0.15 [19], and is presumed to reflect a hydrogen-bonding or solvation-level interaction of the incipient fluoride anion.

2.2.2
Classical General Base Catalysis

General base catalysis is readily observed for the hydrolysis of acyl-activated esters with poor leaving groups, such as ethyl dichloroacetate. It can also be observed in the hydrolysis of typical carboxylate derivatives by using formates. Stefanidis and Jencks studied a series of formate esters, with alcohol leaving groups with pK_as between 12.4 and 16 [20]. In this paper the mechanism is analyzed in detail, in the light of a comprehensive series of structure–reactivity correlations. Solvent deuterium isotope effects of 3.6–5.3 for the water reaction and 2.5–2.8 for the acetate catalyzed process, and Brönsted β-values of 0.36–0.58 for the reaction catalyzed by a series of substituted acetate anions are all consistent with the classical mechanism (steps 1 and 2 of Scheme 2.5). In the hydrolysis of aryl formates both nucleophilic and general base catalysis by acetate are observed, the balance depending on the leaving group. For general base catalysis of the hydrolysis of more reactive esters both the Brönsted β and the solvent deuterium isotope effect fall, as the transition state changes in the direction of hydrogen-bonding catalysis [20].

The same change, from classical general base to nucleophilic catalysis by selected nucleophilic bases, has been observed and studied recently for the hydrolysis of activated amides (an example is 1-benzoyl-3-phenyl-1,2,4-triazole **2.2**, Scheme 2.11) [21], and is observed also for esters of various oxyacids of phosphorus. The changeover to general base catalysis of the hydrolysis of aryl dialkyl phosphates **2.3** parallels that of the corresponding aryl carboxylates [22]. Thus $k_{H2O}/_{D2O}$ for ca-

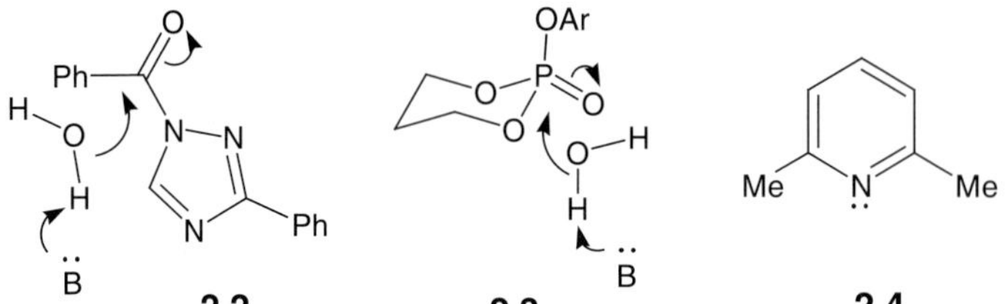

Scheme 2.11

talysis of the hydrolysis of the 2,4-dinitrophenyl triester by 2,6-lutidine **2.4**, for which nucleophilic catalysis is expected to be minimised for steric reasons, is 1.95, consistent with general base catalysis. Catalysis by pyridine, an unhindered nucleophile, is characterized by values of $k_{H2O}/_{D2O}$ ranging from 1.14 to 1.73, as the pK_a of the leaving group is increased, consistent with increasing amounts of general base catalysis: which accounts for some 50% of the reaction for the 4-nitrophenyl ester **2.3**. A recent proton inventory study of the methanolysis of the three triesters $(MeO)_{0\text{-}2}PO(OAr)_{3\text{-}1}$ is consistent with the classical one-proton catalytic bridge model [23]; though the solvent deuterium isotope effects are low (in the region of 1.7) and the distinction from a "generalized solvation effect" is less than clear cut.

It is no coincidence that all the examples described so far involve proton transfer from general acids or bases to or from oxygen. Compared with the OH group a primary or secondary amine is already a strong nucleophile: and thiols RSH, with pK_as in the region of 9, are available as the strongly nucleophilic anions in significant amounts near neutrality. And sulfide sulfur is less basic, so less susceptible to general acid catalysis.

2.2.3
General Base Catalysis of Cyclization Reactions

2.2.3.1 Nucleophilic Substitution

General base catalysis in simple systems is typically a default mechanism, observed in the absence of strong acid or base, or nucleophilic alternatives. It is a relatively inefficient and often slow process, readily observed only with specially designed or activated substrates. The simplest way of increasing reactivity without using "unnatural" activated functional groups is to make reactions intramolecular. Systems where the general base catalysis is itself intramolecular are discussed below, in Section 2.3.5: we discuss here systems where the nucleophilic reaction it supports is intramolecular – that is, a cyclization reaction.

General base catalysis of the S_N2 reaction is not generally observed, for various reasons. Amine nucleophiles do not need it, and hydroxy groups are very weakly nucleophilic towards soft, polarisable centers like sp^3-carbon. The only well-authenticated example of an intermolecular general base catalyzed nucleophilic displacement at sp^3-hybridized carbon is the trifluoroethanolysis of the benzylsulfonium cation **2.5** [24]. This is evidently a very special substrate (the exception that proves the rule?): because the reaction of the corresponding benzyl bromide does not show catalysis, nor do the hydrolysis or ethanolysis of **2.5**. The Brönsted β for catalysis by substituted acetate anions is 0.26, consistent with the classical general base catalysis mechanism shown, but there is no significant solvent deuterium isotope effect, suggesting a mechanism near the minimalist, hydrogen bonding end of the spectrum. (By contrast, general base catalysis, by amines as well as oxyanions, is readily observed for the S_N(Si)-type solvolysis reactions of alkoxy and aryloxysilanes (see Dietze [25] for leading references)).

General base catalysis of S_N2-type reactions of ordinary aliphatic alcohols by oxyanions is observed in the cyclization of 4-chlorobutanol [26], and of the sulfonium cation **2.6** [27] (Scheme 2.12) at 50 °C and 40 °C, respectively. (Amine buffers prefer to demethylate **2.6**.) In all cases (including the reaction of **2.5** discussed above) catalysis by oxyanions shows a low solvent deuterium isotope effect and a Brönsted coefficient β of 0.26 ± 0.1. This reaction may be something of a curiosity, but there is little doubt that it has been properly identified.

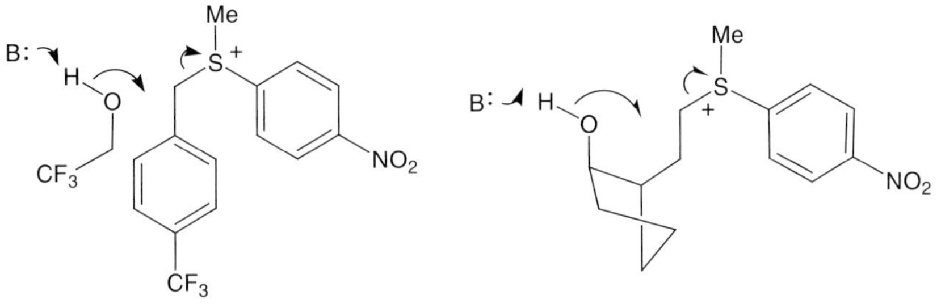

Scheme 2.12

2.2.3.2 Ribonuclease Models

The most interesting, and certainly the best studied general base catalyzed cyclization reaction is the cleavage of RNA and of related model ribonucleotides (Scheme 2.13). Work on this topic designed to shed light on the mechanism of action of

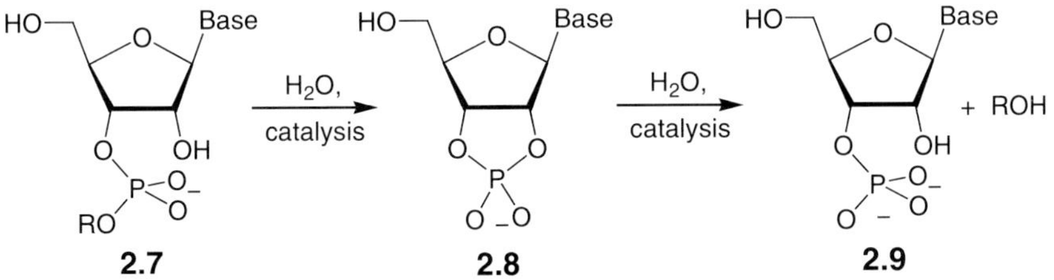

Scheme 2.13

ribonucleases is extensive enough to deserve a review of its own, and several are available [28, 29]. (Of much current interest is the same reaction catalyzed by ribozymes. The mechanisms involved are highly intriguing because the catalysts are themselves RNA molecules, so ill-equipped to support efficient general acid–base catalysis. For recent references to relevant mechanistic work see Kuzmin et al. [30].)

The common nucleophile in ribonuclease enzymes, and thus in relevant models, is the 2′-OH group of the central nucleotide. The work of the Williams group [31] confirmed the mechanism of hydrolysis of uridyl esters (Scheme 2.14, base = U) with good, substituted-phenol leaving groups as a relatively simple process, described by the simple general base catalysis mechanism (Brönsted $\beta = 0.67$), with **2.10** ⇌ **2.11** as the rate determining step (Scheme 2.14), followed by rapid breakdown of the presumed phosphorane (pentacovalent addition) intermediate dianion **2.11** to the reactive cyclic ester **2.8** (Scheme 2.13). Recent evidence for a non-linear Brönsted (leaving group) plot for the alkaline hydrolysis of an extended (to include alcohol leaving groups R = alkyl, base = U) series of the same esters **2.7**, is consistent with a transient intermediate, which rapidly breaks down to reactant and cyclic ester **2.8**. This can only reasonably be the phosphorane **2.11**.

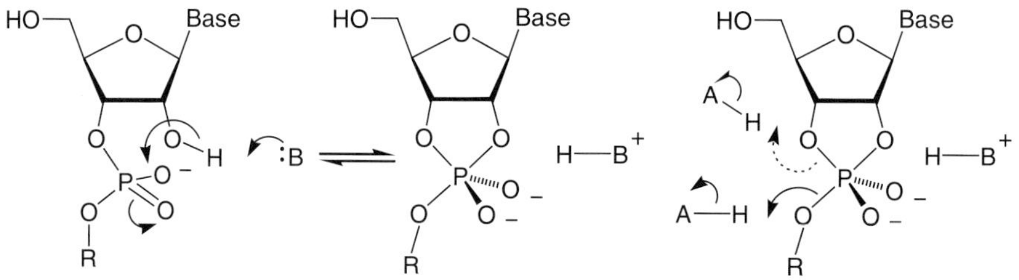

Scheme 2.14

At or near neutral pH, when the leaving group is an alcohol (or a nucleoside or nucleotide) OH, the situation is more complicated. Apart from the protonation state of the initial reactant (relevant because of the extraordinarily low reactivity of phosphodiester anions [32]) the phosphorane **2.11** stands at a mechanistic crossroads (Scheme 2.14). The phosphorane dianion is certainly very short-lived and strongly basic, and can be protonated on any one of the five P–O oxygens. Protonation of O(2′) by BH^+, which certainly starts in the correct position, will regenerate starting material. Alternatively BH^+, or another general acid, could neutralise one of the diastereotopic oxyanions. This opens the way to pseudorotation at the phosphorus center, which makes the 3′-oxygen a potential leaving group: subsequent general acid catalysis of P–O(3′) cleavage (dashed arrow in **2.12**) leads to isomerization to the 2′-phosphodiester [33]. Finally – the route used by ribonuclease enzymes – general acid catalyzed cleavage of the exocyclic P–OR bond gives the cyclic ester **2.8**.

All these processes compete with each other in well-designed model systems, and to establish detailed mechanisms relevant to the situation in natural RNA requires at least oligonucleotide substrates. For example, Beckmann et al. [34] studied the hydrolysis of the bond to the single ribonucleotide in TTUTT (thymidyl-thymidyl-uridyl-thymidyl-thymidine) catalyzed by imidazole buffers, and the Lönnberg group have examined reactivity within longer sequences [35].

2.3 Intramolecular Reactions

There is an enormous gap between the rates of model reactions (which generally have to be studied using activated substrates like *p*-nitrophenyl esters) and those of the same reactions, of natural, unactivated substrates, going on in enzyme active sites. We can go a long way towards bridging this gap by studying intramolecular reactions.

2.3.1 Introduction

Groups held in close proximity on the same molecule can react with each other – depending on the geometry – much faster than the same groups on separate molecules. This is one of the fundamental reasons why enzyme reactions – between groups held in close proximity in the enzyme–substrate complex – can be so fast.

Intramolecular reactions are faster because $\Delta S^\ddagger$ – the entropy of activation (the probability of the reactant groups meeting) – is high: and fastest when the reaction is a cyclisation (corresponding to intramolecular nucleophilic catalysis), which may be particularly favorable enthalpically. The simple measure of efficiency is the effective molarity (EM), the (often hypothetical) concentration of the neighboring group needed to make the corresponding intermolecular process go at the same rate [36]. It is simply measured, as the ratio of the first order rate constant of the intramolecular reaction and the second order rate constant for the (as far as possible identical) intermolecular process. In some convenient cases both reactions can be observed simultaneously, (Scheme 2.15) [37], and EM $= k_1/k_2$ measured di-

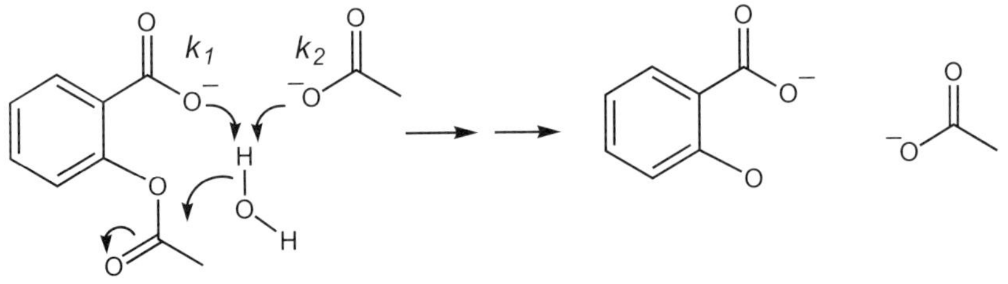

Scheme 2.15

rectly. More often, and always for EM > about 10, the "corresponding" intermolecular reaction is too slow to be measured under the same conditions (or at all), and extrapolations or estimates are needed.

Analysis of the large number of effective molarities available in the literature leads to an important generalisation. For simple cyclisation reactions EMs up to 10^9 M are possible, even in conformationally flexible systems (and can be pushed as high as 10^{13} M by building in ground state strain that is relieved in the transition state, as in the case of the cyclisation of the maleamic acids in Scheme 2.6, above): so the proximity effect could go a long way to explaining the high rates of enzyme reactions (which may involve accelerations of the order of 10^{17-20}). But for intramolecular general acid and general base catalysed reactions, like the aspirin hydrolysis of Scheme 2.15, EMs are typically much lower, usually <10 M [36]. This poses a real problem for attempts to explain the efficiency of enzyme catalysis, since proton transfer is the reaction most often involved in enzyme reactions.

2.3.2 Efficient Intramolecular General Acid–Base Catalysis

A major goal of recent work has been the development of systems showing more efficient intramolecular general acid and general base catalysis. The starting point was the unique known exception to the discouraging generalisation above. Various derivatives of salicylic acid are hydrolyzed with efficient intramolecular general acid catalysis by the COOH group. In the case of acetals, general acid catalysis was first identified as an intramolecular reaction, after it was suggested to contribute to the mechanism of action of lysozyme. Capon [38] showed that the glucoside **3.1** (Scheme 2.16), with the carboxyl group protonated, is hydrolyzed significantly faster than expected for an aryl glycoside; and detailed physical organic studies with the more reactive system **3.2** confirmed the mechanism shown (**3.2**, Scheme

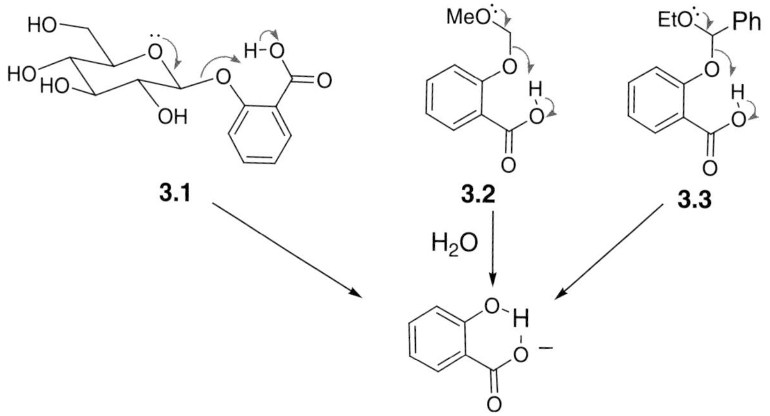

HO HO HO OH O H-O O **3.1**
MeO O H O O **3.2** H_2O
EtO Ph O H O O **3.3**
O H O − O

Scheme 2.16

2.16) in detail [39]. General acid catalysis is easily observed *because* it is highly efficient in salicylic acid derivatives [40]: and Buffet and Lamaty estimated an EM $> 10^4$ M for the reaction of **3.3** [41]. (Possible because there is measurable intermolecular general acid catalysis of the hydrolysis of aryl alkyl acetals of benzaldehyde.)

Detailed studies with several systems derived from salicylic acid suggest that the key to the highly efficient catalysis is the strong intramolecular hydrogen bond in the salicylate anion produced (Scheme 2.16). This is known to raise the pK_a of the phenolic OH group, to 12.95 at 25 °C in water [42], so is worth some 4–5 kcal mol^{-1}: even though the pK_as of the two groups concerned are not closely matched.

New systems designed to test this conclusion confirm the central importance of the intramolecular hydrogen bond. Note that the proton transfer in mechanism **3.2** follows Jencks' libido rule, evolving from strongly unfavorable in the reactant to strongly favorable in the product: so the hydrogen bond could be close to its strongest in the transition state.

Strong intramolecular hydrogen bonds are not common in water because neighboring H-bond acceptor and donor groups are generally solvated separately, but a number of applicable cases are known involving phenol and COOH groups. The most reactive system with this combination of functional groups is the benzisoxazole **3.4** (X = N) (Scheme 2.17). The two acetals **3.4** (X = N and CH) support a closer-to-linear, and therefore stronger, hydrogen bond between carboxyl and leaving group: which are not conjugated with each other – a factor which might have made the salicylate system a special case – because they are in separate rings.

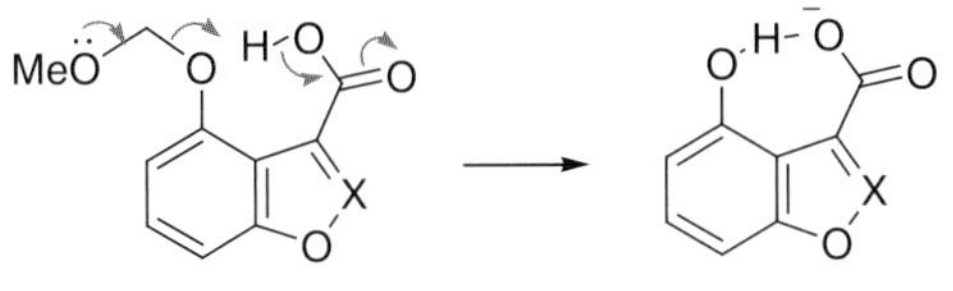

MeO O H O O X O 3.4 → O H O O X O

Scheme 2.17

The benzisoxazole **3.4** (X = N) is hydrolyzed with a half-life of 31 s at 39 °C, compared with 8 min for the salicylate derivative **3.2**. Hydrolysis is faster, at least in part, because of the strength of the general acid (pK_a 1.55 compared with 3.77). The benzofuran **3.4** (X = CH), with pK_a (3.84) close to that of the salicylate derivative **3.2**, has a half-life of 3.3 min. If we assume that the small difference in geometries of the two systems **3.4** is not a factor, this is evidence that the efficiency of catalysis depends on the strength of the general acid. This may seem self-evident, but earlier structure–activity studies on substituted salicylic acid systems [39] indicated that though the rate of hydrolysis depends strongly on the pK_a of the leaving group, it depends not at all on that of the catalytic COOH general acid. The point is

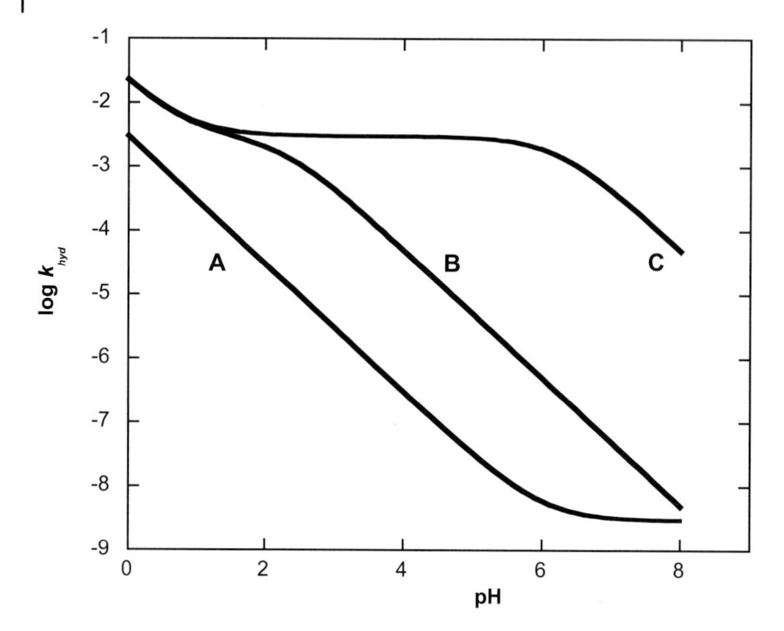

Figure 2.1. pH rate profiles for the intramolecular general acid catalyzed hydrolysis of a typical substrate, for example an acetal (with no other ionizing group). Curve A describes the specific acid catalyzed reaction: at sufficiently high pH reaction becomes pH-independent and a spontaneous, "water-catalyzed" process is observed. Curves B and C show the effect on the rate of adding a catalytic group (for example COOH) ionizing with pK_as of 2.22 and 6.22, respectively, if the efficiency of catalysis is independent of the pK_a.

an important one: if the efficiency of a general acid is independent of its pK_a its rate advantage over specific acid-catalysed hydrolysis, or the spontaneous hydrolysis reaction, increases rapidly with pH – up to its pK_a (Fig. 2.1).

Since systems **3.4** are hydrolyzed faster, by up to an order of magnitude, than similar derivatives of salicylic acid, EMs of up to 10^5 may be estimated. However, it is not generally possible to measure EMs systematically because the necessary "control" – the corresponding intermolecular reaction – is often too slow to be observed above background. The most relevant measure of catalytic efficiency, for comparison with similar reactions in enzyme active sites (where pH is not simply meaningful) is the ratio of the rates of the pH-independent reactions (Fig. 2.1) in the presence and absence of the catalytic group. We use this parameter in the discussion which follows.

Quite different systems, having in common with the salicylate and related systems only a strong intramolecular hydrogen bond, are obtained from 8-dimethylamino-1-naphthol **3.5**, related to proton sponge **3.6** (Scheme 2.18). (The pK_a of the dimethylammonium group of the parent compound **3.5** is normal [43] – as is generally the case for compounds in this series with an exchangeable proton other than the one involved in the intramolecular hydrogen bond.)

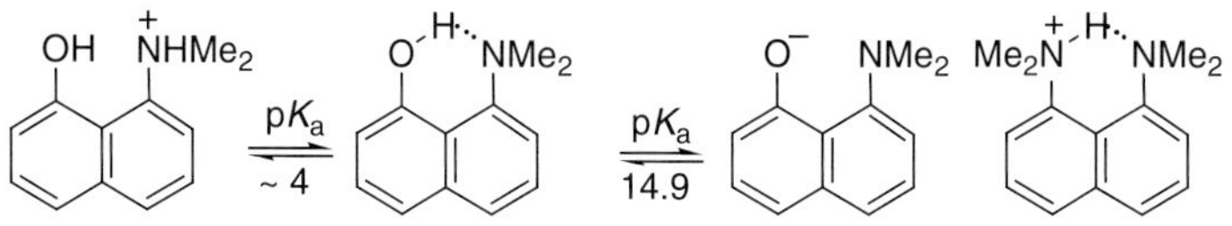

3.5 3.6

Scheme 2.18

The methoxymethyl acetal group of **3.5** is hydrolyzed with efficient catalysis by the neighboring dimethylammonium group (Scheme 2.19). The half-life is 50 min at 65 °C: not obviously fast, but rapid for a methoxymethyl acetal, and corresponding to a rate acceleration of 1900 compared with the expected spontaneous hydrolysis rate for a simple naphthol acetal [44]. Catalytic efficiency is reduced compared with the carboxylic acids discussed above, for two important reasons: (i) The general acid is significantly weaker, by some 3 pK_a units; and (ii) the observed pK_a (7.4) is evidence for a rather strong intramolecular hydrogen bond, which stabilizes the reactant **3.7**. However, an important consequence of this high pK_a is that the spontaneous hydrolysis reaction extends almost to pH 7 (cf. the pH-independent region of curve C of Fig. 2.1).

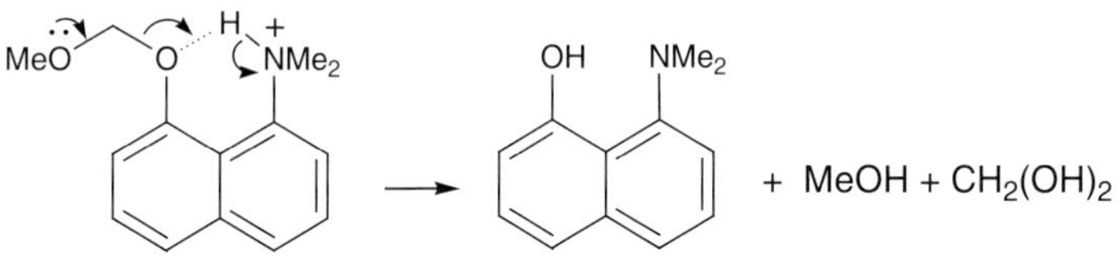

3.7

Scheme 2.19

The acetal cleavage mechanisms sketched out in Schemes 2.15–2.19 have been written – for simplicity – as leading to oxocarbocations. This is likely to be the case for acetals derived from benzaldehyde or tetrahydropyran, but nucleophilic participation by the solvent is undoubtedly involved in the reactions of methoxymethyl acetals, and most likely in those of most glycosides. This aspect is discussed in more detail in Section 2.3.3 below.

2.3.2.1 **Aliphatic Systems**

The systems discussed so far have in common a good, phenolic, leaving group. This is convenient for two reasons: (i) intrinsically unreactive systems react faster with good leaving groups, and (ii) a (changing) aromatic chromophore makes possible continuous monitoring of reactions. An important disadvantage is that natural substrates for known hydrolytic enzymes are almost invariably non-aromatic, with aliphatic alcohols or sugars as typical leaving groups. So model systems derived from alcohols are of special interest. To date they have not been very revealing.

Intramolecular general acid catalysis in aliphatic systems (examples almost always involve the COOH group) is rarely observed, and then rarely convincingly established [40]. Thus the benzaldehyde acetal **3.8** (R = H), expected to be a good candidate, showed none [45], though the enol acetal **3.9** – with similar geometry but a better leaving group (an enol has a pK_a similar to that of a phenol) – is actually slightly more reactive than the salicylic acid derivative **3.2** (Scheme 2.20). (Unpublished work with R. Osborne, cited in Ref. [44]). When the benzaldehyde acetal **3.8** (R = COOH) is deactivated by electron-withdrawing substituents in the aromatic ring some general acid catalysis is observed in 50% aqueous dioxan. A bell-shaped pH–rate profile for hydrolysis between pH 3 and 7 implicates also the second carboxy group in its ionized COO^- form [45].

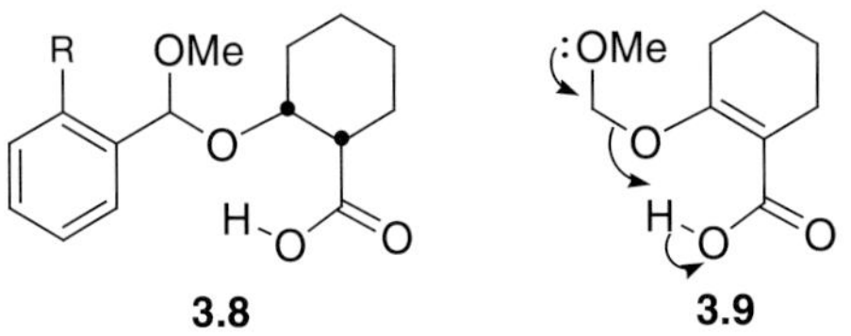

R OMe O H O O **3.8**

:OMe O H O O **3.9**

Scheme 2.20

Brown and Kirby [46] found relatively efficient intramolecular general acid catalysis for the hydrolysis of the three acetals **3.10**, **3.11** and **3.12** (Scheme 2.21). Acetals of benzaldehyde were used to ensure measurable rates of reaction for the much less reactive aliphatic systems. They offered the further advantage that EMs could be estimated, since intermolecular general acid catalysis can be observed for dialkyl acetals of benzaldehyde.

Ph MeO O H O O **3.10**

Ph MeO O H $\overset{+}{N}Me_2$ **3.11**

Ph MeO O H $\overset{+}{N}Me_2$ MeO O O O **3.12**

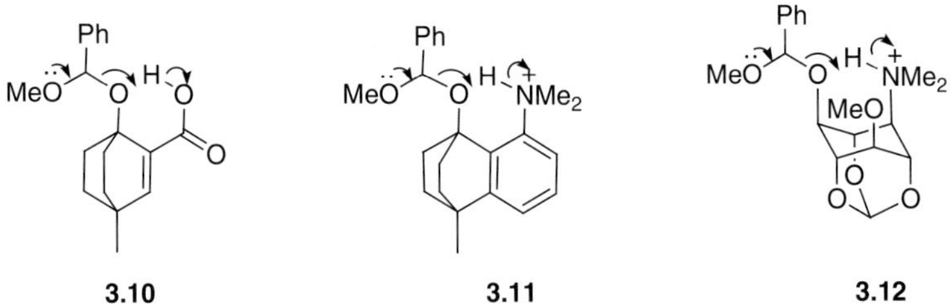

Scheme 2.21

Systems **3.10** and **3.11** are based on the geometries of the salicylic acid and 8-dimethylamino-1-naphthol derivatives, known to show efficient intramolecular general acid catalysis, and all three systems support strong intramolecular hydrogen bonds. Estimated EMs are 3000, 1000 and 10^4 for **3.10**, **3.11** and **3.12**, respectively. Hydrolysis was considered to involve the classical intramolecular general acid catalysis mechanism indicated by the arrows in the scheme: in particular, kinetic solvent deuterium isotope effects k_{H2O}/k_{D2O} of up to 2.2 were observed. System **3.10** based on salicylate has a half-life of <1 s at 25 °C, while **3.12** is some 10^7 times less reactive. General conclusions should not be drawn from such a small sample, but these results suggest strongly that the development of a strong intramolecular hydrogen bond in the transition state supports efficient intramolecular general acid catalysis of the hydrolysis of acetals derived from alcohols as well as from phenols. In this context it is worth noting that acetal **3.11**, which shows the lowest EM, is stabilized (like its aromatic "parent" **3.7**, discussed above) by intramolecular hydrogen bonding in the ground state; as shown by its pK_a of 6.93, (raised by over 2–3 pK_a units compared with a simple naphthylamine).

The most efficient system of this sort – and the only one to show intramolecular general acid catalysis of the hydrolysis of a methoxymethyl acetal derived from an aliphatic alcohol – is compound **3.13**, based on the most reactive aromatic system **3.4**. Compound **3.13** (Scheme 2.22) is hydrolyzed with a half-life of 3.5 h in water at 39 °C, some 10^{10} times faster than expected for the spontaneous hydrolysis of a methoxymethyl acetal of a tertiary alcohol, corresponding to a transition state stabilization of 14 kcal (58 kJ) mol^{-1} [47], and at the rate expected for the methoxymethyl derivative of an alcohol of pK_a about 4. Perhaps significantly, this is almost equal to the pK_a (4.18) of the catalytic COOH group of **3.13**. (Which also confirms that there is no significant intramolecular hydrogen bonding in the reactant.)

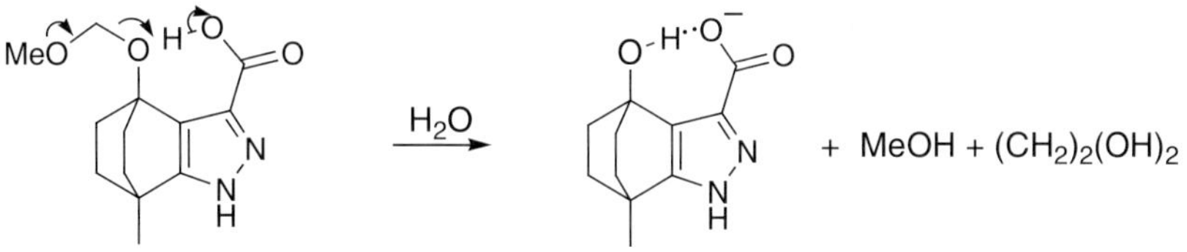

MeO O H O O N N H → H_2O → O H O O N N H + MeOH + $(CH_2)_2(OH)_2$

3.13 **3.14**

Scheme 2.22

2.3.3
Intramolecular General Acid Catalysis of Nucleophilic Catalysis

As discussed above (Section 2.2.2) the hydrolysis reactions of glycosides and methoxymethyl acetals involve nucleophilic participation by water. If water can act as a nucleophile in these systems then reactions with other, better nucleophiles are to be expected. (The model, and the inspiration for most work in this area is the enzyme lysozyme: which – like retaining glycosidases in general – uses two carboxyl groups, one acting as a nucleophile, the other as a general acid, to accomplish (exocyclic) C–O cleavage at the highly unreactive glycosidic center [48].) Much model work, designed to reproduce the two-carboxyl mechanism, has merely underlined how efficient are the enzymes concerned: however, the separate parts of the mechanism can be reproduced fairly convincingly: putting them together remains a challenge.

Nucleophiles attack suitable reactive acetals – with good leaving groups – in well-characterized S_N2 reactions [49, 50]. The methylene center of methoxymethyl acetals is typical, and the reaction easily observed because it is sterically receptive towards nucleophilic attack (**3.15** in Scheme 2.23, Ar = 2,4-dinitrophenyl) [49]. The reaction is highly sensitive to (the pK_a of) the leaving group, but shows very low sensitivity to the nucleophile, with $\beta_{nuc} = 0.1$–0.2. This behavior is consistent with a transition state with weak bonding to both nucleophile and leaving group, and thus a substantial build-up of positive charge at the reaction center. The α-deuterium isotope effect of 1.1–1.2 confirms this picture of an S_N1-like transition structure (**3.16** in Scheme 2.23).

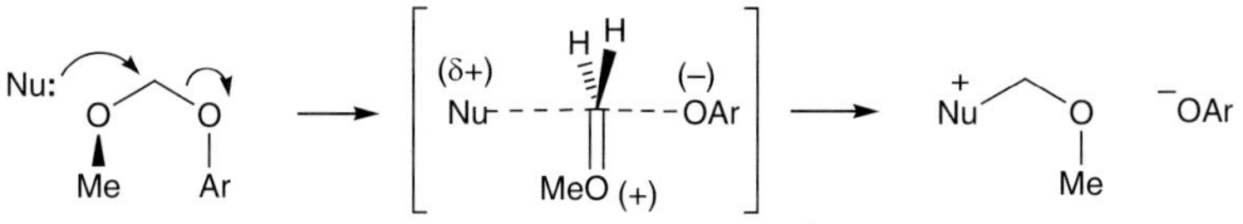

Scheme 2.23

This picture is general for reactions of methoxymethyl acetals and glycosides with any nucleophile, including solvent water. Estimates of the lifetimes of the free oxocarbocations that would be involved in unimolecular processes suggest that they would be too short, or at best borderline, for them to exist in water: and definitely too short in the presence of a better nucleophile than water. (For a succinct review of this topic see Davies et al. [48].)

The good leaving groups of such systems are designed to model the behavior of the partially protonated alcohol or phenol leaving groups of acetals reacting with general acid catalysis. Recent results show that the extrapolation is not a simple one: reactivity depends, unexpectedly, on the charge type of the general acid.

An obvious starting point was to look for general acid catalysis of the attack of nucleophiles on a methyoxymethyl acetal known to be subject to efficient carboxyl-catalyzed hydrolysis. Participation by nucleophiles other than water in the hydrolysis of the salicylic acid derivative **3.17** could not be convincingly distinguished from specific salt effects (the range of nucleophiles is limited by the requirement that the COOH group (pK_a 3.77) be protonated) [49]. On the other hand there is clear involvement of nucleophiles, including carboxylate anions, in the reaction of the dimethylammonium system **3.18** [44] (Scheme 2.24). The difference is presumably simply quantitative.

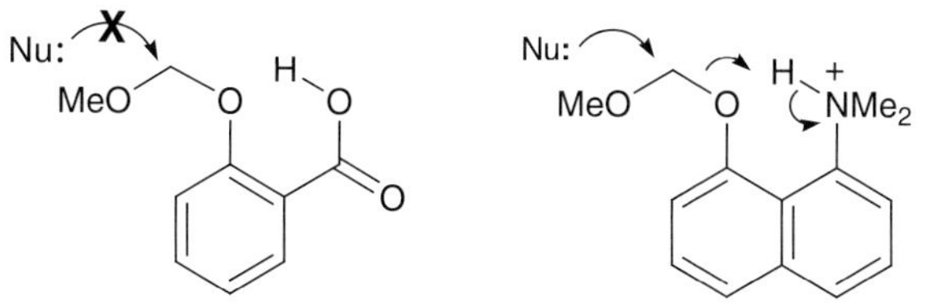

3.17 3.18

Scheme 2.24

An apparently qualitative difference between reactions catalysed by the COOH and Me_2NH^+ groups is observed in the hydrolyses of the phosphate monoesters of salicylic acid and 8-dimethylammonium naphthol. The hydrolysis of the diethyl triester **3.19** (Scheme 2.25) shows efficient intramolecular general acid catalysis of the departure of the naphthol leaving group, assisting, if not concerted with, the attack of nucleophiles. (The reaction is thought to involve the pentacovalent addition intermediate **3.20**: in which case the second step would be rate determining, since the naphthol is a poorer leaving group than the nucleophilic carboxylate.)

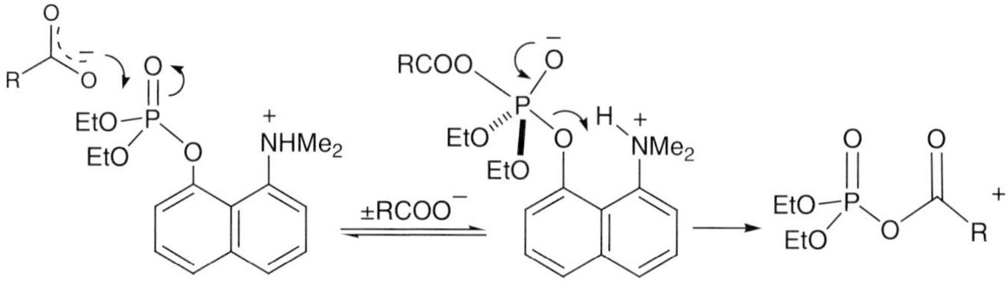

3.19 3.20

Scheme 2.25

On the other hand, phosphorane intermediates are not expected to be involved in the hydrolysis of phosphate monoesters, so the effective observed catalysis by the carboxyl group of salicyl phosphate **3.21** [51] (Scheme 2.26) is presumed to be concerted with nucleophilic attack. (The hydrolysis reaction involves the less abundant tautomer **3.22** of the dianion **3.21**, and the acceleration is $>10^9$-fold relative to the expected rate for the pH-independent hydrolysis of the phosphate monoester dianion of a phenol of pK_a 8.52.) However, this system differs from the methoxymethyl acetals discussed above, in that there is a clear distinction between neutral nucleophiles, which react through an extended transition structure similar to **3.16** in Scheme 2.23, and anions, which do not react at a significant rate, presumably because of electrostatic repulsion. This distinction is well-established for the dianions of phosphate monoesters with good leaving groups (*p*-nitrophenyl [52] and

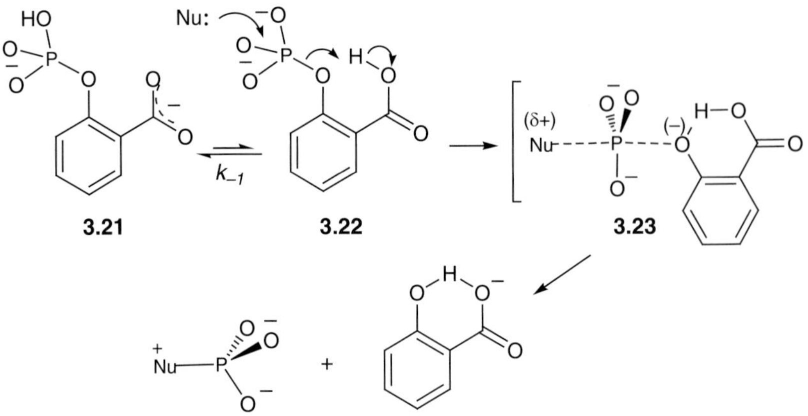

Scheme 2.26

2,4-dinitrophenyl phosphate [53]); and evidently holds also for the reactive tautomer **3.22** of salicyl phosphate (Scheme 2.26).

Structure–activity relationships for the reactions of substituted derivatives of **3.21** are consistent with the extended transition state **3.23** (Nu = H_2O), with the bond to the leaving group largely broken but proton transfer scarcely begun; that is, with the COOH group close to fully protonated, as in the minor tautomer **3.22**. The suggested explanation [51] was that the COOH group is initially rotated out of plane (the observed pK_a of 3.76 refers, of course, to the major tautomer **3.21**, and is consistent with the expected absence of significant intramolecular hydrogen bonding). For the reactive form **3.22** (Scheme 2.26) k_{-1} – a probably intramolecular, thermodynamically favorable proton transfer – is expected to be faster than the rotation of the COOH group into plane, so that the proton transfer step is unlikely to be cleanly rate determining. (There is no significant solvent deuterium isotope effect on the reaction of the dianion.) Water is expected to be (weakly) involved as a nucleophile in the hydrolysis reaction (Scheme 2.26, **3.23**, Nu = H_2O), and reactions with substituted pyridines are catalyzed by the same mechanism (with accelerations of the order of 10^8). A Brönsted $\beta_{nuc} = 0.2$ confirms the weak involvement of the nucleophile.

At first sight the hydrolysis of the phosphate monoester of 8-dimethylammonium naphthyl-1-phosphate **3.24** (Scheme 2.27) looks very similar. However, significant differences have emerged [54]. Most surprising is that in the case of **3.24** there is no discrimination against anionic nucleophiles. Points for oxyanions, fluoride and amine nucleophiles are correlated by the same Brönsted plot, with a slope corresponding to β_{nuc} of 0.2. A second significant difference is that there is a strong intramolecular hydrogen bond in the reactant **3.24**, as indicated by a major shift in the pK_a of the dimethylammonium group to 9.3, compared with an expected, unperturbed value in the region of 4.6 (as observed for the diethyl triester **3.19**).

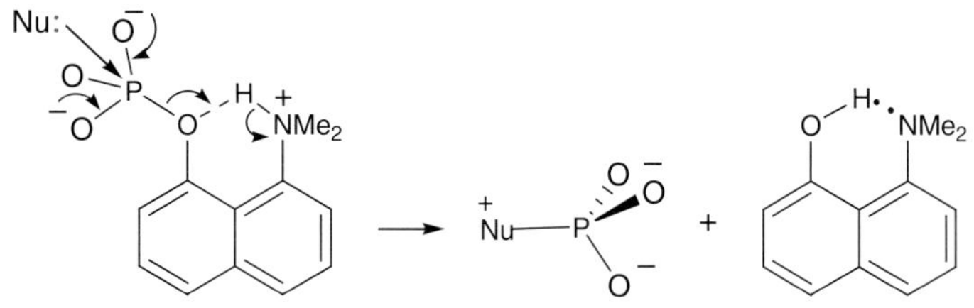

Nu: O⁻ O P O⁻ O H + NMe2 → Nu⁺—P O O⁻ O⁻ + O H NMe2

3.24

Scheme 2.27

It is reasonable to presume that these differences are connected. A strong intramolecular hydrogen bond to the leaving group oxygen will polarize the P–O bond involved. Polarization of this P–O bond will be supported by increased n_O–$\sigma^*_{P\text{–}O}$ interactions (curved arrows in **3.24**: which always play a part in the cleavage of systems X–$PO_3^{=}$). These will reduce the negative charge on, and thus the basicity of, the P–O^- centers. The second pK_a of the phosphate group is reduced accordingly, to 3.94 from the expected value of about 6. It is presumed that this reduction in the negative charge on the three P–O^- centers is sufficient to reduce the electrostatic repulsion to an attacking anionic nucleophile to the levels found for a diester monoanion, or a phosphoramidate, R_3N^+–$PO_3^{=}$, both of which are subject to attack by anionic nucleophiles [54].

Since biological phosphate transfer is often to oxyanions, it is of considerable interest that model reactions are facilitated selectively by cationic (NH^+) general acids: not least because NH^+ centers other than the active general acid are generally present in the enzyme active sites involved. Two model systems illustrate the potential of such effects. In **3.25** [55] (Scheme 2.28) a highly efficient nucleophilic reaction (EM $> 10^{10}$) supports the general acid catalyzed displacement of methoxide as methanol. Primary ammonium cations are well-behaved catalysts, following the Brönsted relationship with $\alpha = -0.33$. Points for neutral, and especially negatively charged, general acids show negative deviations from this line, but diammonium dications show significant positive deviations. The observed rate enhancements depend systematically on the distance between the NH^+ centers, reaching 100-fold for tetramethylethylenediammonium dication. A second stabilizing interaction (circled in **3.27** in Scheme 2.28) at the $PO_2^{=}$ center of the phosphorane (**3.26**, thought to be of borderline stability) is suggested: which could be electrostatic, or involve hydrogen bonding, or both. (An Me_3N^+ group has only a small effect.)

In the second model system **3.28** [56], a proton inventory study indicates that the guanidinium group is involved in hydrolysis as a general acid, rather than simply affording electrostatic stabilization of the phosphorane dianion. The nucleophilic part of the reaction, the cyclization of the 2-hydroxypropyl phosphate, is much less efficient than for **3.25**, and the leaving group is a phenol: so the rate determining step will be the formation of the phosphorane, as shown (**3.28**, Scheme 2.28).

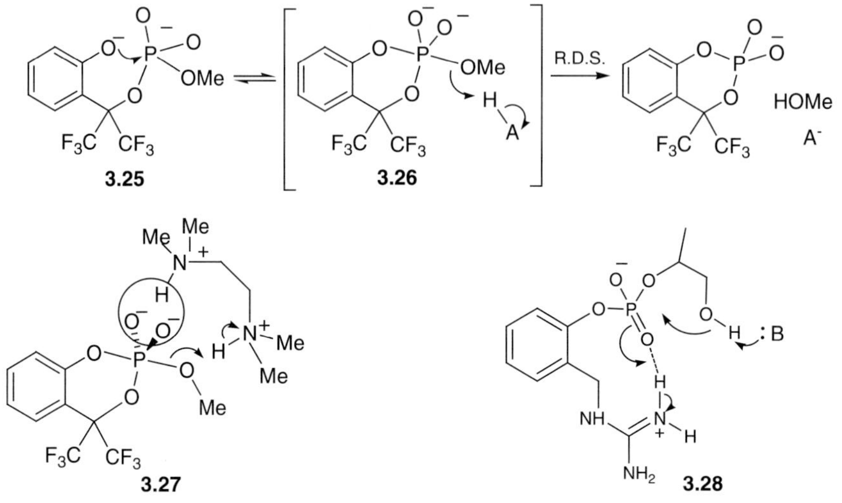

Scheme 2.28

The pK_a of the phosphorane is expected to be higher even than that (ca. 13) of the very weakly acidic guanidinium group, and the proton inventory (in water, pH 10.4 at 50 °C; B = *N*-methyl piperidine) is consistent with two protons in flight in the rate determining step shown. The rate enhancement attributed to assistance by the guanidinium group is 42 [56]: impressive for this flexible system in solvent water, and comparable with the best systems designed to show general acid catalysis of intramolecular nucleophilic catalysis discussed in the following section. (A substantial series of potential intermolecular catalysts for the hydrolysis of RNA based on guanidinium and amidinium groups has been examined by the Göbel group [57, 58].)

2.3.4
Intramolecular General Acid Catalysis of Intramolecular Nucleophilic Catalysis

The logical next step in the construction of intramolecular model reactions based on Scheme 2.5 is to make both nucleophilic and general acid catalysis components intramolecular. This will allow the identification of potential synergy between the two processes when both are optimised. For example, both nucleophilic catalysis and general acid catalysis are generally observed only for acetals with good leaving groups, and both have the effect of turning poorer into better leaving groups. So it is, in principle, possible that an otherwise abortive nucleophilic displacement might be "rescued" by protonation of the under-achieving leaving group by an efficient intramolecular general acid: and vice versa.

Most model systems of this sort have tried to mimic the lysozyme mechanism. Anderson and Fife [59] showed many years ago that benzaldehyde disalicyl acetal **3.29** (Scheme 2.29) is hydrolyzed over 10^9 times faster than its dimethyl ester: but that the contribution of the second, carboxylate group (the potential nucleophile) is

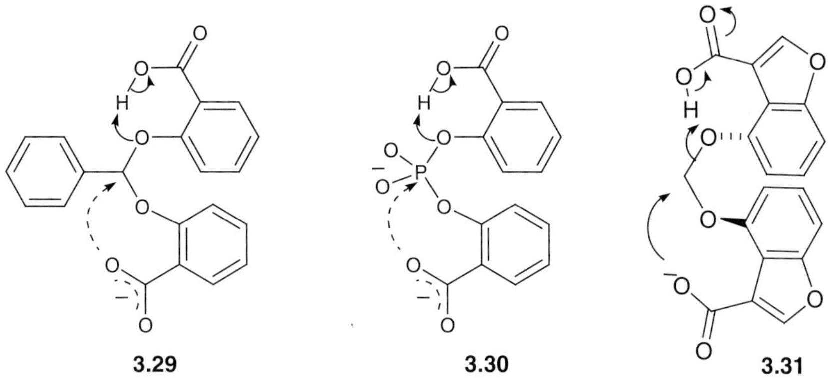

3.29 3.30 3.31

Scheme 2.29

small. The same is true, perhaps for somewhat different reasons, for the rapid hydrolysis of disalicyl phosphate **3.30** (10^{10} times faster than that of diphenyl phosphate): where it is the contribution of the COOH group (the potential general acid) that is small [60].

A less reactive acetal, which nevertheless shows a better balance between the contributions from the two mechanisms, is **3.31**, based on system **3.13** (above, Scheme 2.22), which supports the most efficient intramolecular general acid catalysis (Scheme 2.29). The nucleophilic contribution is less than optimally efficient because it involves the formation of a 7-membered ring, and in this case contributes about 100-fold to the total 10^5-fold acceleration [61]. In all these cases bell-shaped pH–rate profiles are observed, as expected for reactions involving both COOH and CO_2^- groups. But in none of them is there any evidence for synergy between the two mechanisms. (A result consistent, if nothing else, with the broad generalization that catalysis occurs where it is most needed.)

2.3.5 Intramolecular General Base Catalysis

There are no reports of intramolecular general base catalysis of efficiency comparable with those discussed for general acid catalysis in Sections 2.3.2 and 2.3.3 above. Such levels of efficiency must, in principle, be attainable, but there are genuine practical problems. Most significant is the dominance of intramolecular nucleophilic reactions when a general base is brought close to a neighboring electrophilic center. For example, the hydrolysis of aspirin (Scheme 2.15) shows intramolecular general base catalysis only because the much (ca. 10^6-fold) faster nucleophilic mechanism is thermodynamically prohibited [37]. Thus structure–activity studies are only possible in systems, such as the malonate half-esters **3.32** (Scheme 2.30) where the nucleophilic mechanism is specifically disabled (in this case because the formation of a 4-membered ring containing two sp^2-hybridized carbons would be involved). In this system **3.32** the angle of approach α could be varied over a broad range by varying the substituents R [62]. EMs remained <100 over the whole

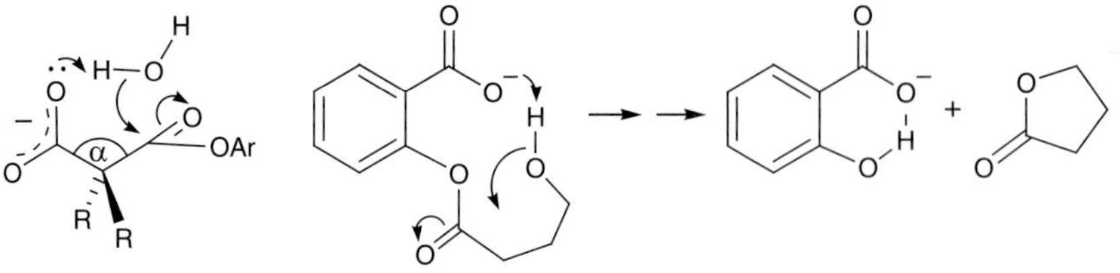

3.32 3.33

Scheme 2.30

series. Also, where intramolecular general base catalysis was coupled to a cyclization reaction, as in **3.33** (Scheme 2.30) the rate enhancements were precisely additive [63].

The most interesting result in this context is a reaction showing "Unusually Rapid Intramolecular General-Base Catalysis" of the reversible cleavage of a C–H bond [64]. The details are discussed below in Section 2.4.2.

2.4
Proton Transfers to and from Carbon

When Y in the general mechanism of Scheme 2.5 is a carbon center, proton transfer from a general acid generates a C–H bond without the possibility of initial hydrogen bond formation. The new bonding interaction involving the proton is to an X–Y bonding orbital (rather than to a lone pair on Y): a process that requires either a strong acid or – for a (relatively weak) general acid – a high energy bonding orbital. Typically this is the π-orbital of an enol or enolate (or an enamine). However, it can be a σ-bonding orbital in special cases. An example is the general acid catalyzed ring-opening of the 1-phenylcyclopropanolate anion **4.1**: here catalysis by tertiary ammonium cations follows a Brönsted correlation, with $\alpha = 0.25$, and the primary deuterium isotope effect k_H/k_D is 1.9: consistent with an early transition state with proton transfer not far advanced, as would be expected for a reaction involving a developing primary carbanion [65]. (The inverted stereochemistry observed in related reactions suggests that the proton is transferred to the tail of the σ_{C-C} bonding orbital, as shown (**4.1**) in Scheme 2.31)

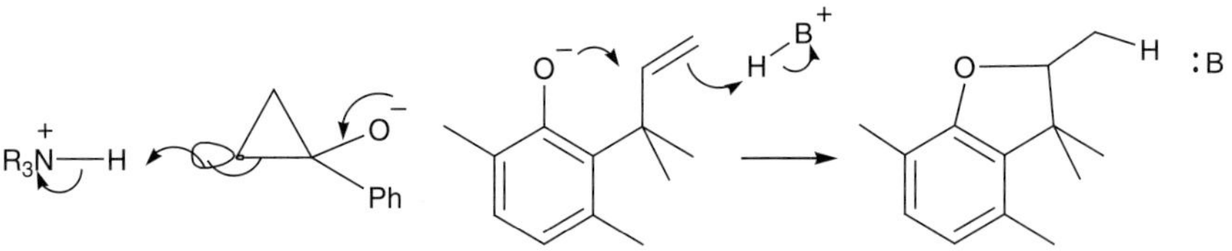

4.1 4.2

Scheme 2.31

Closely related (from the point of view of the general acid) are a series of cyclization reactions in which a carbanion is partially or completely generated in water by the addition of a nucleophile to an alkene or alkyne. The extraordinarily rapid cyclization of the phenolate **4.2** (half-life 50 s at 39 °C, corresponding to an EM of ca. 10^{11}) is described as a preassociation-concerted process (Scheme 2.31) [66]. Catalysis by alkylammonium cations follows the Brönsted equation with $\alpha = 0.06$, and $k_{H2O}/k_{D2O} = 1.6$, consistent again with the late intervention of the general acid before the incipient primary carbanion reverts to reactant.

A (vinyl) carbanion is a full intermediate in the cyclization of the alkyne **4.3** (Scheme 2.32, R = H, half-life 40 s at 39 °C). Cyclization is rate determining and the reaction is not buffer catalyzed) [67]. (The 10^4-fold greater electrophilic reactivity of the alkyne means that the steric acceleration provided by the ortho-methyl groups in **4.2** is not needed in **4.3**.) The solvent deuterium isotope effect k_{H2O}/k_{D2O} on the proton transfer step (derived from the product isotope effect) is 1.6. However, general acid catalysis is observed for the (1375 times slower) cyclization of the C–methyl compound **4.3** (R = CH_3, Scheme 2.32): evidently methylation destabilizes the vinyl carbanion to the point where it no longer has a significant lifetime in water, and the preassociation-concerted mechanism is enforced once more [67].

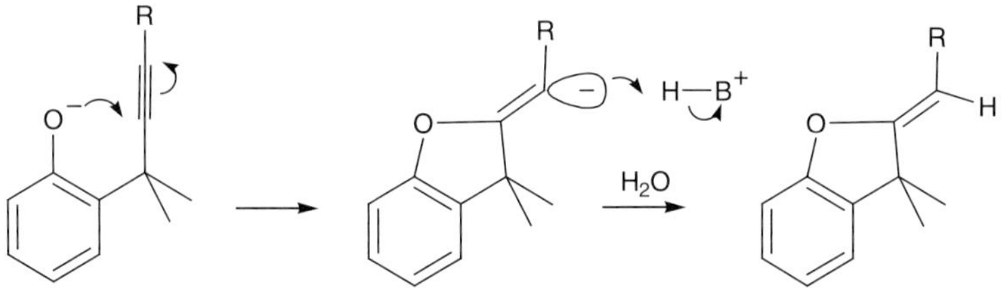

Scheme 2.32

More familiar is proton transfer to the π_{C-C} bonding orbital of an enol or enolate. As discussed in Section 2.1, above, the same rate determining step (proton transfer to the general base, Scheme 2.7) accounts for both general base and general acid catalysed ketonisation reactions of enols. The recent work of the Kresge and Richard groups has extended our knowledge of these reactions to the highly reactive enols and enolates obtained from carboxylic acid derivatives such as esters and amides, and even carboxylate anions. The same mechanisms persist in these extreme cases: thus the ketonization of the enol **4.4** derived from mandelamide is specific base-general acid catalyzed by acetic acid, phosphate and TRIS, and general acid catalyzed by the ammonium ion, all by the mechanism shown (Scheme 2.33): a good Brönsted correlation gives $\alpha = 0.33$ [68]. But eventually proton transfer to such reactive enolates becomes so fast that encounter, or reorganization of solvent water, becomes rate determining [69].

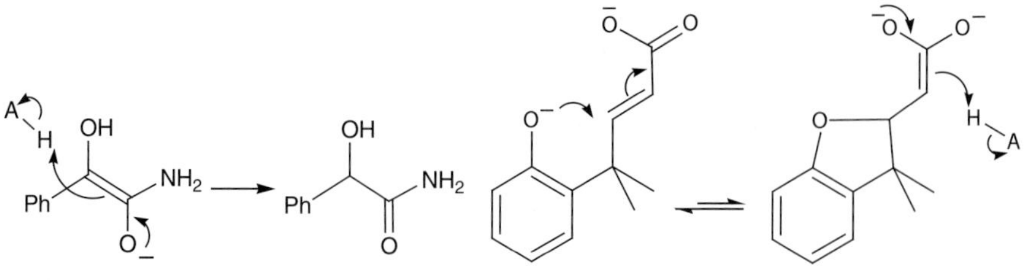

Scheme 2.33

Not surprisingly the dienolate dianion from a carboxylic acid can also be generated by a cyclization reaction of the sort described in Scheme 2.31. Intramolecular Michael addition of the phenolate **4.5** (half-life 60 s at 39 °C) [70] gives the dianion **4.6** as a full intermediate, now stable enough with respect to reversion to reactant that its protonation is rate determining, catalysis by protonated amines following a Brönsted correlation with $\alpha = 0.2$.

The proton transfer step **4.6** behaves remarkably differently depending on the type of general acid involved. Protonation by solvent water gives stereospecific *anti* addition, and shows a solvent isotope effect k_{H2O}/k_{D2O} of 5.1. But catalysis by the quinuclidinium cation gives *syn* addition, and $k_{H2O}/k_{D2O} = 1.34 \pm 0.05$. These results are not fully understood, but suggest a shorter lifetime for the dianion in the presence of the cationic general acid. This is one more example of an increasing number of systems showing apparent electrostatic contributions to general acid catalysis.

2.4.1
Intramolecular General Acid Catalysis

Since the general acid catalyzed enolization of carbonyl compounds actually involves the specific acid–general base catalysis mechanism (Section 2.1.3, above), this would appear to be the wrong reaction to use in a search for intramolecular general acid catalysis. The ketonization of enols, on the other hand, is an example of a well-known range of reactions in which a general acid protonates a reactive π-system. More convenient substrates are the stable enol ethers [71], and several systems have been shown to be subject to general acid catalysis by neighbouring carboxyl groups [34, 72–74]. Of these two show interestingly high catalytic efficiency. Compound **4.7** is modelled on the salicylic acid system, and designed so that the proton transferred from the general acid is involved, as the new C–H bond develops, in an intramolecular hydrogen bond with the carboxylate anion. The geometry is closely similar to that of the salicylate monoanion, known to form a strong intramolecular hydrogen bond (Scheme 2.16, above): and the H–C–C=O^+Me group is certainly strongly acidic (estimated pK_a between −2 and −5) so the libido rule is obeyed. The estimated effective molarities for the two geometrical isomers of **4.7** are about 300 for **4.7E** and 2200 for **4.7Z**. (Catalysis as efficient as this swamps any intermolecular catalysis by buffer general acids, so the comparison is with general acid catalysis of the hydrolysis of the two methyl esters.) Deuterium isotope effects k_{H2O}/k_{D2O} are 3.0 for the reactions of both isomers [73].

Catalysis by the dimethylammonium group of the hydrolysis of the two enol ethers **4.10E** and **4.10Z** (Scheme 2.35) is more efficient still: hydrolysis is faster, with half-lives of the order of 10 s at 39 °C even though the general acids ($pK_a = 4.0$) are slightly weaker. The estimated EM for general acid catalysis by the dimethylammonium group of the reaction of **4.10E** is >60 000 [74]. The mechanism is similar to that of Scheme 2.34, though there are evident differences of detail. In particular, the solvent deuterium isotope effects of 1.86 and 1.76 observed for the reactions of **4.10E** and **4.10Z**, respectively, are the lowest observed for enol ether hydrolysis, and significantly lower than those for compounds **4.7** described above (Scheme 2.34).

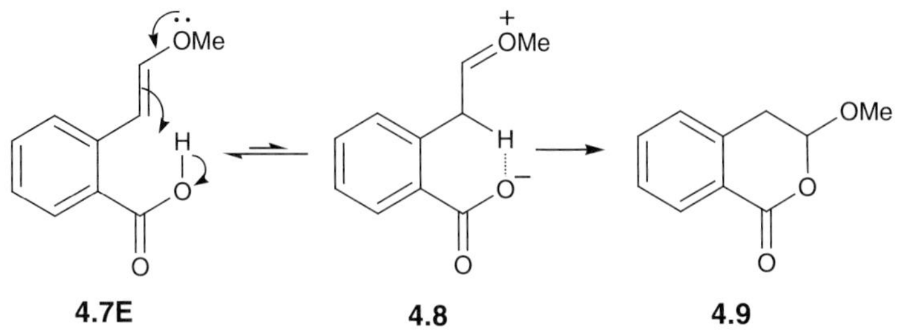

4.7E **4.8** **4.9**

Scheme 2.34

4.10E (⇌ 4.10Z) **4.11** **4.12**

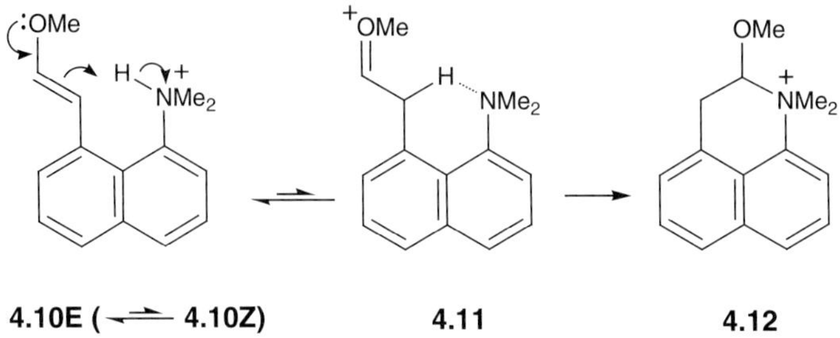

Scheme 2.35

It is suggested that this reflects a mechanism in which the proton transfer (**4.10**) is not exclusively rate determining. If a single step is cleanly rate determining for the overall reaction it is likely to be the one (**4.11**) in which the intramolecular hydrogen bond opens. Equilibrium constants for the opening of strong intramolecular N–H···N and O–H···N hydrogen bonds in similar situations are of the order of 10^{-5}, with correspondingly low rate constants [75, 76]. The rate of the thermodynamically favorable proton jump by which **4.11** reverts to enol ether seems certain to be faster. The effective molarity of the general acid, already much higher than other known values for intramolecular general acid catalyzed reactions, is thus likely to be higher than the initial estimate of 60 000 [74].

2.4.2
Intramolecular General Base Catalysis

As discussed in Section 2.3.5 above, there are no reports of efficient intramolecular general base catalysis of efficiency comparable with those discussed above for general acid catalysis in Sections 2.3.2 and 2.3.3. This generalization applies also to reactions involving proton transfer to and from carbon. For example, early work by Bell and his coworkers [77, 78] found only inefficient catalysis (EMs < 10 [36]) of the enolisation of acetophenone derivatives with ortho COOH or OH substituents. A more detailed study of similar systems by the Anslyn group [79] confirms that the ortho OH group in **4.13**, a potential general acid well placed to support a relatively strong intramolecular hydrogen bond, is responsible for only small (10–100-fold) rate accelerations (Scheme 2.36, B = imidazole). (Note that the adjacent phenol and enol of **4.14** will have similar pK_as.) The system is set up for an early proton transfer, and the authors suggest non-perfect synchronization as an explanation for the weak intramolecular catalysis. They also make the interesting suggestion that proton transfer transfer to the π_{C-O} bonding orbital (a geometry not seen in hydrogen bonding interactions in typical crystal structures [80]) may be a more efficient way to activate the carbonyl group.

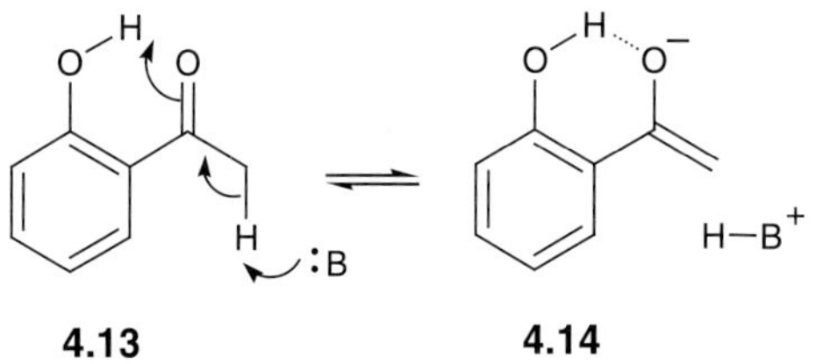

4.13 **4.14**

Scheme 2.36

In the light of the successful identification of efficient intramolecular general acid catalysis in the 8-dimethylaminonaphthol system (Scheme 2.35), this system would appear to offer the best chance of observing efficient intramolecular general base catalysis, by setting up the microscopic reverse process involving the same functional groups in the same geometry. The chosen system, conveniently also a plausible model for the mandelate racemase reaction (in which amine general bases remove a C–H proton from the position alpha to the mandelate carboxyl group [81]), were esters **4.15** (Scheme 2.37). The enolate **4.16**, and thus the transition state leading to it, reproduces in detail the geometry of the enol ether **4.17**,

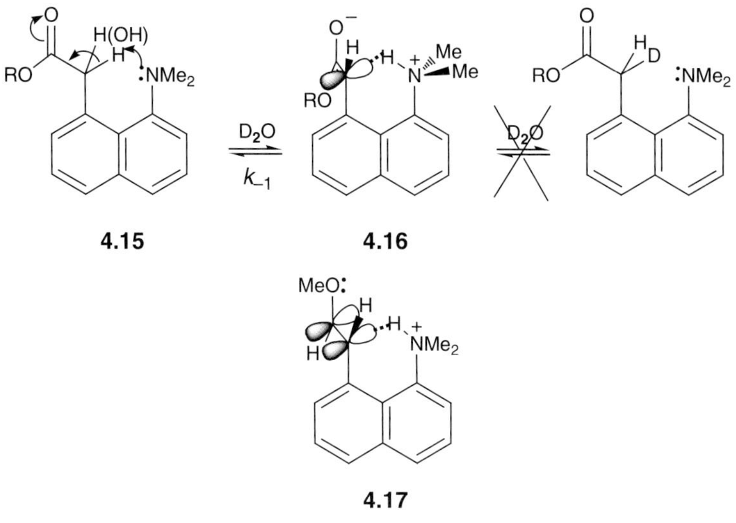

Scheme 2.37

shown to support very efficient proton transfer to the π_{C-C} bonding orbital (see Scheme 2.35, above).

When esters **4.15** were incubated as the free bases in D_2O, standard conditions for detecting small amounts of enolization by the exchange of solvent deuterium into the α-positions of esters [82], no trace of exchange was observed, even after many days under vigorous conditions [83]. A working explanation is that the enolate intermediate **4.16**, specifically designed to support a strong intramolecular hydrogen bond between the NH^+ group and the high energy π_{C-C} orbital of the enolate, reverts to reactant ester (k_{-1} in Scheme 2.37) much faster than the cyclic H-bond opens – a necessary preliminary to exchange with solvent deuterium. Proton transfers within such intramolecular hydrogen bonds have been shown in related systems to be much faster than ring opening [84], and it has already been suggested that ring-opening is at least partially rate determining in the efficient general acid catalyzed hydrolysis of the enol ether **4.10** (Scheme 2.35, above), a proton transfer with the same geometrical requirements (**4.17**).

In this situation the report mentioned above (Section 2.3.5) of the reversible cleavage of a C–H bond **4.18** (Scheme 2.38) showing "unusually rapid" intramolec-

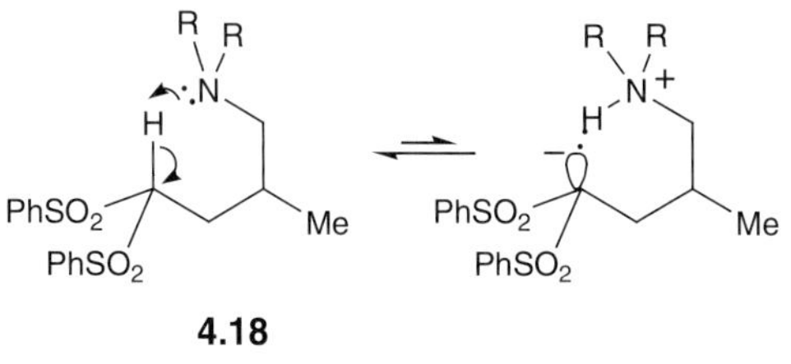

Scheme 2.38

ular general base catalysis [64] is of particular interest. The authors conclude that intramolecular site-exchange of the proton proceeds at least 10^4–10^5 times faster than the intermolecular reaction, so has an EM comparable with the best observed values for intramolecular general acid catalysis. Considerable extrapolations are involved, and the rates were measured by NMR line broadening, in toluene-d_8 or $CDCl_3$ (so the conditions differ from those for all the reactions discussed above): and the work has not been followed up. (Rapid deuterium exchange should be observed, for example, in the ND_2 compound.) But this reaction suggests, as a minimum, an alternative starting point for work on efficiency in such systems.

2.4.3
Simple Enzyme Models

The model systems discussed so far in this chapter have been designed and studied specifically to investigate mechanisms of bond making and breaking. Any serious enzyme model must also incorporate a substrate binding step, and much effort has gone into this area in recent years. Not surprisingly, in the light of the evidence from small molecule systems discussed in Section 2.3, above, efficient general acid–base catalysis has not been observed in systems designed to bind specific substrates, and the synthetic effort involved in adding properly positioned catalytic functionality to such systems is a serious deterrent [85].

One model reaction that has proved a popular and appropriate probe for general base catalysis by potential binders of hydrophobic substrates is the Kemp elimination, the simple, single transition state ring-opening elimination of benzisoxazoles **4.19** (Scheme 2.39).

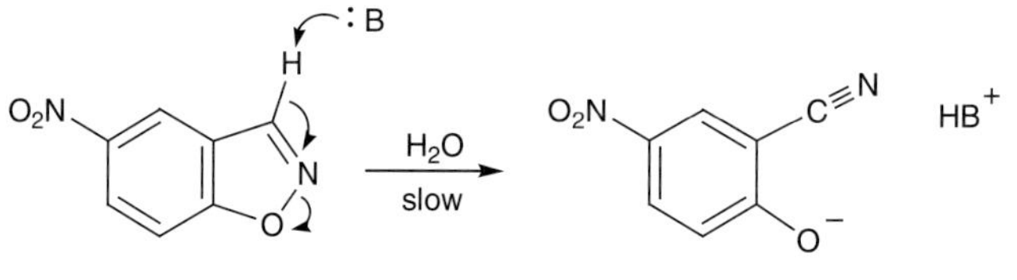

:B
H
O_2N
N
O
4.19
H_2O
slow
O_2N
C≡N
O^-
HB^+

Scheme 2.39

The reaction is a well-characterized E2-type process involving the general base catalyzed removal of the 3-proton (k_H/k_D 4–6, depending on the ring substituents) [86], which is extraordinarily sensitive to the medium when the general base is an acetate anion; though much less so for catalysis by amine general bases. It is catalyzed apparently efficiently by catalytic antibodies raised against carefully designed haptens [87], but it is not a simple matter to distinguish medium effects from the contributions of general base catalysis by binding site carboxylate groups. The finding that serum albumens, non-specific binders of hydrophobic substrates, catalyze the reaction almost as efficiently using local lysine amino groups as general bases

[88], suggests that the observed accelerations result from a combination of medium effects and functional group catalysis; and that general base catalysis is efficient but not extraordinarily so [89]. The Kemp elimination has also been shown to be catalyzed by polyethylenimine modified to bind hydrophobic substrates [90], by micelles and vesicles [91], and by various more or less specific binders with general bases attached [91–93]. The general conclusion remains the same.

2.5
Hydrogen Bonding, Mechanism and Reactivity

Recent thinking on general acid–base catalysis has been dominated by discussions of the contribution of "short, strong" or "low-barrier" hydrogen bonds, and – for proton transfers to and from carbon – the possible contribution of tunneling. Tunneling is discussed in detail by other contributors to this Handbook, but decisive results with model systems have been conspicuous by their absence. The hydrogen bond, on the other hand, is of central importance to any proton transfer reaction, and all but ubiquitous in discussions of the topic. It figures, explicitly or implicitly, in all the mechanistic schemes above, and has been suggested to make key contributions to the high efficiency of enzyme catalysis.

The experimental evidence is clear, up to a point. Proton transfers between electronegative centers involve the transient formation of hydrogen bonds, along which the proton is transferred. The same mechanism applies to proton transfer to carbon, but only in special, overlapping, cases: if the carbanion concerned lives long enough for solvent equilibration, and if the C–H bond is strongly acidic. Weakly acidic C–H, and probably also π-bonding orbitals, are not expected to make significant hydrogen bonds to solvent or to general bases in solution in protic solvents (though examples are well established in the solid state [94], and are of potential importance in non-polar solvents and enzyme active sites). Strong hydrogen bonding in protic solvents is observed only between groups held or brought together in close and appropriate proximity: it is generally thermodynamically preferable for strong hydrogen bonding groups to be solvated separately.

General acid–base catalysis is typically inefficient, compared with nucleophilic catalysis, and this is particularly well documented for intramolecular reactions, as discussed above (Section 2.3.1). The reasons for this disparity have been discussed in terms of broad generalizations, citing most often the "looseness," and thus relatively low entropy, of the transition state for a general acid–base catalyzed reaction compared with a cyclization process in which ring formation is complete apart from one partial covalent bond. (Compare, for example, the observed general base catalyzed hydrolysis **5.1** and the abortive but 10^6 times faster nucleophilic reaction **5.2** (Scheme 2.40) reactions of aspirin (Section 2.3.5, above) [36].) However, it is not immediately obvious why, for example, the transition state for intramolecular general base catalyzed enolization **5.3** (EM 56 M [36]) should be significantly looser than **5.2**. Though it seems clear that the presence of the proton in the cyclic transition structure is the key. (The entropy of activation is not: no third molecule is in-

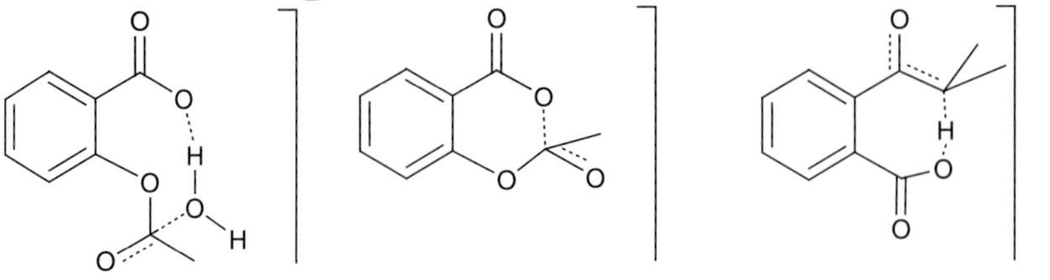

5.1 5.2 5.3

Scheme 2.40

volved in **5.3**, and favorable $T\Delta S^{\ddagger}$ values are typically far too small to explain the high effective molarities of nucleophilic reactions.)

Pascal [95] has argued that, because proton transfer goes by way of a hydrogen-bond, this already makes an intermolecular reaction effectively intramolecular: thus explaining the generally low efficiency of intramolecular general acid–base catalysis. This would be true if – and only if – the H-bonded complex were the major species in solution: it is not. A polar substrate in water is hydrogen-bonded predominantly to the solvent not to the general acid or base, so that the state of the reactant – on which free energy estimates must be based – retains the degrees of freedom which are lost in the transition state and the complex leading to it. (A corollary is that intramolecular proton transfers involving C–H bonds should not suffer from this disadvantage, but there is no evidence from the new generation of systems showing efficient general acid catalysis (Section 2.3.2, above) that proton transfer to carbon is specifically favored.)

The low efficiency of intramolecular general acid–base catalysis, compared with intramolecular nucleophilic catalysis, is general but not universal. The exceptions, systems showing general acid catalysis with EMs > 1000, have in common a strong intramolecular hydrogen bond. So the simple (though not especially informative!) generalisation is that the efficiency of catalysis depends on the free energy of formation of the cyclic transition state. For nucleophilic catalysis this is known to mirror the equilibrium free energy change for the overall cyclization reaction [36]; so the effect on efficiency must be primarily thermodynamic rather than kinetic.

A specifically kinetic effect may derive from the nature of the transition state. For a cyclization reaction the transition state is a ring with one partial covalent bond (for example, **5.2**). This may be formed almost completely, as in some lactonisations; moderately, as in S_N2-type processes; or to a very minor extent, as perhaps in the hard-to-detect cyclisation reactions of acetals (see **3.29**, above). Observed EMs for cyclization reactions fall in this order [36]. For a general base catalyzed reaction, by contrast, there is no formal equilibrium free energy of cyclization, because the product is not a ring: rather – at best – a cycle interrupted by an intramolecular hydrogen bond, set up for the reverse, intramolecular general acid catalyzed reaction (**5.5**, Scheme 2.41). In this situation it is clear that stronger

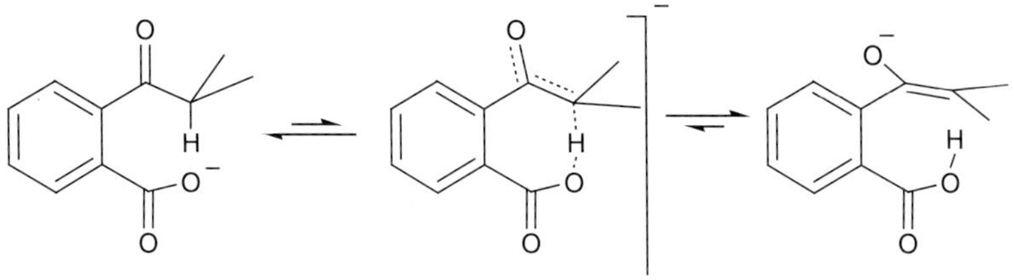

5.4 5.3 5.5

Scheme 2.41

hydrogen-bonding can make ring formation more thermodynamically favorable, and thus stronger in the transition state than in either reactant or product, for example because the pK_as of the groups concerned can be better matched.

The systems described above (Section 2.3.2) showing highly efficient intramolecular general acid catalysis are more or less rigid, and movement of the heavy atoms directly concerned in the proton transfer process minimal. (The Principle of Least Motion may apply [96].) Schowen et al. [97] have argued convincingly that the proton "in general catalysis bridges between oxygen, nitrogen and sulfur" will be in a stable potential, with the bond order conserved at unity, and will shift towards a residue that becomes more basic: that is, in the direction indicated by the curved arrows used in the schemes. In this picture the reaction coordinate is defined exclusively by the motions of the heavy atoms involved, uncoupled from the bridging proton, which is in a stable potential at the transition state (for example, **3.7** in Scheme 2.42). This would explain why the deuterium kinetic isotope effects for these reactions are particularly low. For the highly efficient general acid catalyzed reactions of Schemes 2.16–2.19 (Section 2.3.2, above) $k_H/k_D = 1.5 \pm 0.2$. This compares with higher – but still low – values in the region of 2–3 for more typical general acid–base catalyzed reactions.

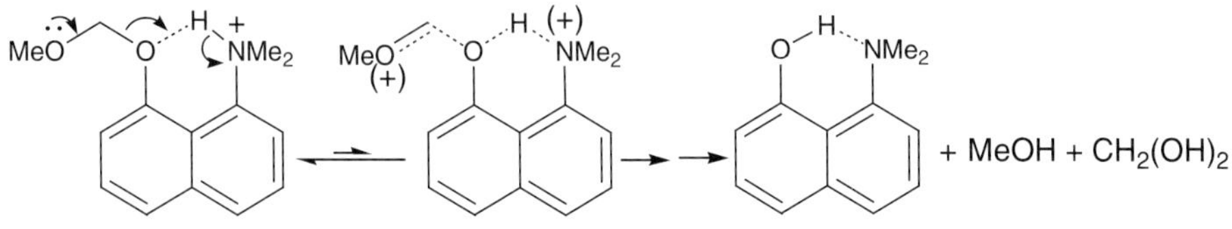

MeO O H + NMe2 MeO (+) O H (+) NMe2 O H NMe2 + MeOH + $CH_2(OH)_2$

3.7

Scheme 2.42

There has been intensive recent discussion, summarized elsewhere in this Handbook, about the potential contribution of strong hydrogen bonding to the catalytic efficiency of enzymes. A key question is conveniently summarized by

Chen et al. [98] "Is it possible that some hydrogen bonds … in the active sites of enzymes have energies in the range 10–20 kcal mol^{-1}?" (Guthrie suggested that the maximum reasonable value is 10 kcal mol^{-1} [99].) The accumulating evidence from model systems suggests that the answer is positive, at least for transition state hydrogen bonds. (Note that hydrogen bonds in a product complex offer no rate advantage, and in a reactant complex are disadvantageous.) The rate enhancement from intramolecular general acid catalysis by the carboxyl group of **3.13** (Scheme 2.22), a system designed for optimal hydrogen bonding in the product, is worth 14 kcal mol^{-1}, in water at 39 °C [47]. In this system there is little or no intramolecular hydrogen bonding in the reactant. Systems like **3.7** (Sections 2.3.2 and 2.3.3) are less efficient in part because significant intramolecular hydrogen bonding stabilizes the reactant when the leaving group oxygen is sufficiently basic.

All these rather subtle requirements are a challenge to the ingenuity of the designer of simple model systems: and a reminder of the extraordinary flexibility that is built in to enzymes, many of which have evolved to catalyze so efficiently complex and intrinsically slow reactions which may have many separate steps, with different geometrical requirements, in a single active site.

References

1. H. Maskill, *The Physical Basis Of Organic Chemistry*, Oxford University Press, Oxford, New York, 1985.
2. C. W. Wharton, in *Comprehensive Biological Catalysis*, Sinnott, M. L. (Ed.), Academic Press, London and San Diego, 1998, pp. 345–379.
3. M. F. Aldersley, A. J. Kirby, P. W. Lancaster, R. S. McDonald, C. R. Smith, *J. Chem. Soc., Perkin Trans 2* (1974) 1487–1495.
4. A. S. Baynham, F. Hibbert, M. A. Malana, *J. Chem. Soc., Perkin Trans. 2* (1993) 1711–1715.
5. R. A. Bednar, W. P. Jencks, *J. Am. Chem. Soc.* 107 (1985) 7117–7126.
6. M. W. Washabaugh, J. T. Stivers, K. A. Hickey, *J. Am. Chem. Soc.* 116 (1994) 7094–7097.
7. C. A. Coleman, C. J. Murray, *J. Am. Chem. Soc.* 113 (1991) 1677–1684.
8. A. R. Butler, V. Gold, *J. Chem. Soc.* (1961) 2305–2310.
9. W. P. Jencks, *J. Am. Chem. Soc.* 94 (1972) 4731–4732.
10. W. P. Jencks, *Chem. Rev.* 72 (1972) 705–718.
11. E. H. Cordes, H. G. Bull, *Chem. Rev.* 74 (1974) 581–603.
12. E. H. Cordes, H. G. Bull, in *Transition States of Biolochemical Processes*, R. D. Gandour, R. L. Schowen (Eds.), 1978, pp. 429–465.
13. T. H. Fife, L. H. Brod, *J. Am. Chem. Soc.* 92 (1970) 1681–1684.
14. E. Anderson, T. H. Fife, *J. Am. Chem. Soc.* 91 (1969) 7163–7166.
15. A. T. N. Belarmino, S. Froehner, D. Zanette, J. P. S. Farah, C. A. Bunton, L. S. Romsted, *J. Org. Chem.* 68 (2003) 706–717.
16. J. L. Jensen, L. R. Herold, P. A. Lenz, S. Trusty, V. Sergi, K. Bell, P. Rogers, *J. Am. Chem. Soc.* 101 (1979) 4672–4677.
17. J. P. Richard, W. P. Jencks, *J. Am. Chem. Soc.* 106 (1984) 1396–1401.
18. M. M. Toteva, J. P. Richard, *J. Am. Chem. Soc.* 124 (2002) 9798–9805.
19. N. S. Banait, W. P. Jencks, *J. Am. Chem. Soc.* 113 (1991) 7958–7963.
20. D. Stefanidis, W. P. Jencks, *J. Am. Chem. Soc.* 115 (1993) 6045–6050.
21. N. J. Buurma, M. J. Blandamer, J. Engberts, *J. Phys. Org. Chem.* 16 (2003) 438–449.
22. S. A. Khan, A. J. Kirby, *J. Chem. Soc. B* (1970) 1172–1182.

23 C. D. Bryan, K. B. Schowen, R. L. Schowen, *Can. J. Chem.* 74 (1996) 931–938.
24 P. E. Dietze, W. P. Jencks, *J. Am. Chem. Soc.* 111 (1989) 340–344.
25 P. E. Dietze, Y. Y. Xu, *J. Org. Chem.* 59 (1994) 5010–5016.
26 T. H. Cromartie, C. G. Swain, *J. Am. Chem. Soc.* 97 (1975) 232–233.
27 J. O. Knipe, J. A. Coward, *J. Am. Chem. Soc.* 101 (1979) 4339–4348.
28 D. M. Perreault, E. V. Anslyn, *Angew. Chem., Intl. Ed. Engl.* 36 (1997) 432–450.
29 M. Oivanen, S. Kuusela, H. Lonnberg, *Chem. Rev.* 98 (1998) 961–990.
30 Y. I. Kuzmin, C. P. Da Costa, M. J. Fedor, *J. Mol. Biol.* 340 (2004) 233–251.
31 A. M. Davis, A. D. Hall, A. Williams, *J. Am. Chem. Soc.* 110 (1988) 5105–5108.
32 N. H. Williams, *Biochim. Biophys. Acta* 1697 (2004) 279–287.
33 E. Maki, M. Oivanen, P. Poijarvi, H. Lonnberg, *J. Chem. Soc., Perkin Trans. 2* (1999) 2493–2499.
34 C. Beckmann, A. J. Kirby, S. Kuusela, D. C. Tickle, *J. Chem. Soc., Perkin Trans. 2* (1998) 573–581.
35 S. Mikkola, M. Kosonen, H. Lonnberg, *Curr. Org. Chem.* 6 (2002) 523–538.
36 A. J. Kirby, *Adv. Phys. Org. Chem.* 17 (1980) 183–278.
37 A. R. Fersht, A. J. Kirby, *J. Am. Chem. Soc.* 89 (1967) 4857–4863.
38 B. Capon, M. C. Smith, E. Anderson, R. H. Dahm, G. H. Sankey, *J. Chem. Soc., B* (1969) 1038–1047.
39 G.-A. Craze and A. J. Kirby, *J. Chem. Soc., Perkin Trans. 2* (1974) 61–66.
40 A. J. Kirby, A. R. Fersht, in *Advances in Bioorganic Mechanisms, Vol. 1*, Kaiser, E. T., Kezdy, F. J. (Eds.), Wiley, New York 1971, pp. 1–82.
41 C. Buffet, G. Lamaty, *Rec. Trav. Chim.* 95 (1976) 1–30.
42 J. Hermans, S. J. Leach, H. A. Scheraga, *J. Am. Chem. Soc.* 85 (1963) 1390–1395.
43 F. H. Hibbert, personal communication.
44 A. J. Kirby, J. M. Percy, *J. Chem. Soc., Perkin Trans. 2* (1989) 907–912.
45 T. H. Fife, T. J. Przystas, *J. Am. Chem. Soc.* 101 (1979) 1202–1210.
46 C. J. Brown, A. J. Kirby, *J. Chem. Soc., Perkin Trans. 2* (1997) 1081–1093.
47 E. Hartwell, D. R. W. Hodgson, A. J. Kirby, *J. Am. Chem. Soc.* 122 (2000) 9326–9327.
48 G. Davies, M. L. Sinnott, S. G. Withers, in *Comprehensive Biological Catalysis*, Sinnott, M. L. (Ed.), Academic Press, London and San Diego, 1998, pp. 119–207.
49 G.-A. Craze, A. J. Kirby, R. Osborne, *J. Chem. Soc., Perkin Trans. 2* (1978) 357–368.
50 B. L. Knier, W. P. Jencks, *J. Am. Chem. Soc.* 102 (1980) 6789–6798.
51 R. H. Bromilow, A. J. Kirby, *J. Chem. Soc., B* (1972) 149.
52 A. J. Kirby, W. P. Jencks, *J. Am. Chem. Soc.* 87 (1965) 3209.
53 A. J. Kirby, A. G. Varvoglis, *J. Chem. Soc., B* (1968) 135.
54 A. J. Kirby, M. F. Lima, D. da Silva, F. Nome, *J. Am. Chem. Soc.* 126 (2004) 1350–1351.
55 K. N. Dalby, A. J. Kirby, F. Hollfelder, *J. Chem. Soc., Perkin Trans. 2* (1993) 1269.
56 A. M. Piatek, M. Gray, E. V. Anslyn, *J. Am. Chem. Soc.* 126 (2004) 9878–9879.
57 M.-S. Muche, M. W. Göbel, *Angew. Chem., Intl. Ed. Engl.* 35 (1996) 2126–2129.
58 S. Pitsch, S. Scheffer, M. Hey, A. Strick, M. W. Göbel, *Helv. Chim Acta* 86 (2003) 3740–3752.
59 E. Anderson, T. H. Fife, *J. Am. Chem. Soc.* 95 (1973) 6437–6441.
60 K. W. Y. Abell, A. J. Kirby, *J. Chem. Soc., Perkin Trans. 2* (1983) 1171–1174.
61 K. E. S. Dean, A. J. Kirby, *J. Chem. Soc., Perkin Trans. 2* (2002) 428–432.
62 A. J. Kirby, G. J. Lloyd, *J. Chem. Soc., Perkin Trans. 2* (1976) 1753–1761.
63 A. J. Kirby, G. J. Lloyd, *J. Chem. Soc., Perkin Trans. 2* (1974) 637–642.
64 F. M. Menger, K. Gabrielson, *J. Am. Chem. Soc.* 114 (1992) 3574–3575.

65 A. Thibblin, W. P. Jencks, *J. Am. Chem. Soc.* 101 (1979) 4963–4973.
66 C. M. Evans, A. J. Kirby, *J. Chem. Soc., Perkin Trans. 2.* (1984) 1259–1268.
67 C. M. Evans, A. J. Kirby, *J. Chem. Soc., Perkin Trans. 2* (1984) 1269–1275.
68 Y. Chiang, H. X. Guo, A. J. Kresge, J. P. Richard, K. Toth, *J. Am. Chem. Soc.* 125 (2003) 187–194.
69 J. P. Richard, G. Williams, A. C. O'Donoghue, T. L. Amyes, *J. Am. Chem. Soc.* 124 (2002) 2957–2968.
70 T. L. Amyes, A. J. Kirby, *J. Am. Chem. Soc.* 110 (1988) 6505–6514.
71 A. J. Kresge, *Accts. Chem. Res.* 20 (1987) 364–370.
72 A. J. Kirby, N. H. Williams, *J. Chem. Soc., Chem. Commun.* (1991) 1644–5.
73 A. J. Kirby, N. H. Williams, *J. Chem. Soc., Perkin Trans. 2* (1994) 643–648.
74 A. J. Kirby, F. O'Carroll, *J. Chem. Soc., Perkin Trans. 2* (1994) 649–655.
75 F. Hibbert, *J. Chem. Soc., Perkin Trans. 2* (1974) 1862–1866.
76 A. Awwal, F. Hibbert, *J. Chem. Soc., Perkin Trans. 2* (1977) 152–156.
77 R. P. Bell, D. W. Earls, J. B. Henshall, *J. Chem. Soc., Perkin Trans. 2* (1976) 39–44.
78 R. P. Bell and D. W. Earls, *J. Chem. Soc., Perkin Trans. 2* (1976) 45–46.
79 Z. L. Zhong, T. S. Snowden, M. D. Best, E. V. Anslyn, *J. Am. Chem. Soc.* 126 (2004) 3488–3495.
80 A. J. Kirby, *The Anomeric Effect and Related Stereoelectronic Effects at Oxygen*. Springer-Verlag, Berlin and Heidelberg, 1983, pp. 43–44.
81 S. L. Bearne, R. Wolfenden, *Biochemistry* 36 (1997) 1646–1656.
82 T. L. Amyes, J. P. Richard, *J. Am. Chem. Soc.* 118 (1996) 3129–3141.
83 F. Hollfelder, Thesis, Cambridge, 1997.
84 F. Hibbert, *Acc. Chem. Res.* 17 (1984) 115–120.
85 A. J. Kirby, *Angew. Chem., Intl. Ed. Engl.* **35** (1996) 707–724.
86 M. L. Casey, D. S. Kemp, K. G. Paul, D. D. Cox, *J. Org. Chem.* 38 (1973) 2294–2301.
87 S. N. Thorn, R. G. Daniels, M. T. M. Auditor, D. Hilvert, *Nature (London)* 373 (1995) 228–230.
88 F. Hollfelder, A. J. Kirby, D. S. Tawfik, *Nature (London)* 383 (1996) 60–63.
89 F. Hollfelder, A. J. Kirby, D. S. Tawfik, K. Kikuchi, D. Hilvert, *J. Am. Chem. Soc.* 124 (2000) 1022–1029.
90 F. Hollfelder, A. J. Kirby, D. S. Tawfik, *J. Am. Chem. Soc.* 119 (1997) 9578–9579.
91 J. E. Klijn, J. Engberts, *Org. Biomol. Chem.* (2004) 1789–1799.
92 X. C. Liu, K. Mosbach, *Macromol. Rapid Commun.* 19 (1998) 671–674.
93 P. G. McCracken, C. G. Ferguson, D. Vizitiu, C. S. Walkinshaw, Y. Wang, G. R. J. Thatcher, *J. Chem. Soc., Perkin Trans. 2* (1999) 911–912.
94 T. Steiner, *Angew. Chem., Intl. Ed. Engl.* 41 (2002) 48–76.
95 R. Pascal, *J. Phys. Org. Chem.* 15 (2002) 566–569.
96 M. L. Sinnott, *Adv. Phys. Org. Chem.* 24 (1988) 113–204.
97 K. B. Schowen, H. H. Limbach, G. S. Denisov, R. L. Schowen, *Biochim. Biophys. Acta-Bioenergetics* 1458 (2000) 43–62.
98 J. G. Chen, M. A. McAllister, J. K. Lee, K. N. Houk, *J. Org. Chem.* 63 (1998) 4611–4619.
99 J. P. Guthrie, *Chem. Biol.* 3 (1996) 163–170.

3
Hydrogen Atom Transfer in Model Reactions

Christian Schöneich

3.1
Introduction

Hydrogen transfer reactions are of fundamental importance in synthetic [1], environmental [2] and biological [3] processes. This chapter focuses on the experimental and theoretical treatment of model reactions designed to understand the mechanistic details of and the parameters controlling hydrogen transfer processes. Specific emphasis is placed on hydrogen transfer reactions of oxygen-, nitrogen- and sulfur-centered radicals relevant to oxidation mechanisms of amino acids, peptides and proteins. After much progress in deciphering the human genome, it was realized that insights into pathologic processes would require the analysis of the protein complement, the "proteome". Such analysis must include post-translational protein modifications, and a thorough knowledge of the nature, abundance and location of oxidative post-translational modifications, such as are prevalent in many disease states and biological aging, requires a mechanistic and structural understanding of the reactions leading to the accumulation of specific oxidation products. The hydrogen transfer reactions described in this chapter comprise a subset of the reactions leading to the ultimate formation of oxidized proteins *in vivo*.

3.2
Oxygen-centered Radicals

Oxygen-centered radicals represent the most abundant class of radicals in biological systems. Several recent reviews have dealt with the reactions of oxygen-centered amino acid, peptide and protein radicals [4, 5]. Therefore, we will only give a brief review of the reactions of common oxygen-centered radicals, and especially those of amino acids and peptides, before focusing on some selected mechanistic aspects of biologically relevant hydrogen transfer reactions of specific oxygen-centered radicals.

The smallest, but most reactive, oxygen-centered radical is the hydroxyl radical

Hydrogen-Transfer Reactions. Edited by J. T. Hynes, J. P. Klinman, H.-H. Limbach, and R. L. Schowen

ISBN: 978-3-527-30777-7

($HO^•$). The $HO^•$ radical reacts efficiently with most amino acid and peptide C–H bonds [6], where $k_1 \approx 2 \times 10^7$–$2 \times 10^9$ $M^{-1}s^{-1}$ for free amino amino acids and small peptides at pH ca. 7 [7].

$$HO^• + P–H \rightarrow H_2O + P^• \tag{3.1}$$

The product carbon-centered radicals add molecular oxygen in a diffusion-controlled process ($k_2 \approx 2 \times 10^9$ $M^{-1}s^{-1}$ [8]). Pulse radiolysis studies on the reactions of peroxyl radicals from cyclic model dipeptides (diketopiperazines) have demonstrated a base-catalyzed elimination of superoxide [8]. In proteins, the resultant peroxyl radicals can abstract additional hydrogen atoms (Reaction (3.3)), initiating chain reactions, which ultimately lead to the accumulation of protein-bound hydroperoxides [9–11].

$$P^• + O_2 \rightarrow P–O–O^• \tag{3.2}$$

$$P–O–O^• + P'–H \rightarrow P–O–O–H + P'^• \tag{3.3}$$

Evidence for these pathways *in vivo*, and the initial involvement of $HO^•$ radicals, has been presented [12]. An important feature of these protein-bound hydroperoxides is their sensitivity to reductive cleavage by transition metals, generating highly reactive alkoxyl radicals (Reaction (3.4)) [13].

$$P–O–O–H + M^{n+} \rightarrow P–O^• + HO^- + M^{(n+1)+} \tag{3.4}$$

Recent theoretical studies have dealt with the potential structures of amino acid and protein peroxyl radicals and hydroperoxides at the $^{\alpha}C$ position, i.e. $A^{\alpha}C–O–O^•$, $A^{\alpha}C–O–O–H$, $P^{\alpha}C–O–O^•$ and $P^{\alpha}C–O–O–H$, respectively [14]. The amino acid hydroperoxides are stabilized through hydrogen bonding to the acyl oxygen, whereas the protein hydroperoxides show a hydrogen bond to the acyl oxygen of the $i-1$ residue. No hydrogen bonding is apparent for the amino acid peroxyl radicals, but for the protein peroxyl radicals hydrogen bonding to the amide group of the $i+1$ residue appears likely.

The amino acid or peptide alkoxyl radicals formed in Reaction (3.4) eventually undergo an α–β fragmentation [15, 16], a 1,2-H-shift (*vide infra*), or react via hydrogen- or electron transfer [17]. The latter processes will ultimately generate hydroxy amino acids, which have been observed *in vivo* as a consequence of oxidative stress [12]. However, not all endogenous antioxidants will react efficiently with these alkoxyl radicals. For example, pulse radiolysis experiments in aqueous solution revealed a surprisingly low reactivity of model *tert*-butoxyl radicals ($^{tert}BuO^•$) with glutathione (GSH), where $k < 4 \times 10^7$ M^{-1} s^{-1}, while hydrogen abstraction from other antioxidants proceeded ca. two orders of magnitude faster, i.e., $k = 1.6 \times 10^9$ M^{-1} s^{-1} and 1.1×10^9 M^{-1} s^{-1} for the reaction of $^{tert}BuO^•$ with ascorbate and the water-soluble vitamin E analog trolox C, respectively [17]. Even the reaction of $^{tert}BuO^•$ radicals with polyunsaturated fatty acids occurred faster than that with GSH, where $k = 1.3, 1.6$ and $1.8 \times 10^8\ \mathrm{M^{-1}\ s^{-1}}$ for linoleate, linolenate and arachidonate, respectively [17]. For the α–β fragmentation of amino acid and protein $^{\alpha}$C-alkoxyl radicals, theoretical data [14] predict a clear preference for scission of the $^{\alpha}$C–CO bond, based on the free energies of activation. Such fragmentation has been observed experimentally in aqueous solution [15]. However, experiments with model alkoxyl radicals have shown that the nature of the solvent has a significant influence on both the α–β fragmentation [18] and the 1,2-H-shift [19]. Especially for proteins, that means that alkoxyl radicals could react differently depending on whether they are located in hydrophilic or hydrophobic protein regions. Recent studies on the 1,2-H-shift in benzyloxyl radicals suggest that this reaction is catalyzed by nucleophilic solvents, which contain hydroxy groups, such as water (Reaction (3.5)) and alcohols [19]. The mechanism of catalysis was suggested to involve two hydrogen-bonds, one between the benzylic hydrogen and the solvent oxygen and one between the exchangeable solvent proton and the alkoxyl radical oxygen.

$$\mathrm{PhCH_2O^{\bullet} + H_2O \rightarrow PhC^{\bullet}HOH + H_2O} \tag{3.5}$$

Pulse radiolysis studies on the 1,2-H-shift in ethyloxyl radicals in water indicate a kinetic isotope effect of $k_H/k_D \approx 50$ (for the intramolecular reaction of $\mathrm{CH_3CH_2O^{\bullet}}$ vs. $\mathrm{CD_3CD_2O^{\bullet}}$), indicating the potential participation of tunneling [20]. The latter observation suggests that the mechanism of the 1,2-H-shift is actually more complex than anticipated.

Of particular interest in biology are redox processes of the amino acid tyrosine

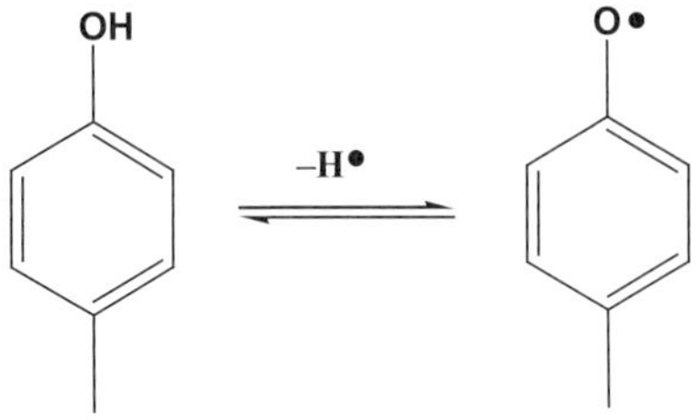

OH –H• O• (3.6)

(Equilibrium (3.6)), where hydrogen transfer (or proton-coupled electron transfer, PCET) generates an aromatic alkoxyl (tyrosyl) radical. In a series of recent publications ([21] and references therein), Ingold and coworkers have pointed out the importance of hydrogen bonding for the rate constants of such hydrogen transfer reactions. This is summarized in Scheme 3.1, where hydrogen transfer to an attacking radical $\mathrm{Y^{\bullet}}$ requires dissociation of any hydrogen bond between the hydrogen donor XH and the solvent S.

A single empirical equation (I) was developed [21], which allows the prediction of rate constants k^S for hydrogen transfer between the hydrogen donor XH and any radical $\mathrm{Y^{\bullet}}$ in any solvent S based on the following parameters: (i) the rate constant k^0 of hydrogen transfer from XH in a reference solvent (a saturated hydrocarbon with no hydrogen bond acceptor properties), (ii) a parameter $\alpha_2{}^{H}$, which

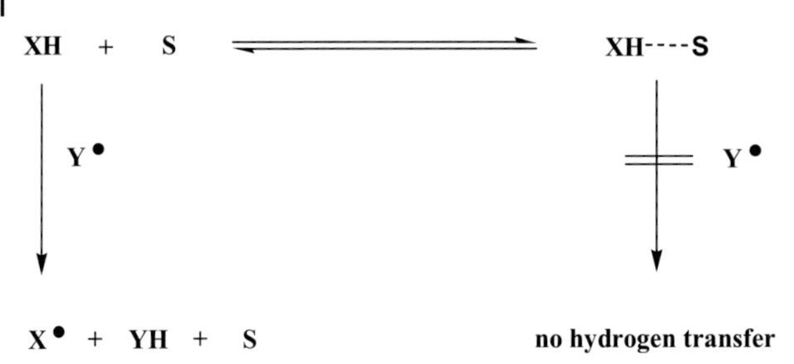

Scheme 3.1. Hydrogen bonding between substrate and solvent affects the efficiency of hydrogen transfer reaction to an attacking radical.

describes the ability of XH to act as a hydrogen bond donor, and (iii) a parameter ${\beta_2}^H$, which describes the ability of the solvent to act as a hydrogen bond acceptor.

$$\log(k^S/M^{-1}\ s^{-1}) = \log(k^0/M^{-1}\ s^{-1}) - 8.3{\alpha_2}^H{\beta_2}^H \quad (I)$$

It is important to note that Scheme 3.1 and Eq. (I) are valid for hydrogen transfer and, possibly, proton transfer processes [21]. However, it must be realized that while hydrogen bonding between the phenolic hydroxy group and a hydrogen acceptor may prevent hydrogen transfer, electron transfer processes can still proceed. Evidence is mounting that, especially in β-sheet structures, proteins form ${}^{\alpha}$C–H⋯O=C hydrogen bonds [22, 23]. These bonds display association enthalpies of $\Delta H^{298} \approx -3.0 \pm 0.5$ kcal mol^{-1} [22, 23], supported by theoretical and NMR spectroscopic data. It remains to be shown experimentally whether the formation of such ${}^{\alpha}$C–H⋯O=C hydrogen bonds in peptides and proteins may affect hydrogen transfer kinetics of the ${}^{\alpha}$C–H bond (*vide infra*).

Tyrosyl radicals are important intermediates utilized by enzymes, such as, for example, the ribonucleotide reductase class I (RNR1) [3, 24–27] or prostaglandin H synthase [28–30]. In RNR1, a tyrosyl radical ultimately oxidizes a Cys residue to produce a Cys thiyl (cysteinyl) radical, which subsequently attacks the C3′ C–H bond of a ribonucleotide substrate [3, 24–27] (*vide infra*). Prostaglandin H synthase utilizes a tyrosyl radical at position Tyr^{385} to abstract a hydrogen atom from arachidonic acid to yield a pentadienyl radical in the first step of the cyclooxygenase reaction [28–30]. Multiple additional *inter*- and *intra*molecular reactions may occur in proteins when tyrosyl radicals are produced in an uncontrolled manner during conditions of oxidative stress. The results of Foti et al. [31] suggest that tyrosyl radicals could actually be significantly more reactive than peroxyl radicals, at least in a nonaqueous environment. In other words, protein tyrosyl radicals could be efficient initiators of chain processes leading to protein hydroperoxides (*vide supra*). The hydrogen transfer reactivities of phenoxyl and peroxyl radicals towards a series of reductants were analyzed by laser flash photolysis experiments in organic solvents [31]. Generally, the phenoxyl radicals reacted ca. 100-fold faster than the peroxyl radicals. In benzene, the rate constant for the hydrogen transfer reaction between the phenoxyl radical and α-tocopherol (vitamin E), $k = 1.1 \times 10^9\ M^{-1}\ s^{-1}$, indicates a diffusion-controlled process. The large differences between the rate constants for phenoxyl and peroxyl radicals were rationalized in terms of hydrogen-bonded complexes and transition state structures. It was suggested that peroxyl radicals form tightly associated hydrogen-bonded complexes, which are, however, incorrectly oriented for a hydrogen transfer reaction. In contrast, the phenoxyl radical associates weakly with the substrate phenol, enabling successful hydrogen transfer.

3.3 Nitrogen-dentered Radicals

3.3.1 Generation of Aminyl and Amidyl Radicals

Aminyl and amidyl radicals are conveniently generated from the homolytic or reductive cleavage of chloramines and chloramides [32–39]. The latter form under inflammatory conditions when amino acids and/or peptides are exposed, for example, to hypochlorous acid (HOCl). *In vivo*, the reduction of chloramines and chloramides may proceed through the action of superoxide, eventually catalyzed by redox-active transition metals, M^{n+}, where M may be Fe and/or Cu (Reactions (3.7) and (3.8)) [38, 39]. Nitrogen-centered protein radicals were detected by EPR-spin trapping after the exposure of isolated proteins and plasma as well as red blood cells to HOCl (and HOBr) [32–35].

$$R\text{–}NH\text{–}Cl + M^{n+} \rightarrow R\text{–}NH^{\bullet} + M^{(n+1)+} + Cl^{-} \tag{3.7}$$

$$R'\text{–}CO\text{–}N(Cl)R + M^{n+} \rightarrow R'\text{–}CO\text{–}NR^{\bullet} + M^{(n+1)+} + Cl^{-} \tag{3.8}$$

Aminyl radicals have also been detected indirectly during the reaction of hydroxyl radicals ($HO^{\bullet}$) or their conjugated base ($^{\bullet}O^{-}$) with the free amino group of amino acids (Reactions (3.9) and (3.10)) [40–43], and identified by time-resolved EPR experiments [44]. Similar reactions may be expected for peptides. While Reactions (3.9) and (3.10) show a net hydrogen transfer, they likely proceed via a stepwise electron-transfer and proton-transfer (Reaction (3.11)), a reaction commonly referred to as proton-coupled electron transfer (PCET). Proton transfer from the aminium radical cation to the base (OH^{-}) will likely occur within the solvent cage.

$$HO^{\bullet} + R\text{–}NH_2 \rightarrow H_2O + R\text{–}NH^{\bullet} \tag{3.9}$$

$$^{\bullet}O^{-} + R\text{–}NH_2 \rightarrow HO^{-} + R\text{–}NH^{\bullet} \tag{3.10}$$

$$HO^{\bullet} + R\text{–}NH_2\ [HO^{-}\ {}^{+\bullet}H_2N\text{–}R] \rightarrow H_2O + R\text{–}NH^{\bullet} \tag{3.11}$$

Aminyl and amidyl radicals are electrophilic, oxidizing radicals (cf. the oxidation of hydroquinone by aminyl radicals from Gly at pH 11; $k = 7.4 \times 10^7\ M^{-1}\ s^{-1}$ [40]).

Moreover, they involve several fragmentation reactions, hydrogen transfer and protonation equilibria of potential biological significance (*vide infra*).

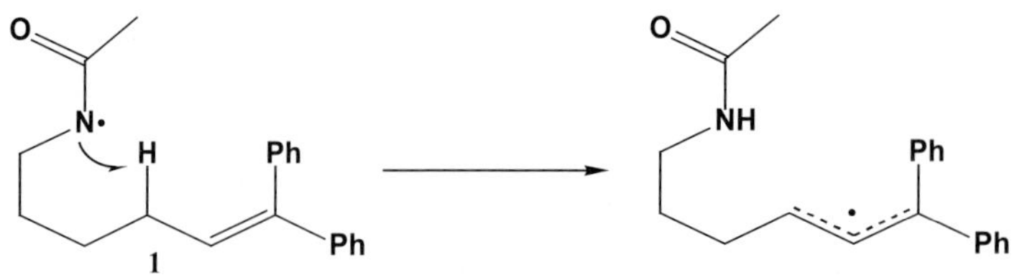

(3.12)

3.3.2 Reactions of Aminyl and Amidyl Radicals

Specifically *intra*molecular hydrogen transfer reactions of aminyl and amidyl radicals have been described, e.g., the 1,5-H-shift in the Hoffmann–Löffler–Freytag reaction, involving protonated aminyl radicals (aminium radical cations; reviewed in Ref. [45]). Amidyl radicals do not require protonation for hydrogen transfer. Representative kinetic data were obtained by laser flash photolysis for the *intra*molecular hydrogen transfer (1,5-H-shift) in the *N*-(6,6-diphenyl-5-hexenyl)acetamidyl radical **1** (Reaction (3.12); $k_{12} = 5.5 \times 10^6\ s^{-1}$ [46]). Here, hydrogen transfer is facilitated by formation of the highly conjugated product radical. In addition, the *inter*molecular reaction of amidyl radicals with thiophenol (PhSH) proceeds with $k = 9 \times 10^7$ $M^{-1}\ s^{-1}$ [46].

Davies and coworkers examined the reactivities of amidyl radicals derived from glucosamines, where evidence for both 1,5- and 1,2-H-shift processes was obtained (Scheme 3.2, Reactions (3.13) and (3.14)) [38, 39].

In aminyl radicals from amino acids and amidyl radicals from peptides, such a 1,2-H-shift (Reaction (3.15)) was considered feasible based on (i) the analogy to the well-known, solvent-assisted 1,2-H-shift within alkoxy radicals [19] and (ii) the exothermicity based on the homolytic bond dissociation energies (BDEs) of the N–H (406 kJ mol^{-1}) and the $^{\alpha}C$–H bond (363 kJ mol^{-1}) (representative values for the Gly anion [47]). However, both pulse radiolysis and γ-radiolysis experiments concluded that the 1,2-H-shift in aminyl and amidyl radicals derived from amino acids and peptides must be rather slow ($k_{15} \approx 1.2 \times 10^3\ s^{-1}$) [37, 40].

$$^{\bullet}HN{-}CH_2{-}CO_2^- \rightarrow H_2N{-}C^{\bullet}H{-}CO_2^- \quad (3.15)$$

Some evidence for a reverse 1,2-H-shift was also obtained. In their studies on the radical-induced decarboxylation of Gly anion, Bonifačić et al. observed a proton-catalyzed decarboxylation of $H_2N{-}C^{\bullet}H{-}CO_2^-$, which may proceed via Reactions (3.16)–(3.18) [40, 41].

$$H_2N{-}C^{\bullet}H{-}CO_2^- \rightarrow {}^{+}H_3N{-}C^{\bullet}H{-}CO_2^- \quad (3.16)$$

$$^{+}H_3N{-}C^{\bullet}H{-}CO_2^- \rightarrow {}^{+\bullet}H_2N{-}CH_2{-}CO_2^- \quad (3.17)$$

$$^{+\bullet}H_2N{-}CH_2{-}CO_2^- \rightarrow H^+ + H_2N{-}CH_2^{\bullet} + CO_2 \quad (3.18)$$

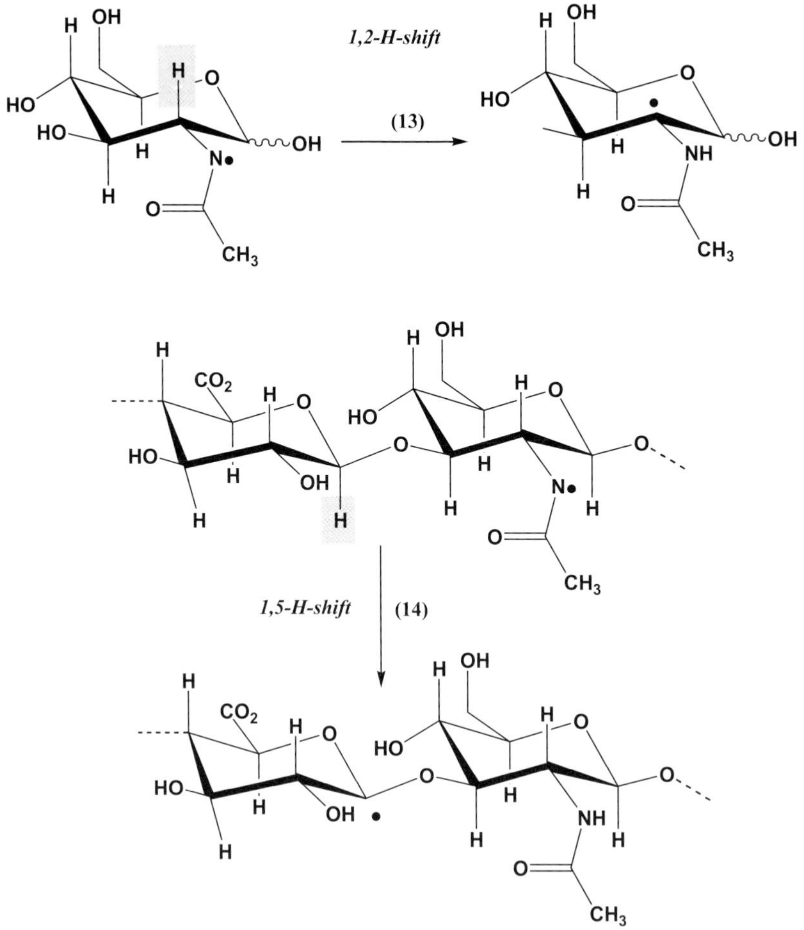

Scheme 3.2. 1,2- and 1,5-H-shift in amidyl radicals of glucosamine moieties.

The latter mechanism would also serve to explain the decarboxylation via the reaction of Gly anion with methyl radicals, $^{\bullet}CH_3$, and isopropyl radicals, $(CH_3)_2C^{\bullet}OH$, namely via initial hydrogen abstraction by the carbon-centered radicals at the weak Gly anion $^{\alpha}C$–H bond, followed by Reactions (3.16)–(3.18).

3.4 Sulfur-centered Radicals

This section will mainly focus on two biologically relevant sulfur-centered radical species, thiyl radicals ($RS^{\bullet}$) and sulfide radical cations ($R_2S^{\bullet +}$).

3.4.1
Thiols and Thiyl Radicals

3.4.1.1 Hydrogen Transfer from Thiols

Thiols play an important role in the maintenance of the cellular redox state, the structural and functional integrity of proteins, and redox signaling. Moreover, specific enzymes utilize the one-electron oxidation products of thiols, thiyl radicals, for substrate turnover, e.g. the ribonucleotide reductases [3, 24–27], pyruvate formate lyase [3, 48–50], and benzylsuccinate synthase [51, 52]. Radiation chemical experiments have documented the potency of endogenous and exogenous thiols to protect cells against ionizing radiation [53–55]. Part of this protection is due to the direct reaction of primary and secondary radicals with thiols (the "repair reaction") according to the general reactions (3.19) and (3.20).

$$X^{\bullet} + RSH \rightarrow XH + RS^{\bullet} \tag{3.19}$$

$$X^{\bullet} + RS^{-} \rightarrow X^{-} + RS^{\bullet} \tag{3.20}$$

The repair of carbohydrate radicals within polynucleotides can proceed with remarkable stereoselectivity, as demonstrated for hydrogen transfer from both 2-mercaptoethanol and dithiothreitol to deoxyuridin-1′-yl radials within single- and double-stranded oligonucleotides (Scheme 3.3, Reactions (3.21α) and (3.21β)) [56].

Reactions (3.21α) and (3.21β) yield a ca. 4-fold excess of β- over α-doxyuridine in single-stranded oligonucleotides, which increases to a ca. (7–8)-fold excess for double-stranded oligonucleotides. To date, numerous reports have identified thiyl radical formation *in vitro* and *in vivo* as a consequence of enzyme turnover [3, 24–27, 48–52, 57], oxidative stress [58] and drug metabolism [59]. Among the biologically relevant species generating thiyl radicals during conditions of oxidative stress are the nitrogen monoxide ($NO^{\bullet}$) metabolites nitrogen dioxide ($^{\bullet}NO_2$) [60] and peroxynitrite/peroxynitrous acid ($ONOO^-$/ONOOH) [58, 61, 62], the oxygen-centered hydroxyl ($HO^{\bullet}$) [53], alkoxyl ($RO^{\bullet}$) [17], peroxyl ($ROO^{\bullet}$) [63] and phenoxyl ($ArO^{\bullet}$) [64] radicals, and carbon-centered radicals [53].

The hydrogen transfer reaction between carbon-centered radicals and thiols has been especially the focus of intense theoretical and mechanistic investigation. Central to these studies is the controversy about the structure of the transition state, affecting the heights of the activation barriers. Zavitsas and coworkers utilize the four canonical structures I–IV, given below, to describe the transition state [65–67].

$$X{\uparrow}{\downarrow}H\ Y{\uparrow}(I) \quad {\uparrow}X\ H{\downarrow}{\uparrow}Y(II) \quad X{\uparrow}\ H{\downarrow}\ Y{\uparrow}(III) \quad [X\ H\ Y]^{\bullet}(IV)$$

Structure III represents an antibonding, triplet repulsion between the atoms or groups transferring the hydrogen atom (here, a carbon-centered and a thiyl radical), which has a significant influence on the activation barrier. They conclude that polar transition state structures are not necessary to rationalize experimental re-

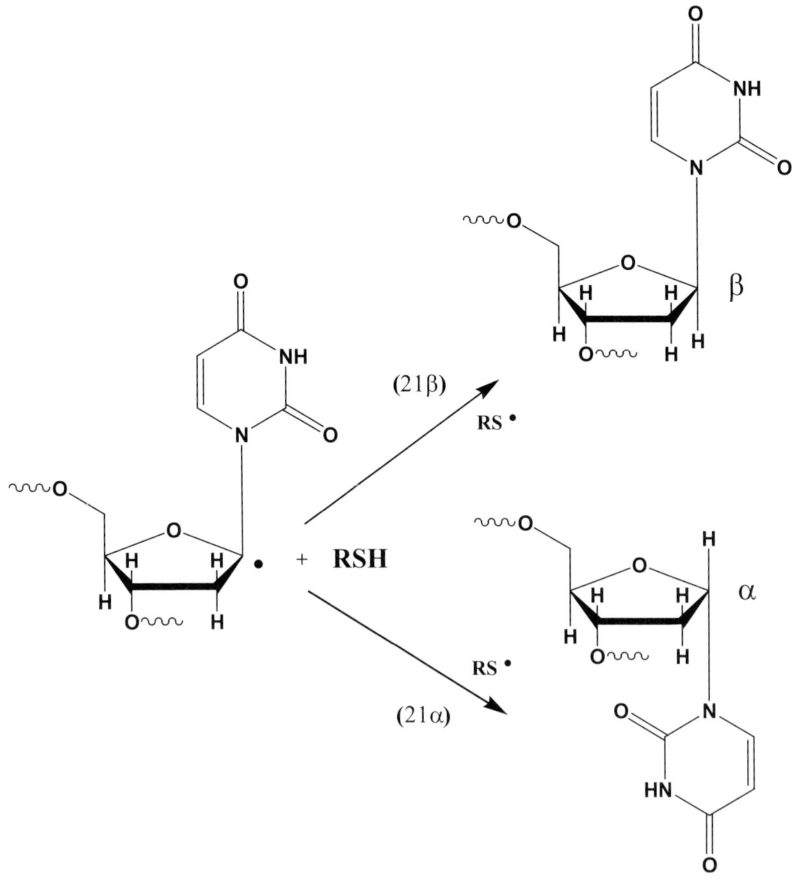

Scheme 3.3. Reactions of deoxyuridin-1′-yl radials with thiols.

sults (and may even lead to predictions of reactivity which strongly deviate from experimental results). In contrast, Roberts and coworkers have devised the concept of "polarity reversal catalysis" (PRC), where polar transition states are dominant features explaining the catalytic affect of thiols and thiyl radicals on hydrogen transfer reactions between carbon-centered radicals and several hydrogen donors [68–70]. Theoretical support for polar transition states in the hydrogen transfer from thiols to carbon-centered radicals comes from recent work of Beare and Coote [71], who provide strong evidence for the involvement of Structures V and VI, given below.

$$C^{+}H^{\bullet}S^{-}(V) \quad C^{-}H^{\bullet}S^{+}(VI)$$

Reid and coworkers experimentally re-determined the rate constants for hydrogen transfer between a model thiol, 1,4-dithiothreitol, and several carbon-centered radicals [72]. The substrate 1,4-dithiothreitol (structure **2**) represents an elegant model system for the direct time-resolved measurement of absolute rate constants for hydrogen transfer according to Reactions (3.22) and (3.23).

Here, the product thiyl radical **3** undergoes spontaneous cyclization and deprotonation ($pK_{a,2} = 5.2$) to the radical anion **4**, which has a strong UV absorption with $\lambda_{max} = 390$ nm. In this sequence, concentrations can be adjusted such that Reaction (3.22) is rate-determining. The reactivity of carbon-centered radicals with 1,4-dithiothreitol increased in the following order: $^{\bullet}CH_2C(CH_3)_2OH < {}^{\bullet}CH_3 < {}^{\bullet}CH_2OH < {}^{\bullet}CH(CH_3)OH < {}^{\bullet}C(CH_3)_2OH$. Among these radicals, Reaction (3.24) generates the product with the lowest C–H bond energy (isopropanol), but, nevertheless, shows the highest rate constant for hydrogen transfer.

$$R^{\bullet} + \mathbf{2} \xrightarrow{(3.22)} RH + \mathbf{3} \underset{(3.23)}{\overset{-H^+}{\rightleftharpoons}} \mathbf{4}$$

$$^{\bullet}C(CH_3)_2OH + RSH \rightarrow H\text{–}C(CH_3)_2OH + RS^{\bullet} \qquad (3.24)$$

Parallel theoretical studies (with CH_3SH instead of 1,4-dithiothreitol) gave clear evidence for polar transition state structures of the general type V, where, in the alcohol-derived radicals, the positive charge developing on the carbon forming the new C–H bond is stabilized by the α-OH substituent. Importantly, increasing charge separation in the transition state was observed within the series $^{\bullet}CH_3 < {}^{\bullet}CH_2OH < {}^{\bullet}C(CH_3)_2OH$, consistent with the introduction of additional α-substituents. These results are in agreement with the data of Beare and Coote [71]. The results of Reid et al. [72] are also discussed in terms of orbital interaction theory, where the isopropyl radical, $^{\bullet}C(CH_3)_2OH$, is characterized by the highest SOMO energy, supporting interaction with the LUMO of CH_3SH, i.e. the σ^* orbital of the H–S bond.

Several enzymes utilize thiyl radicals for substrate conversion. In the ribonucleotide reductase (RNR) class III, pyruvate formate lyase and benzylsuccinate synthase, cysteine thiyl radicals are generated via hydrogen transfer from cysteine to glycine radicals (Reaction (3.25)) [3, 24–27, 48–52].

$$\text{—HN—}\overset{\bullet}{C}H\text{—}C(=O)\text{—NH—} + RSH \underset{(-3.25)}{\overset{(3.25)}{\rightleftharpoons}} \text{—HN—}CH_2\text{—}C(=O)\text{—NH—} + RS^{\bullet} \qquad (3.25)$$

The efficiency of Reaction (3.25) depends on the conformational properties of glycine and the product glycyl radical, respectively, within the protein. If the glycyl radical can adopt an ideal planar structure, hydrogen transfer from Cys is endothermic and the reverse reaction, i.e. hydrogen transfer from glycine to cysteinyl radicals (Reaction (−3.25)), will prevail (*vide infra*). On the other hand, if the glycyl radical adopts a pyramidal structure, Reaction (3.25) would be exothermic. Here, the protein has the opportunity to fine-tune hydrogen transfer equilibria through conformational properties. Calculations have shown that the C–H homolytic bond dissociation energy (BDE) of glycine depends on the secondary strucure, in which the amino acid is located, i.e. $BDE(C_\alpha\text{–}H) = 402, 404$, and 361 kJ mol^{-1} for Gly within an α-helix, parallel β-sheet and antiparallel β-sheet, respectively [73]. In contrast, $BDE(C_\alpha\text{–}H) = 330\text{–}370$ kJ mol^{-1} for Gly within linear, relaxed peptide structures [73]. Hence, glycyl radicals within α-helical or parallel β-sheet conformation should especially rapidly abstract hydrogen from protein cysteine residues, which show S–H BDE values of the order of 370 kJ mol^{-1}. In contrast, cysteinyl radicals may abstract hydrogen atoms from Gly located in antiparallel β-sheets or relaxed peptide conformations.

In ribonucleotide reductasc class II, thiyl radicals are generated via hydrogen transfer from Cys to the primary carbon-centered radical generated from 5′-deoxy-5′-adenosylcobalamin [3]. In contrast, ribonucleotide reductase class I generates cysteinyl radicals via long-range electron and/or proton-coupled electron transfer involving an ultimate hydrogen transfer from Cys to a tyrosyl radical [3].

3.4.1.2 **Hydrogen Abstraction by Thiyl Radicals**

The calculated bond energies (*vide supra*) provide one rationale for the propensity of thiyl radicals to abstract hydrogen atoms from a variety of biological substrates (Reaction (3.26)). Hence, with suitable substrates the "repair reaction" may actually proceed in the reverse direction [74].

$$RS^\bullet + YH \rightarrow RSH + Y^\bullet \qquad (3.26)$$

In all three classes of ribonucleotide reductases, a cysteinyl radical (in the *E. coli* RNR1 sequence at position Cys^{439}) abstracts a hydrogen atom from the C3′ position of the carbohydrate moiety of the ribonucleotide substrate [3]. Biomimetic model studies of this enzymatic process were designed, achieving *intra*molecular hydrogen transfer within a tetrahydrofurane-appended thiyl radical (Scheme 3.4; Reactions (3.27) and (3.28) [75, 76].

In kinetic NMR experiments, rate constants for the *inter*molecular hydrogen transfer from several carbohydrates to cysteinyl radicals were found to be of the order of $k_{29} = (1\text{–}3) \times 10^4\ M^{-1}\ s^{-1}$ at 37 °C [77]. These values agree with previous, pulse radiolytically determined rate constants for thiyl radical-mediated hydrogen abstraction from various model alcohols and ethers [74, 78, 79]. In contrast, the reverse reaction, hydrogen transfer from thiols to carbohydrate radicals, proceeds with $k_{-29} > 10^6\ M^{-1}\ s^{-1}$ [80, 81], indicating that equilibrium (3.29) is normally located far to the left-hand side.

$$RS^\bullet + (\text{Carbohydrate})C\text{–}H \underset{k_{-29}}{\overset{k_{29}}{\rightleftharpoons}} RSH + (\text{Carbohydrate})C^\bullet \qquad (3.29)$$

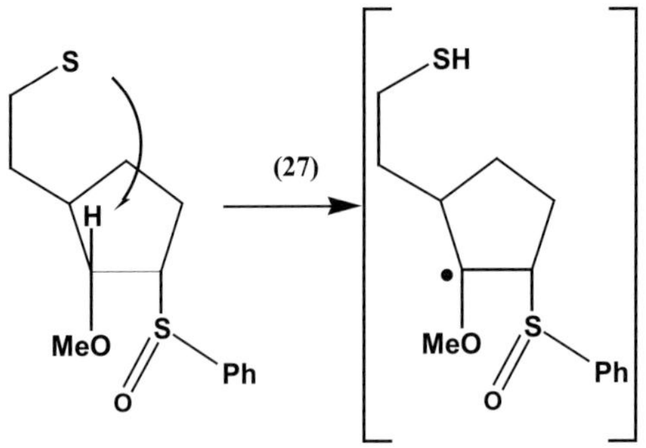

Scheme 3.4. Biomimetic model reaction displaying a 1,5-H-shift in thiyl radicals containing a tetrahydrofuran substituent.

However, efficient water elimination [82] from the resulting carbohydrate C3′ radical of the original ribonucleotide will shift equilibrium (3.29) to the right-hand side. Moreover, in RNR1, the activation barrier for hydrogen transfer to the cysteinyl radical may be lowered through hydrogen bonding of the C2′OH group to Glu^{441} (model systems were calculated, in which the ribose moiety was replaced by ethylene glycol or *cis*-tetrahydrofurane-2,3-diol, and Glu^{441} was replaced by formate, acetate and acetamide) [83].

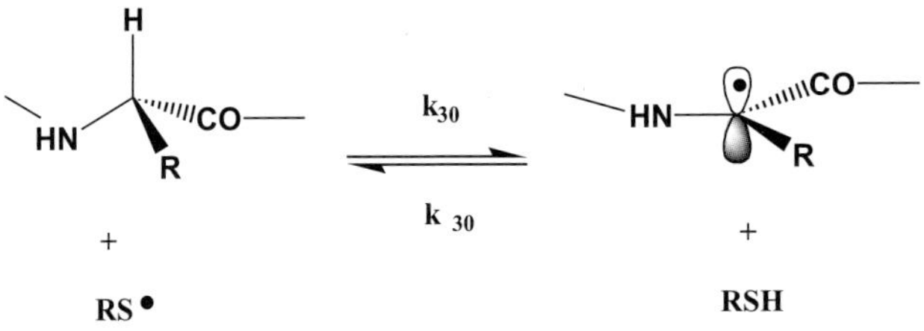

(3.30)

While equilibrium (3.29) for the uncatalyzed reaction of thiyl radicals with carbohydrates is located far to the left, the analogous equilibrium (3.30) with peptide and protein substrates may actually shift to the right-hand side. In equilibrium (3.30), the $^{\alpha}C^{\bullet}$ radical of the amino acid moiety is displayed in a planar conformation; in reality, this conformation may be approached in linear peptides by glycine (R=H) or cyclic peptide models, such as the diketopiperazines, but less likely by amino acid moieties different from glycine.

Table 3.1 displays rate constants for the hydrogen abstraction by cysteamine thiyl radicals from several *N*-acetyl-amino acid amides and diketopiperazines [84].

These rate constants were obtained through competition kinetics in D_2O at pD 3.0–3.4, using isopropanol as a competitor. Several important features are noted: (i) the cyclic diketopiperazines show generally higher rate constants per $^{\alpha}C$–H bond compared to their linear analogs (cf., GlyA and N-Ac-Gly-NH_2), (ii) the trend of calculated $^{\alpha}C$–H bond energies does not match the trend of experimental rate

Table 3.1. Rate constants for the reaction of cysteamine thiyl radicals with model peptides in D_2O, pD 3.0–3.4 at 37 °C (adapted from Ref. [84]; abbreviations for the diketopiperazines: SarcA = sarcosine anhydride, Gly = glycine anhydride).

Substrate	k_{30}, 10^4 M^{-1} s^{-1}	k_{30} per $^{\alpha}$C–H bond, 10^4 M^{-1} s^{-1}	BDE of $^{\alpha}$C–H[a], kJ mol^{-1}
SarcA	40 ± 8	10	–
GlyA	32 ± 16	8.0	350 (340[b])
N-Ac-Gly-NH_2	6.4 ± 2.8	3.2	350
N-Ac-Ala-NH_2	1.0 ± 0.3	1.0	345
N-Ac-Asp-NH_2	0.44 ± 0.16	0.44	332
N-Ac-Gln-NH_2	0.19 ± 0.06	0.19	334
N-Ac-Pro-NH_2	0.18 ± 0.06	0.18	358 (cis) 369 (trans)

[a] Ab initio calculated values [85]; [b] experimental value [86].

constants, i.e. decreasing $^{\alpha}$C–H bond energies are paralleled by decreasing rate constants. This effect can be rationalized by steric constraints induced by the increasing bulkiness of the amino acid side chain, which prevents the generated peptide $^{\alpha}C^{\bullet}$ radical from approaching the ideal planar conformation for maximal captodative stabilization of the radical. Importantly, pulse radiolysis and steady-state radiolysis experiments failed to measure a rate-constant for the reaction of $^{\alpha}C^{\bullet}$ radicals from glycine anhydride with thiols, suggesting that for GlyA, $k_{-30} \leq 10^5$ M^{-1} s^{-1}. Here $k = 10^5$ M^{-1} s^{-1} reflects the lower limit of second order rate constants measurable by the pulse radiolysis technique, except for a few cases where very high substrate concentrations (>0.1 M) can be adjusted. Hence, K_{30} $(= k_{30}/k_{-30}) \geq 3.0$, indicating that equilibrium (3.30) for the diketopiperazines is located more on the right-hand side. This conclusion is well-supported by the deuterium NMR studies of Anderson and coworkers [87–89], who have exposed amino acids and peptides in D_2O to radiation chemically generated hydroxyl radicals, and monitored repair of the amino acid radicals by deuterated dithiothreitol. In all their experiments, Gly showed the lowest efficiency of deuterium incorporation, indicating that the reaction of glycyl radicals with dithiothreitol is of low efficiency.

A low value for k_{-30} is also in accord with similar findings for the reaction of thiyl radicals with polyunsaturated fatty acids, where hydrogen abstraction from bisallylic methylene groups occurs with $k_{31} \geq 3 \times 10^6$ M^{-1} s^{-1} [90] (k_{31} depends on the number of bisallylic methylene groups within the fatty acid chain), generating a stable pentadienyl radical. No experimental evidence for the reverse reaction was obtained, i.e. $k_{-31} < 10^5$ M^{-1} s^{-1}. Analogous pulse radiolysis experiments detected no measurable reaction of cyclohexadienyl radicals with thiols [81].

Table 3.2. Rate constants for the reaction of cysteamine thiyl radicals with selected amino acid substrates containing reactive side chains; in D_2O, pD 3.0–3.4 at 37 °C (adapted from Ref. [91]).

Substrate	k_{32}, 10^4 M^{-1} s^{-1}	$k_{side\ chain}$, 10^4 M^{-1} s^{-1}
$AlaNH_2$	0.4 ± 0.1	n.d.
$GlyNH_2$	0.7 ± 0.4	–
$HisNH_2$	0.4 ± 0.1	n.d.
$MetNH_2$	1.8 ± 0.5	0.9 ± 0.6
$PheNH_2$	1.5 ± 0.2	1.3 ± 0.6
$SerNH_2$	23 ± 7	16 ± 4
$ThrNH_2$	10 ± 5	4.4 ± 0.8
$ValNH_2$	0.8 ± 0.3	<0.3

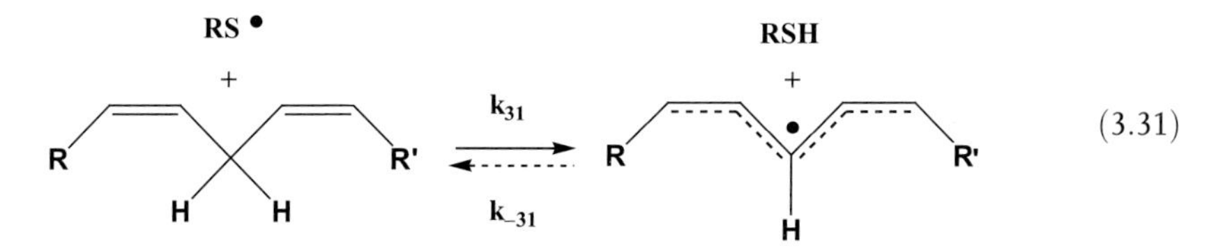

The kinetic NMR method permitted also the determination of rate constants k_{32} for hydrogen transfer to cysteamine thiyl radicals from selected amino acids containing reactive side chains [91]. A summary of these rate constants is given in Table 3.2. Here, the rate constant k_{32} represents the sum of the individual rate constants for hydrogen transfer from $^{\alpha}$C–H (k_{30}) and from the side chain C–H bonds (k_{sc}), i.e., $k_{32} = k_{30} + k_{sc}$.

RS• + NH–C(H)(R)–CO — $\xrightarrow[-RSH]{k_{32}}$ NH–C(H)(•R(-H))–CO — + HN–•C(R)–CO — (3.32)

These rate constants were obtained with amino acid amides at pD 3.0–3.4. The low pD value ensures full protonation (deuteration) of the amino group while the use of the respective amides avoids any possible interference from zwitterion formation, which could be a problem with the free amino acids. Full protonation of the amino group has a decelerating effect on hydrogen abstraction from $^{\alpha}$C–H due to the lack of possible captodative stabilization [92]. Hence, especially in the amino acid amides at low pD, hydrogen transfer from the amino acid side chains can compete with hydrogen transfer from $^{\alpha}$C–H. Noteworthy in Table 3.2 are the high rate constants k_{32} and k_{sc} for $SerNH_2$, which are significantly larger than the comparable rate constants for $ThrNH_2$. These data are in clear contrast to rate constants for hydrogen transfer from ethanol ($k = 6 \times 10^3$ M^{-1} s^{-1} at 20 °C) and isopropanol ($k = 2 \times 10^4$ M^{-1} s^{-1} at 20 °C) [78, 79]. Moreover, these show the opposite trend with the higher value for the higher substituted substrate, consistent with a lower C–H BDE. At this point we can only speculate about a possible rationale for this observation, such as a potential stabilization of a polar transition state by the C-terminal amide (theoretical studies demonstrate that polar transition states play an important role in the hydrogen abstraction from amino acid moieties by thiyl radicals [93]). An analogous effect has been postulated by Easton and Merett for anchimeric assistance in the hydrogen abstraction from Phe derivatives [94].

$$RS^{\bullet} + O_2 \underset{(-3.33)}{\overset{(3.33)}{\rightleftharpoons}} RSOO^{\bullet} \xrightarrow{(3.34)} R{-}S^{\bullet}(=O)_2$$

Model calculations allow the predictions that, especially, the hydrogen transfer from $^{\alpha}$C–H to thiyl radicals may be of physiological significance. A detailed summary of these calculations is given elsewhere [91] and will not be repeated here. Important for these considerations is the fact that any of the carbon-centered amino acid/peptide radicals will react efficiently with molecular oxygen ($k \approx 2 \times 10^9$ M^{-1} s^{-1} [8]). While thiyl radicals also react with oxygen, this reaction is reversible ($k_{33} = 2.2 \times 10^9$ M^{-1} s^{-1}; $k_{-33} = 6 \times 10^5$ s^{-1}) [95] and an efficient elimination of thiyl radicals from equilibrium (3.30) is only possible through the irreversible rearrangement of thiyl peroxyl radicals to sulfonyl radicals, where $k_{34} \approx 2 \times 10^3$ s^{-1} at 37 °C [95].

To date no rate constants have been published for the intramolecular hydrogen transfer between amino acid moieties and cysteinyl radicals in peptides. Preliminary data exist for the reversible hydrogen transfer in radicals of the peptide *N*-acetyl-Cys-$(Gly)_6$, where glycyl radicals abstract the thiol hydrogen from Cys with $k \approx 10^6$ s^{-1} and the reverse reaction occurs with $k \approx 2 \times 10^5$ s^{-1} [96]. The high reactivity of the glycyl radicals in this linear peptide (vs. the low reactivity of glycyl radicals in GlyA) may be rationalized by the propensity of polyGly peptides to adopt secondary structures such as helical and β-sheet conformations [97, 98] (possibly raising the $^{\alpha}$C–H BDE of the glycyl residues). A specific case of *intra*-molecular hydrogen transfer was reported for thiyl radicals of the peptide glutathione [99, 100]. Here, thiyl radicals abstract the hydrogen atom from the *N*-terminal γ-glutamyl residue, where the forming carbon-centered radical is stabilized captodatively by both a free amino group and a free carboxylate group.

Besides peptides, proteins and polyunsaturated fatty acids, DNA and RNA represent a third large class of biomolecules sensitive to covalent modification under conditions of oxidative stress. Few studies have focused on the potential reaction of thiyl radicals with polynucleotides. The active site cysteinyl radical required for deoxynribonucleotide synthesis by the ribonucleotide synthases, and model studies with carbohydrates, have shown that DNA damage by thiyl radicals is, in principle, feasible. Obviously, thiyl radical reactions with DNA have to compete with thiyl radical reactions with endogenous antioxidants such as ascorbate and glutathione. However, charge repulsion excludes especially, negatively charged molecules [54, 55] (such as ascorbate and glutathione at physiological pH) from the immediate environment of the DNA strand. Hence, there will be a concentration gradient, where the concentration of these negatively charged antioxidant molecules close to the DNA strand may be lower than that in the "bulk solution". Jain et al. studied the degradation of DNA by glutathione and Cu(II) under anaerobic conditions, suggesting the involvement of glutathione thiyl radicals [101]. Recently, evidence for a reaction of thiyl radicals with nucleobases has been provided. Carter et al. designed model experiments, which confirmed that thiyl radicals would add to pyrimidine bases [102]. Moreover, our own experiments provided rate constants for hydrogen abstraction at the C5–CH_3 group of both thymine (Scheme 3.5; $k_{35a} = 1.2 \times 10^4$ M^{-1} s^{-1}) and thymidine-5′-monophosphate ($k_{35b} = 0.9 \times 10^4$ M^{-1} s^{-1}) [103]. These rate constants are comparable to that for hydrogen abstraction from toluenesulfonate (tosylate) ($k = 2.7 \times 10^4$ M^{-1} s^{-1}) [103] and also to that

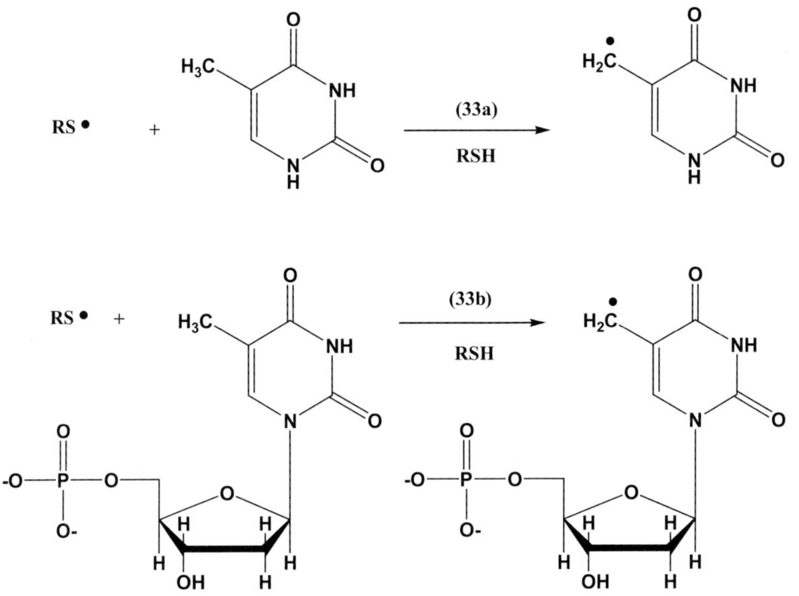

Scheme 3.5. Thiyl radicals react with the C5–CH_3 group of thymine and thymidine-5′-monophosphate.

for the reaction of thiyl radicals with the benzylic $^{\beta}$C H bond in PheNH$_2$ ($k_{sc} = 1.3 \times 10^4$ M^{-1} s^{-1}; see Table 3.2) [91].

3.4.2
Sulfide Radical Cations

The one-electron oxidation of organic sulfides yields sulfide radicals cations (Reaction (3.36)) [104].

$$>S + Ox \rightarrow >S^{\bullet +} + Ox^{\bullet -} \quad (3.36)$$

In general, sulfide radical cations would either deprotonate in the α-position to the sulfur, yielding α-(alkylthio)alkyl radicals, or engage in the one-electron oxidation of additional substrates. However, recent hypothesis and results have focused on a possible role of methionine sulfide radical cations in hydrogen abstraction reactions within the Alzheimer's disease β-amyloid peptide (βAP) [73]. These mechanisms will be discussed here in some detail.

βAP represents a 39–42 amino acid peptide, released from the amyloid precursor protein, with a pronounced tendency to form low and high molecular weight aggregates [105, 106]. The NMR structure of βAP recorded in aqueous micelles shows a helical conformation around Met^{35} in the C-terminus [107]. This Met^{35} residue appears to have a critical function as electron donor during the reduction of Cu(II) [108], which complexes via three His residues to the *N*-terminal part of βAP [106]. Model calculations have pointed to the possible role of the helical conformation in stabilizing Met sulfide radical cations through three-electron bond formation with the carbonyl oxygen of the peptide bond C-terminal to Ile^{31} [109]. In fact, substitution of Ile^{31} by the helix-breaking Pro^{31} lowers the propensity of βAP to reduce Cu(II) [110]. A schematic representation of the three-electron bonded Met sulfide radical cation is given in structure **5** (Scheme 3.6).

Theoretical studies by Rauk and coworkers [73] predict that such a sulfur–oxygen bonded radical cation complex can abstract a hydrogen atom from Gly, ultimately yielding a proton and reducing the Met sulfide radical cation back to Met (Scheme 3.6, Reactions (3.37) and (3.38)). Such a mechanism would combine the known redox activity of βAP with the generation of a carbon-centered radical at a specific site, which could serve as an origin for βAP cross-linking and formation of insoluble aggregates. This is an attractive hypothesis, and subsequent studies by Butterfield and coworkers showed that βAP containing the Gly^{33}Val mutation was less neurotoxic [111]. When dissociated from the carbonyl complex (structure **5**), the sulfide radical cation of Met would probably meet the conditions defined by Parker and coworkers [112], restricting hydrogen abstraction by radical cations to those species, where the odd electron is located in a nonbonding orbital. However, several facts and observations must be considered, which may limit the necessity or importance of Reactions (3.37) and (3.38) in βAP. First, the deprotonation of Met sulfide radical cations to α-(alkylthio)alkyl radicals generates carbon-centered radicals, even in the absence of any potential hydrogen transfer from Gly residues

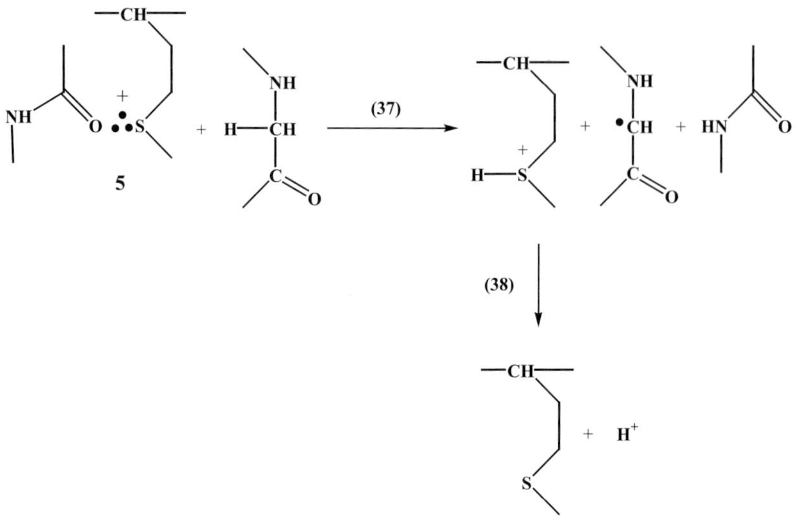

Scheme 3.6. The proposed hydrogen abstraction by (S∴O) three-electron-bonded Met sulfide radical cation from Gly.

[113]. These α-(alkylthio)alkyl radicals could lead to covalent aggregation. Moreover, the addition of oxygen would yield peroxyl radicals, species, which have been identified by electron paramagnetic resonance (EPR) spectroscopy during the incubation of βAP in buffer. Second, pulse radiolysis experiments, in which Met sulfide radical cations were generated in linear, flexible *N*-Ac-$(Gly)_n$Met$(Gly)_n$ model peptides ($n = 1$ and 3) did not reveal any hydrogen transfer from either of the Gly residues to the sulfide radical cations [114]. In these peptides the $Gly^{\bullet}$ radical would have been easily detected on the basis of their strong absorbance around 320 nm. Hence, it must be concluded that the theoretically predicted hydrogen transfer from Gly to Met sulfide radical cations in βAP may be restricted to highly organized assemblies such as β-sheet structures. However, mechanistic studies suggest that redox reactions of βAP, involving Met oxidation, may occur predominantly in low molecular weight aggregates, which may not contain the β-sheet structure.

Our recent results suggest an alternative potential pathway for the generation of ${}^{\alpha}C^{\bullet}$ radicals in Met-containing peptides [114], originating from sulfide radical cations and/or their intramolecular complexes with functionalities of the peptide bond. Pulse radiolysis experiments with model peptides provided the kinetics and yields for sulfur–oxygen three-electron bond formation between Met sulfide radical cations and peptide bond carbonyl groups. Supported by time-resolved UV and conductivity studies, a pH-dependent conversion of these sulfur–oxygen-bonded radical cations into sulfide–nitrogen-bonded radicals was observed, displayed in Scheme 3.7, Reactions (3.39)–(3.41) (control experiments with the model substrate *N*-acetylmethionineethyl ester confirmed that this mechanism can occur with the peptide bond *N*-terminal of the Met residue).

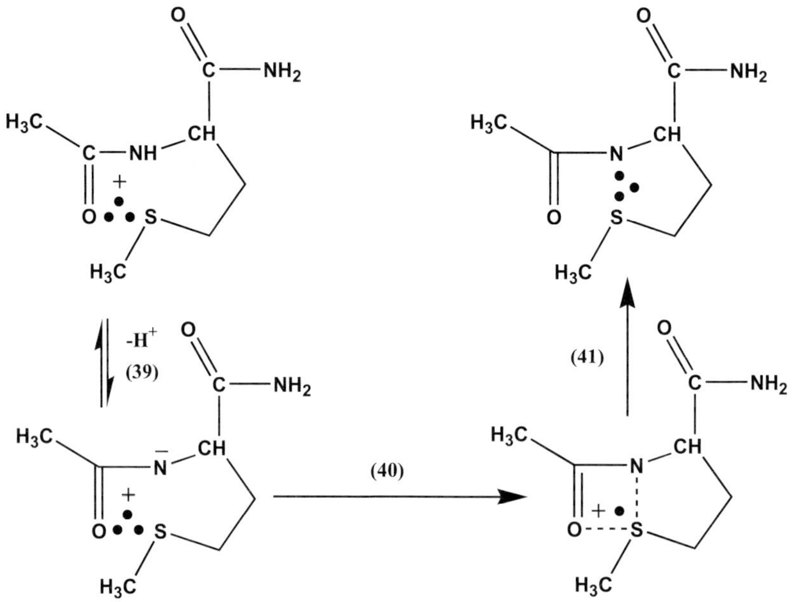

Scheme 3.7. The pH-dependent rearrangement of (S∴O) to (S∴N) three-electron-bonded Met sulfide radical cation involving the amide bond *N*-terminal of the Met residue.

A fraction of these sulfur–nitrogen bonded radical complexes ultimately converted into $^{\alpha}C^{\bullet}$ radicals, identified through their characteristic absorbance with $\lambda_{max} = 350$ nm. Scheme 3.8, Reactions (3.42)–(3.45), displays a tentative mechanism for the formation of these $^{\alpha}C^{\bullet}$ radicals, involving the formal 1,2-H-shift of

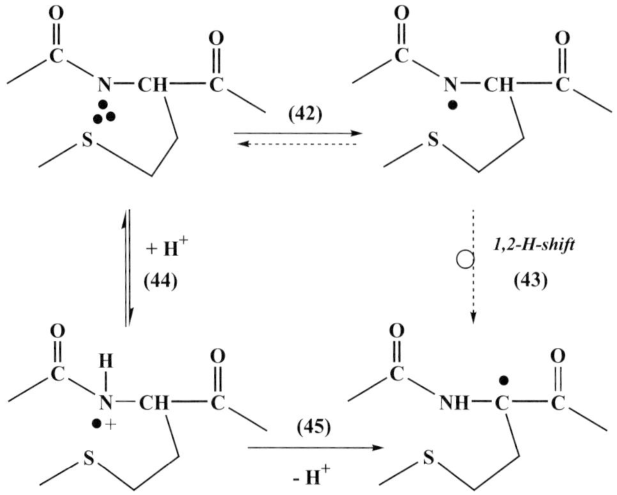

Scheme 3.8. Conversion of *N*-terminal (S∴N) three-electron bonded Met sulfide radical cation into the $^{\alpha}C^{\bullet}$ radical.

an intermediary amidyl radical (analogous to the reactions described in the section on N-centered radicals; *vide supra*). Alternatively, amide radical protonation followed by $^{\alpha}$C–H deprotonation (Reactions (3.44) and (3.45)) of the intermediary nitrogen-centered radical cation would lead to the same product.

3.5 Conclusion

This article summarizes the mechanisms and kinetics of selected biologically relevant hydrogen transfer reactions of oxygen-, nitrogen- and sulfur-centered radicals. Special emphasis has been placed on hydrogen transfer reactions involving amino acids and peptides. Many of the rate constants known to date have been measured with small organic model compounds. These results provide a reasonable basis for an approximate extrapolation onto proteins. However, the higher order structure and molecular dynamics of proteins will affect both the mechanisms and kinetics of *inter-* and *intra*molecular hydrogen transfer reactions. Therefore, these reactions need to be monitored directly at some point. Hence, future research should focus on the design of experiments, instruments and software to support the direct kinetic measurements of hydrogen transfer reactions in proteins.

Acknowledgment

Support by the NIH (PO1AG12993) is gratefully acknowledged.

References

1 *Radicals in Organic Synthesis*, P. Renaud, M. P. Sibi (Eds.), Vol. 1 and 2, Wiley-VCH, Weinheim, 2001.
2 I. W. M. Smith, A. R. Ravishankara, *J. Phys. Chem. A* **2002**, *106*, 4798–4807.
3 J. Stubbe, W. A. van der Donk, *Chem. Rev.* **1998**, *98*, 705–762.
4 R. T. Dean, S. Fu, R. Stocker, M. J. Davies, *Biochem. J.* **1997**, *324*, 1–18.
5 C. L. Hawkins, M. J. Davies, *Biochim. Biophys. Acta* **2001**, *1504*, 196–219.
6 C. L. Hawkins, M. J. Davies, *J. Chem. Soc., Perkin Trans. 2* **1998**, 2617–2622.
7 G. V. Buxton, C. L. Greenstock, W. P. Helman, A. B. Ross, *J. Phys. Chem. Ref. Data* **1988**, *17*, 513–886.
8 O. J. Mieden, M. N. Schuchmann, C. von Sonntag, *J. Phys. Chem.* **1993**, *97*, 3783–3790.
9 S. Gebicki, J. M. Gebicki, *Biochem. J.* **1993**, *289*, 743–749.
10 J. Neužil, J. M. Gebicki, R. Stocker, *Biochem. J.* **1993**, *293*, 601–606.
11 S. Gieseg, S. Duggan, J. M. Gebicki, *Biochem. J.* **2000**, *350*, 215–218.
12 S. Fu, R. Dean, M. Southan, R. Truscott, *J. Biol. Chem.* **1998**, *273*, 28603–28609.
13 M. J. Davies, S. Fu, R. T. Dean, *Biochem. J.* **1995**, *305*, 643–649.
14 M. L. Huang, A. Rauk, *J. Phys. Org. Chem.* **2004**, *17*, 777–786.
15 M. J. Davies, *Arch. Biochem. Biophys.* **1996**, *336*, 163–172.

16 H. A. Headlam, A. Mortimer, C. J. Easton, M. J. Davies, *Chem. Res. Toxicol.* **2000**, *13*, 1087–1095.
17 M. Erben-Russ, C. Michel, W. Bors, M. Saran, *J. Phys. Chem.* **1987**, *91*, 2362–2365.
18 D. V. Avila, C. E. Brown, K. U. Ingold, J. Lusztyk, *J. Am. Chem. Soc.* **1993**, *115*, 466–470.
19 K. G. Konya, T. Paul, S. Lin, J. Lusztyk, K. U. Ingold, *J. Am. Chem. Soc.* **2000**, *122*, 7518–7527.
20 M. Bonifačić, D. A. Armstrong, I. Štefanić, K.-D. Asmus, *J. Phys. Chem. B* **2003**, *107*, 7268–7276.
21 D. W. Snelgrove, J. Lusztyk, J. T. Banks, P. Mulder, K. U. Ingold, *J. Am. Chem. Soc.* **2001**, *123*, 469–477.
22 R. Vargas, J. Garza, D. A. Dixon, B. P. Hay, *J. Am. Chem. Soc.* **2000**, *122*, 4750–4755.
23 F. Cordier, M. Barfield, S. Grzesiek, *J. Am. Chem. Soc.* **2003**, *125*, 15750–15751.
24 J. Stubbe, J. Ge, C. S. Yee, *Trends Biochem. Sci.* **2001**, *26*, 93–99.
25 J. Stubbe, D. G. Nocera, C. S. Yee, M. C. Y. Chang, *Chem. Rev.* **2003**, *103*, 2167–2202.
26 F. Himo, P. E. M. Siegbahn, *Chem. Rev.* **2003**, *103*, 2421–2456.
27 L. Noodleman, T. Lovell, W.-G. Han, J. Li, F. Himo, *Chem. Rev.* **2004**, *104*, 459–508.
28 A. Tsai, R. J. Kulmacz, G. Palmer, *J. Biol. Chem.* **1995**, *270*, 10503–10508.
29 S. Peng, N. M. Okeley, A. Tsai, G. Wu, R. J. Kulmacz, W. A. van der Donk, *J. Am. Chem. Soc.* **2001**, *123*, 3609–3610.
30 S. Peng, N. M. Okeley, A. Tsai, G. Wu, R. Kulmacz, W. A. van der Donk, *J. Am. Chem. Soc.* **2002**, *124*, 10785–10796.
31 M. Foti, K. U. Ingold, J. Lusztyk, *J. Am. Chem. Soc.* **1994**, *116*, 9440–9447.
32 C. L. Hawkins, M. J. Davies, *J. Chem. Soc., Perkin Trans. 2* **1998**, 1937–1945.
33 C. L. Hawkins, M. J. Davies, *Biochem. J.* **1998**, *332*, 617–625.
34 C. L. Hawkins, M. J. Davies, *Biochem. J.* **1999**, *340*, 539–548.
35 C. L. Hawkins, B. E. Brown, M. J. Davies, *Arch. Biochem. Biophys.* **2001**, *395*, 137–145.
36 C. L. Hawkins, D. I. Pattison, M. J. Davies, *Biochem. J.* **2002**, *365*, 605–615.
37 D. I. Pattison, M. J. Davies, K.-D. Asmus, *J. Chem. Soc., Perkin Trans. 2* **2002**, 1461–1467.
38 M. D. Rees, C. L. Hawkins, M. J. Davies, *J. Am. Chem. Soc.* **2003**, *125*, 13719–13733.
39 M. D. Rees, C. L. Hawkins, M. J. Davies, *Biochem. J.* **2004**, *381*, 175–184.
40 M. Bonifačić, I. Štefanić, G. L. Hug, D. A. Armstrong, K.-D. Asmus, *J. Am. Chem. Soc.* **1998**, *120*, 9930–9940.
41 M. Bonifačić, D. A. Armstrong, I. Carmichael, K.-D. Asmus, *J. Phys. Chem. B* **2000**, *104*, 643–649.
42 I. Štefanić, M. Bonifačić, K.-D. Asmus, D. A. Armstrong, *J. Phys. Chem. A* **2001**, *105*, 8681–8690.
43 D. A. Armstrong, K.-D. Asmus, M. Bonifačić, *J. Phys. Chem. A* **2004**, *108*, 2238–2246.
44 P. Wisniowski, I. Carmichael, R. W. Fessenden, G. L. Hug, *J. Phys. Chem. A* **2002**, *106*, 4573–4580.
45 L. Feray, N. Kuznetsov, P. Renaud, in *Radicals in Organic Synthesis*, P. Renaud, M. P. Sibi (Eds.), Vol. 2, Wiley-VCH, Weinheim, 2001, pp. 246–278.
46 J. H. Horner, O. M. Musa, A. Bouvier, M. Newcomb, *J. Am. Chem. Soc.* **1998**, *120*, 7738–7748.
47 D. Yu, A. Rauk, D. A. Armstrong, *J. Am. Chem. Soc.* **1995**, *117*, 1789–1796.
48 F. Himo, L. A. Eriksson, *J. Am. Chem. Soc.* **1998**, *120*, 11449–11455.
49 A. Becker, K. Fritz-Wolf, W. Kabsch, J. Knappe, S. Schultz, A. F. Volkner Wagner, *Nature Struct. Biol.* **1999**, *6*, 969–975.
50 A. Becker, W. Kabsch, *J. Biol. Chem.* **2002**, *277*, 40036–40042.
51 C. J. Krieger, W. Rooseboom, S. P. J. Albracht, A. M. Spormann, *J. Biol. Chem.* **2001**, *276*, 12924–12927.
52 F. Himo, *J. Phys. Chem. B* **2002**, *106*, 7688–7692.

53 C. von Sonntag, *The Chemical Basis of Radiation Biology*, Taylor & Francis, London, 1987.
54 S. Zheng, G. L. Newton, G. Gonick, R. C. Fahey, J. F. Ward, *Radiat. Res.* **1988**, *114*, 11–27.
55 J. A. Aguilera, G. L. Newton, R. C. Fahey, J. F. Ward, *Radiat. Res.* **1992**, *130*, 194–204.
56 J.-T. Hwang, M. M. Greenberg, *J. Am. Chem. Soc.* **1999**, *121*, 4311–4315.
57 C. Mottley, R. P. Mason, *J. Biol. Chem.* **2001**, *276*, 42677–42683.
58 J. Vasquez-Vivar, A. M. Santos, V. B. Junqueira, O. Augusto, *Biochem. J.* **1996**, *314*, 869–876.
59 I. Wilson, P. Wardman, G. M. Cohen, M. d'Arcy Doherty, *Biochem. Pharmacol.* **1986**, *35*, 21–22.
60 E. Ford, M. N. Hughes, P. Wardman, *Free Radical Biol. Med.* **2002**, *32*, 1314–1323.
61 R. M. Gatti, R. Radi, O. Augusto, *FEBS Lett.* **1994**, *348*, 287–290.
62 X. Shi, Y. Rojanasakul, P. Gannett, K. Liu, Y. Mao, L. N. Daniel, N. Ahmed, U Saffiotti, *J. Inorg. Biochem.* **1994**, *56*, 77–86.
63 K. Hildenbrand, D. Schulte-Frohlinde, *Int. J. Radiat. Biol.* **1997**, *71*, 377–385.
64 W. A. Prütz, in *Sulfur-Centered Reactive Intermediates in Chemistry and Biology*, C. Chatgilialoglu, K.-D. Asmus (Eds.), NATO ASI Series A, Vol. 197, 1990, pp. 389–399.
65 A. A. Zavitsas, J. A. Pinto, *J. Am. Chem. Soc.* **1972**, *94*, 7390–7396.
66 A. A. Zavitsas, A. A. Melikian, *J. Am. Chem. Soc.* **1975**, *97*, 2757–2763.
67 A. A. Zavitsas, C. Chatgilialoglu, *J. Am. Chem. Soc.* **1995**, *117*, 10645–10654.
68 B. P. Roberts, *Chem. Soc. Rev.* **1999**, *28*, 25–35.
69 Y. Cai, B. P. Roberts, *J. Chem. Soc., Perkin Trans. 2*, **2002**, 1858–1868.
70 H.-S. Dang, B. P. Roberts, D. A. Tocher, *J. Chem. Soc., Perkin Trans. 1* **2001**, 2452–2461.
71 K. D. Beare, M. L. Coote, *J. Phys. Chem. A* **2004**, *108*, 7211–7221.
72 D. L. Reid, G. V. Shustov, D. A. Armstrong, A. Rauk, M. N. Schuchmann, M. S. Akhlaq, C. von Sonntag, *Phys. Chem. Chem. Phys.* **2002**, *4*, 2965–2974.
73 A. Rauk, D. A. Armstrong, D. P. Fairlie, *J. Am. Chem. Soc.* **2000**, *122*, 9761–9767.
74 M. S. Akhlaq, H.-P. Schuchmann, C. Von Sonntag, *Int. J. Radiat. Biol.* **1987**, *51*, 91–102.
75 M. J. Robins, G. J. Ewing, *J. Am. Chem. Soc.* **1999**, *121*, 5823–5824.
76 Z. Guo, M. C. Samano, J. W. Krzykawski, S. F. Wnuk, G. J. Ewing, M. J. Robins, *Tetrahedron* **1999**, *55*, 5705–5718.
77 D. Pogocki, Ch. Schöneich, *Free Radical Biol. Med.* **2001**, *31*, 98–107.
78 Ch. Schöneich, M. Bonifačić, K.-D. Asmus, *Free Radical Res. Commun.* **1989**, *6*, 393–394.
79 Ch. Schöneich, K.-D. Asmus, M. Bonifačić, *J. Chem. Soc., Faraday Trans.* **1995**, *91*, 1923–1930.
80 M. Z. Baker, R. Badiello, M. Tamba, M. Quintiliani, G. Gorin, *Int. J. Radiat. Biol. Relat. Stud. Phys. Chem. Med.* **1982**, *41*, 595–602.
81 C. von Sonntag, in *Sulfur-Centered Reactive Intermediates in Chemistry and Biology*, C. Chatgilialoglu, K.-D. Asmus (Eds.), NATO ASI Series A, Vol. 197, 1990, pp. 359–366.
82 R. Lenz, B. Giese, *J. Am. Chem. Soc.* **1997**, *119*, 2784–2794.
83 H. Zipse, *Org. Biomol. Chem.* **2003**, *1*, 692–699.
84 T. Nauser, Ch. Schöneich, *J. Am. Chem. Soc.* **2003**, *125*, 2042–2043.
85 A. Rauk, D. Yu, J. Taylor, G. V. Shustov, D. A. Block, D. A. Armstrong, *Biochemisty* **1999**, *38*, 9089–9096.
86 M. Jonsson, D. D. M. Wayner, D. A. Armstrong, D. Yu, A. Rauk, *J. Chem. Soc., Perkin Trans. 2* **1998**, 1967–1972.
87 M. B. Goshe, V. E. Anderson, *Radiat. Res.* **1999**, *151*, 50–58.
88 M. B. Goshe, Y. H. Chen, V. E. Anderson, *Biochemistry* **2000**, *39*, 1761–1770.
89 B. N. Nukuna, M. B. Goshe, V. E. Anderson, *J. Am. Chem. Soc.* **2001**, *123*, 1208–1214.

90 Ch. Schöneich, U. Dillinger, F. von Bruchhausen, K.-D. Asmus, *Arch. Biochem. Biophys.* **1992**, *292*, 456–467.

91 T. Nauser, J. Pelling, Ch. Schöneich, *Chem. Res. Toxicol.* **2004**, in press.

92 C. J. Easton, in *Radicals in Organic Synthesis*, P. Renaud, M. P. Sibi (Eds.), Vol. 2, Wiley-VCH, Weinheim, 2001, pp. 505–525.

93 D. L. Reid, D. A. Armstrong, A. Rauk, C. von Sonntag, *Phys. Chem. Chem. Phys.* **2003**, *5*, 3994–3999.

94 C. J. Easton, M. C. Merrett, *J. Am. Chem. Soc.* **1996**, *118*, 3035–3036.

95 X. Zhang, N. Zhang, H.-P. Schuchmann, C. Von Sonntag, *J. Phys. Chem.* **1994**, *98*, 6541–6547.

96 T. Nauser, G. Casi, W. H. Koppenol, Ch. Schöneich, unpublished results.

97 T. E. Creighton, *Proteins*, W. H. Freeman, New York, 1993.

98 R. Schweitzer-Stenner, F. Eker, Q. Huang, K. Griebenow, *J. Am. Chem. Soc.* **2001**, *123*, 9628–9633.

99 L. Grierson, K. Hildenbrand, E. Bothe, *Int. J. Radiat. Biol.* **1992**, *62*, 265–277.

100 R. Zhao, J. Lind, G. Merényi, T. E. Eriksen, *J. Chem. Soc., Perkin Trans. 2*, **1997**, 569–574.

101 A. Jain, N. K. Alvi, J. H. Parish, S. M. Hadi, *Mutat. Res.* **1996**, *357*, 83–88.

102 K. N. Carter, T. Taverner, C. H. Schiesser, M. M. Greenberg, *J. Org. Chem.* **2000**, *65*, 8375–8378.

103 T. Nauser, Ch. Schöneich, *Chem. Res. Toxicol.* **2003**, *16*, 1056–1061.

104 Ch. Schöneich, *Arch. Biochem. Biophys.* **2002**, *397*, 370–376.

105 A. I. Bush, W. H. Pettingell, G. Multhaup, M. Paradis, J.-P. Vonsattel, J. F. Gusella, K. Beyreuther, C. L. Masters, R. E. Tanzi, *Science* **1994**, *265*, 1464–1467.

106 C. C. Curtain, F. Ali, I. Volitakis, R. A. Cherny, R. S. Norton, K. Beyreuther, C. J. Barrow, C. L. Masters, A. I. Bush, K. J. Barnham, *J. Biol. Chem.* **2001**, *276*, 20466–20473.

107 M. Coles, W. Bicknell, A. A. Watson, D. P. Fairlie, D. J. Craik, *Biochemistry* **1998**, *34*, 11064–11077.

108 S. Varadarajan, J. Kanski, M. Aksenova, C. Lauderback, D. A. Butterfield, *J. Am. Chem. Soc.* **2001**, *123*, 5625–5631.

109 D. Pogocki, Ch. Schöneich, *Chem. Res. Toxicol.* **2002**, *15*, 408–418.

110 J. Kanski, M. Aksenova, Ch. Schöneich, D. A. Butterfield, *Free Radical Biol. Med.* **2002**, *32*, 1205–1211.

111 J. Kanski, S. Varadarajan, M. Aksenova, D. A. Butterfield, *Biochim. Biophys. Acta* **2001**, *1586*, 190–198.

112 K. L. Handoo, J.-P. Cheng, V. D. Parker, *J. Chem. Soc., Perkin Trans. 2* **2001**, 1476–1480.

113 K.-O. Hiller, K.-D. Asmus, *Int. J. Radiat. Biol.* **1981**, *40*, 597–604.

114 Ch. Schöneich, D. Pogocki, G. L. Hug, K. Bobrowski, *J. Am. Chem. Soc.* **2003**, *125*, 13700–13713.

4
Model Studies of Hydride-transfer Reactions

Richard L. Schowen

4.1
Introduction [1]

Enzymes catalyze many oxidation–reduction reactions in which the equivalent of dihydrogen is added or removed from a substrate molecule.

$$S + H_2 \text{ (equivalent to } 2H^+ + 2e^-) \rightleftharpoons SH_2 \tag{4.1}$$

As Eq. (4.1) emphasizes, dihydrogen is the equivalent of two protons and two electrons. If the reaction is conceived mechanistically as consisting of one proton and two electrons moving as a unit (the *hydride ion*, $H:^-$) and the second proton being moved separately (or omitted entirely), then the process may with justice be called a *hydride-transfer reaction.*

For example, in the case of the reduction of acetaldehyde to ethanol (as catalyzed by alcohol dehydrogenases) the following sequence of events, where Donor-H represents a hydride donor, is composed of a hydride-transfer mechanism for the step in Eq. (4.2a), a proton-transfer mechanism for the step in Eq. (4.2b), and a hydride-transfer/proton-transfer mechanism for the overall reaction formed by Eqs. (4.2a) and (4.2b):

$$\text{Donor-H} + CH_3CHO \rightarrow CH_3CH_2O^- + \text{Donor}^+ \text{ (hydride transfer)} \tag{4.2a}$$

$$CH_3CH_2O^- + H^+ \rightarrow CH_3CH_2OH \text{ (proton transfer)} \tag{4.2b}$$

Other mechanistic variants decouple the transfers of one or both electrons from the transfer of the proton so that, for example, the reaction of Eq. (4.2a) might be accomplished by an initial proton transfer followed by a sequence of two one-electron transfers or the transfer of an electron succeeded by the transfer of a hydrogen atom. These processes or combinations of steps within them may be called *formal hydride-transfer reactions.*

This chapter is concerned with mechanistic observations on hydride-transfer processes in non-enzymic systems, but under conditions and with structures such that the observations are considered relevant to enzyme-catalyzed reactions. Even

Hydrogen-Transfer Reactions. Edited by J. T. Hynes, J. P. Klinman, H.-H. Limbach, and R. L. Schowen

ISBN: 978-3-527-30777-7

within this narrow compass, no effort at a comprehensive survey of the vast available literature has been made, but instead publications are described that are judged particularly relevant to current biochemical concerns. Their extensive lists of references should be consulted if comprehensive information is wanted.

The terms "coenzyme" and "cofactor" are used as synonyms here, as seems to be the case in most textbooks (see Ref. [1] as an example). Duine (2004) has suggested specific distinctions among these and related terms but the distinctions appeared not to be crucial for the discussions in this chapter, which deals with nonenzymic reactions.

4.1.1 Nicotinamide Coenzymes: Basic Features

Nicotinamide coenzymes have large complex structures (Fig. 4.1) that divide conceptually into two parts, the nicotinamide nucleus and the dinucleotide-derived part which may either be simple (NAD^+/NADH) or phosporylated at the

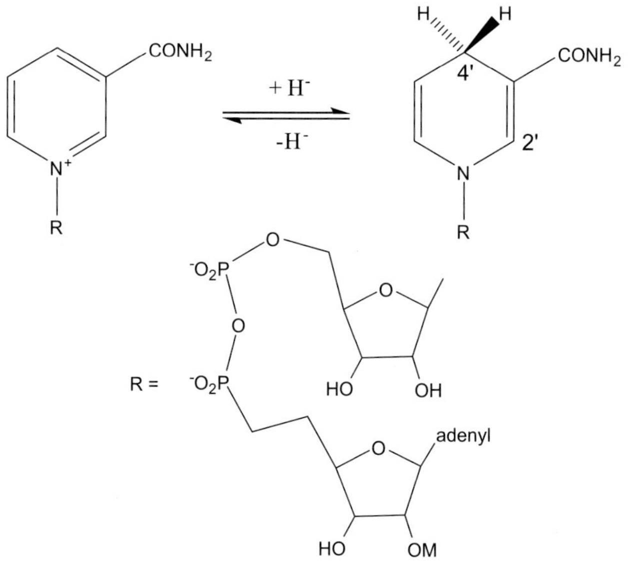

Figure 4.1. The nicotinamide coenzymes, nicotinamide adenine dinucleotide (NAD^+ or $NADP^+$) and dihydronicotinamide adenine dinucleotide (NADH or NADPH). The locations of C(4′) and C(2′) are indicated. The groups R and M are important only for binding and orientation in complexes of the cofactor with enzymes and do not participate directly in the redox chemistry. The oxidized coenzymes bear an electrical charge one unit more positive than the reduced coenzymes and the interconversion is a formal hydride transfer.

adenosine-2′-position ($NADP^+$/NADPH). The nicotinamide nucleus can exist either in the pyridinium form (oxidized form, with a net charge positive by one unit over the charge on phosphate ester moieties, NAD^+/$NADP^+$) or the 1,4-dihydropyridine form (reduced form, with a net charge equal to the charge on the phosphate ester moieties, NADH/NADPH). The redox interconversion then is formally a hydride-transfer reaction. The remaining part of the structure does not undergo chemical reactions in the course of coenzyme action, and serves instead to bind the entire coenzyme in some specific orientation to the active site of the host redox enzymes. The two sets of nicotinamide coenzymes, NAD^+/NADH and $NADP^+$/NADPH, differ from each other only in the structural region that is not a participant in the redox chemistry, so that a given model reaction can suffice to describe features of both.

4.1.2 Flavin Coenzymes: Basic Features

In a manner similar to the structures of the nicotinamide cofactors, the flavin cofactors FMN/$FMNH_2$ and FAD/$FADH_2$ also have structures that resolve into a chemically reactive unit, the flavin nucleus, and a large ancillary structure that has the function of binding the cofactor in a specific orientation to the host enzyme, as emerges from Fig. 4.2. Here again the flavin nucleus and thus the strictly chemical properties are common to the two coenzymes, which differ in the ancillary part of the structure.

4.1.3 Quinone Coenzymes: Basic Features

In relatively recent times, a number of biologically novel structures that function as cofactors in redox reactions, some of them as agents of hydride transfer, have been discovered to be present in enzyme or other protein structures. They are often covalently bound and are formed in post-translational reactions from the side-chains of normal, genetically encoded amino-acid residues. The structures are shown in Fig. 4.3, taken from Mure's excellent review [2]. The important chemical functionality of these coenzymes is the quinone ring, an *ortho*-quinone in most cases, both a *para*-quinone and an *ortho*-quinone in TPQ.

4.1.4 Matters Not Treated in This Chapter

This chapter describes model studies of hydride transfer entirely with respect to nicotinamide coenzymes, flavin coenzymes and quinone coenzymes. Other coenzymes/cofactors may be alluded to but are not reviewed in detail. Some coenzymes involved either in hydride transfer or the transfer of other hydrogen species have been treated elsewhere in these volumes (thiamin diphosphate is treated by Hübner et al., pyridoxal phosphate by Spies and Toney, folic acid by Benkovic

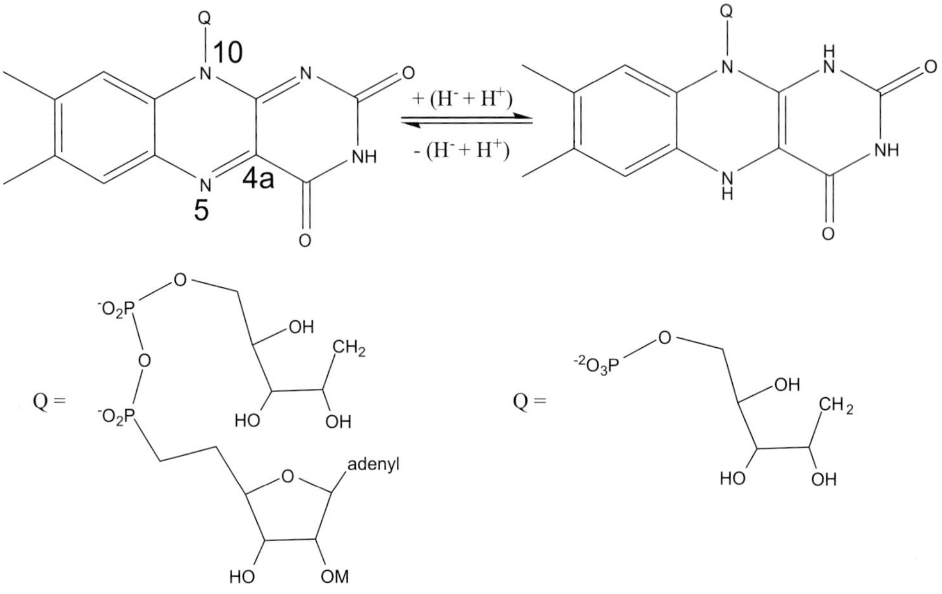

$FAD/FADH_2$ $FMN/FMNH_2$

Figure 4.2. The flavin coenzymes, flavine adenine dinucleotide/dihydroflavine adenine dinucleotide ($FAD/FADH_2$), and flavin mononucleotide/dihydroflavin mononucleotide (FMN/FMN). The locations of N(5), N(10), and C(4a) are indicated. The groups Q are important only for binding and orientation in complexes of the cofactor with enzymes and do not participate directly in the redox chemistry. The oxidized coenzymes bear an electrical charge equal to that of the reduced coenzymes and the interconversion is formally a dihydrogen-addition reaction, equivalent to the transfer of one hydride ion and one proton.

and Hammes-Schiffer, and cobalamin by Banerjee et al.) and articles on electron transfer and proton-coupled electron transfer may treat the role of metal ions, metal clusters, hemes, and related structures.

4.2 The Design of Suitable Model Reactions

The term, "model reactions," can mean several things. Some studies that may be thought of as involving models focus on *biomimetic reactions*. This name is customarily applied to processes in which structures similar in some sense to those in-

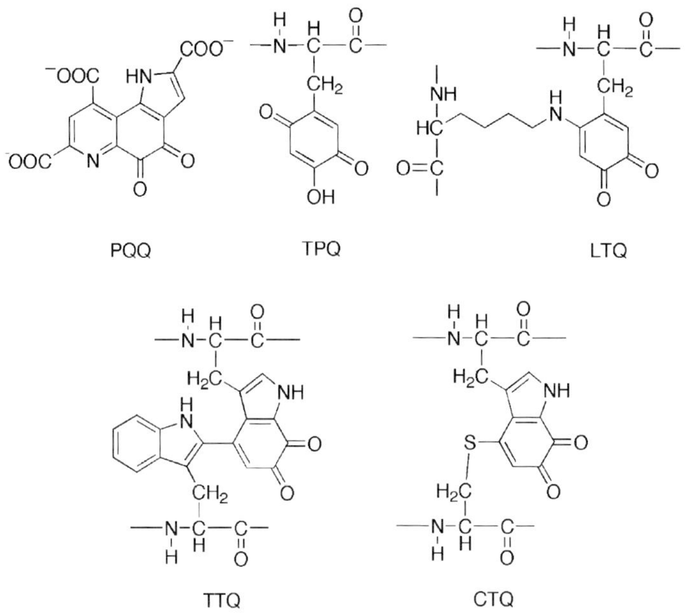

PQQ TPQ LTQ

TTQ CTQ

Figure 4.3. Structures of quinone cofactors. PQQ: pyrolloquinoline quinone; TPQ: 2,4,5-trihydroxyphenylalanine quinone; LTQ: lysine tyrosylquinone; TTQ: tryptophan tryptophyl quinone; CTQ: cysteine tryptophyl quinone. Taken with permission from Mure [2].

volved in biology are put to work in chemical or technological synthesis or analysis: for example, the use of NADH analogs as chemical reducing agents. There is a vast and valuable literature on this subject, which will be passed over here without comment. Sometimes, theoretical studies, such as the construction of potential surfaces for reactions important in biology, are referred to as "modeling" of the biological reactions. This sort of work will be brought in as needed.

For our purposes, "model studies" or studies of "model reactions" will refer to investigations in which enzymes or other features of the biological environment are omitted so as to provide a system that is easier to construct and control than the true biological system and where the influences of biological agents such as enzymes are omitted. Often the structures most intimately involved are themselves decreased in complexity, as discussed below. The rates and mechanisms of the reactions thus simplified, which are the focus of model studies, then provide a kind of baseline information in which one hopes the biological influences have been removed, leaving only strictly chemical factors to determine the observed behavior. Comparison with the biological system can then illuminate what biological evolution has used from the basic chemical system, what it has invented, and what basic features it has enhanced or diminished.

Previous writers have provided reviews of very high quality of parts or all of the studies reviewed here. Some of these are mentioned below at appropriate points. Here, two reviews by one of the great modern originators and practitioners of chemistry applied to biology, F.H. Westheimer, should be mentioned. His account [3] of the discoveries of enzyme mechanisms in the period 1947 to 1963 and an extraordinary critical review of model studies of nicotinamide (and to some extent flavin) coenzymes [4] are of great value and should be consulted by every reader of the present article.

Earlier writers have also expressed useful views about proper characteristics of model reactions. In particular, Kosower, in a work that broke new ground in chemical biology ([5], pp. 276–277), suggested the difficulty of achieving the *duplication* of enzyme mechanisms with model compounds but noted that mechanistic *parallels* between enzyme and model reactions can nevertheless lead to informative results, culminating in what he denoted *congruency* between enzyme and model reactions, i.e., a very strong resemblance in terms of reactant structures and of the nature and sequential order of mechanistic events.

4.2.1 The Anchor Principle of Jencks

The structural complexity of coenzymes, in contrast to the chemical simplicity of the reactions they are involved in, has always given chemists pause for thought when they contemplate biology. Thus NADH serves as the carrier of a hydride ion (three elementary particles, with a mass slightly over 1 Da) yet has a total molecular weight around 664 Da. At first glance, this may seem quite extravagant of Nature, to which Newton [6] famously atttributed the opposite virtue ("To this purpose the philosophers say that Nature does nothing in vain, and more is in vain when less will serve ..."). Something closer to the borohydride ion might seem to have been more advisable.

It was William Jencks who put most clearly the idea that, for example, the effective delivery of three elementary particles to a specific atomic location at a high rate in the biological context can adhere entirely to the Newtonian principle of parsimony while requiring a substantial array of molecular superstructure. The concept is formulated this way in Jencks's words [7]: *Energy from the specific-binding interactions between an enzyme and a substrate or coenzyme is required to bring about the (highly improbable) positioning of reacting groups in the optimum manner and such binding requires both a high degree of three-dimensional structure and a large interaction area. Thus the binding interaction of the adenine-ribose-phosphate-phosphate-ribose moiety of NAD^+* [see Fig. 4.1] *with a dehydrogenase provides the binding energy that anchors the coenzyme in the correct position, so that only an internal rotation of the C–N bond to the nicotinamide ring ... need be frozen in order to bring the 4 position into the correct relationship for reduction to occur ...* Jencks also notes more generally: *This anchoring effect immediately provides a qualitative rationale for the large sizes of enzymes, coenzymes, and some substrates.*

For model studies that target the chemical mechanistic features of the enzyme-catalyzed reactions, the anchor principle is of great practical value, for it allows the researcher to discard large and inconvenient fragments of the biologically active species, once these fragments have been shown to have an anchor function only. In such a case, a model compound need contain only the chemically significant features of the non-anchor portion of the natural species. One may then enjoy the convenience of structures that are smaller, easier to synthesize and modify, and cheaper. Figure 4.4 illustrates some of the ways in which these opportunities have been seized in model studies for the nicotinamide cofactors.

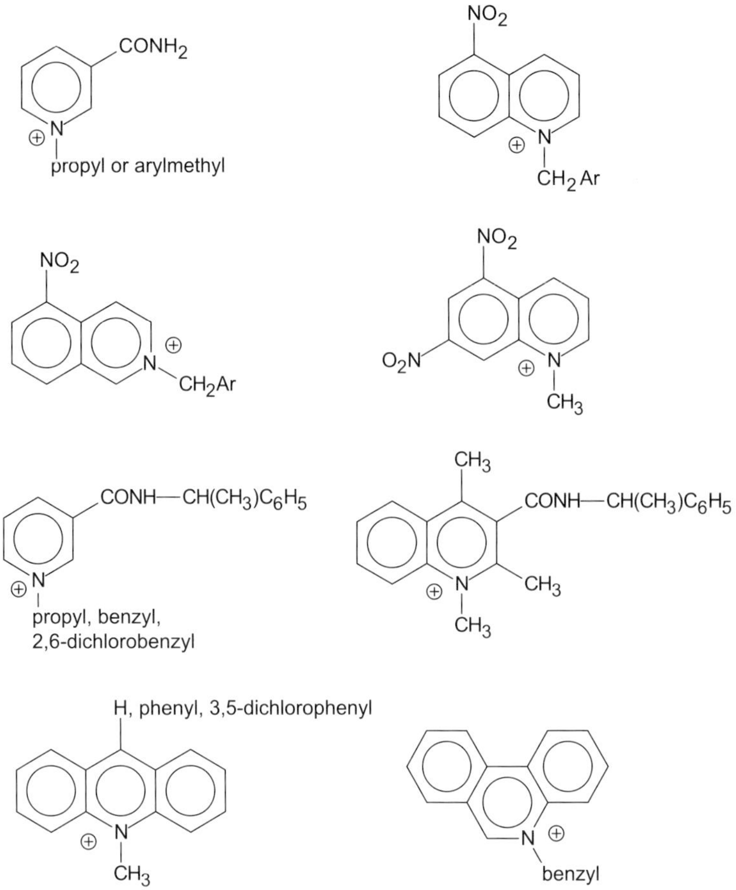

Figure 4.4. Illustrative examples of the liberty of design permitted the student of model reactions for the nicotinamide cofactors. From upper left, descriptions and references may be found in Bunting and Sindhuatmadja [8]; Bunting and Norris [9]; Bunting and Norris [9]; Bunting and Norris [9]; Ohnishi et al. [10]; Ohno et al. [11]; Lee et al. [12], Lee et al. [12]. It is easy to see the role of convenience in selection of compounds with chromophoric or solubility properties that lighten the burden of kinetics experiments or with substitution patterns that alleviate the pain of synthetic procedures.

4.2.2
The Proximity Effect of Bruice

Two species near each other have a higher probability of colliding (the first prerequisite for reaction) than when they are not near each other: the idea that reaction of two species simultaneously bound to an enzyme active site might therefore react more rapidly with each other than when both are free in solution is thus a venerable idea in enzymology. It is an idea that has been fruitfully examined in non-enzymic intramolecular reactions.

The fundamental development of the concept in quantitative terms owes much to Bruice and his coworkers. By the thorough study of intramolecular reactions in which the reacting groups were required by synthetic creativity to adopt specific distances and orientations, Bruice showed that the rate effects of appropriate relative locations were potentially many decades in magnitude and thus beyond doubt significant for an understanding of enzyme catalysis. Initially, Bruice and Benkovic [13] used the term "propinquity catalysis" to describe such accelerations. Later, others employed various terms for the same idea but today probably the most commonly used name is "proximity effect" and a deeply painstaking documentation and thorough analysis has been provided by Kirby [14].

As the power of computational science has grown, Bruice and his coworkers [15, 16] have further developed the concept by defining, through the use of experiential mechanistic information, specific molecular configurations in which the distances between reacting centers are less than a value slightly larger than a bond distance and the orientations of electronic orbitals are close to the final orientations appropriate for bond formation or fission. These configurations, which need not be, and generally are not, stationary points on the potential-energy surface, are called "near-attack conformations" or "NACs." If the reacting-center separations and the orbital orientations are suitably chosen, then the NAC can be considered a structure especially likely to lead on to the transition state for the reaction.

One carries out a molecular-dynamics simulation for the reactant-state assembly at a temperature of interest for a particular example, and counts the number of configurations attained within a specific time period that are within the chosen NAC limits. Very commonly, those species for which large numbers of NACs are observed are found experimentally to undergo reaction between the proximate centers more rapidly than is true for species with a smaller population of NACs. Such studies are capable of developing a reasonably reliable catalog of the distance and orientational requirements for a broad range of reactions and thus defining some of the requirements for enzymes to promote reactions by the adjustment of distances and orientation.

The proper design of intramolecular reactions that simulate desired features of biological reactions is thus a major line of approach in using model reactions to investigate biochemical processes in general and hydride-transfer reactions in particular.

4.2.3
Environmental Considerations

It has again been long realized that the chemical environments within enzymes can have an enormous range of properties, because of the diversity of the chemical structures of the natural amino-acid side chains in terms of polarity and lack thereof, electrostatic features, and the capacity to donate or accept hydrogen bonds, as well as the potential diversity of the secondary, tertiary and quaternary levels of protein structure in modulating these properties and producing others [17]. It is equally well-known that chemical reactions are very powerfully affected by the medium in which they occur [18]. It is one of the legitimate aims of model reactions to explore the ways in which environmental conditions within enzyme active sites may aid in controlling the rates and mechanisms of such processes as hydride-transfer reactions.

Three main approaches have been taken in the general case of model reactions for all biochemical processes. On the one hand, model reactions have been studied in solvents with various characteristics that simulate particular features of a protein environment. In this approach, such macroscopic properties as dielectric constant or such microscopic properties as the propensities for donating and accepting hydrogen bonds are examined. Second, a more direct simulation of some of the properties of intact proteins has been attempted by the synthesis of small-peptide analogs with highly defined and thoroughly controlled structures. This approach to hydrogen-transfer reactions is described by Lars Baltzer in Chapter 5. Third, intramolecular reactions, particularly those within host–guest complexes have been exploited in the simulation of features of enzymic reactions, including environmental effects. Such studies have been reviewed recently from several viewpoints by Rebek [19–21].

4.3
The Role of Model Reactions in Mechanistic Enzymology

4.3.1
Kinetic Baselines for Estimations of Enzyme Catalytic Power

Enzymes are catalysts and the question of their quality as catalysts is irresistible to the chemist, although biologists may now and then regard the question as idle or eccentric. However, the magnitudes of enzymic catalytic acceleration factors certainly have a biological value, because the catalytic power of a modern enzyme represents the effect of molecular evolution in developing, from some primitive enzyme of low quality, the modern enzyme of today, which is often impressive in its catalytic power [22]. The catalytic power can be quantitatively measured, in principle, by a catalytic acceleration factor for any enzyme, which might logically be taken as the ratio of the rate of the enzyme-catalyzed reaction under chosen conditions to the rate of the same reaction in the absence of the enzyme under the same conditions. By such a definition, one requires, for calculation of enzyme catalytic power, a knowledge of the kinetics – ideally the rate constants – for both the enzyme-catalyzed reaction and the non-enzymic reaction.

The kinetic constants for the enzymic reaction, in this day of cloned enzymes and high-throughput kinetics, are frequently available at minimum investment, but the same may not be true for the non-enzymic reaction, particularly if the ratio of rates is large. Indeed it is frequently large, consistent with the idea that the basic biospheric strategy is to select intrinsically slow reactions as chemical components of physiological networks, then evolve powerfully catalytic enzymes, with the overall result that organismic chemistry exhibits a high ratio of signal (enzymic reaction) to noise (non-enzymic reaction). This means that determination of the kinetics of a non-enzymic reaction can present an experimental challenge of daunting proportions, a matter that has been addressed with great skill and determination by Wolfenden and his coworkers [23].

A much easier problem is posed by the fact that even the simplest enzymic reaction has two kinetic parameters (a second-order rate constant commonly denoted k_{cat}/K_M and a first-order rate constant known as k_{cat}; the notation may seem confusing to non-enzymologists but these two quantities are quite independent of each other) while the simplest non-enzymic reaction will have one rate constant (call it k_{unc} and imagine it, for the sake of argument, to be a first-order rate constant). There are therefore two possible measures of catalytic power, $[(k_{cat}/K_M)/k_{unc}]$ and $[k_{cat}/k_{unc}]$, so which is correct?

Radzicka and Wolfenden [24] elegantly show that both are correct and that they yield different and valuable information about the effects of the enzyme. The relevant equations have been gathered into Chart 4.1. The quantity $[(k_{cat}/K_M)/k_{unc}]$, given the name *catalytic proficiency*, measures the equilibrium constant for binding of the transition state for the uncatalyzed reaction to the unoccupied active site of the enzyme (see Eq. (iv) in Chart 4.1), and thus the total stabilization of that transition state by the enzyme. The quantity $[k_{cat}/k_{unc}]$ is called the *rate enhancement* and measures (Eq. (v) of Chart 4.1) the equilibrium constant for expulsion of the reactant-state substrate molecule from the active site and its replacement in the active site by the transition state for the uncatalyzed reaction. The quantity there-

Chart 4.1. Measures of enzyme catalytic power for a unireactant enzyme.

Catalytic Power of a Unireactant Enzyme with Unimolecular Non-enzymic Reaction

$E + S \rightarrow E{:}TS_{k/K}$	$K = \exp[-(\Delta G^*_{k/K})/RT] = (k_{cat}/K_M)/(k_B T/h)$	(i)
$ES \rightarrow E{:}TS_k$	$K = \exp[-(\Delta G^*_{k})/RT] = (k_{cat})/(k_B T/h)$	(ii)
$S \rightarrow TS_{unc}$	$K = \exp[-(\Delta G^*_{unc})/RT] = (k_{unc})/(k_B T/h)$	(iii)
$E + TS_{unc} \rightarrow E{:}TS_{k/K}$	$K = (k_{cat}/K_M)/(k_{unc})$ = catalytic proficiency	(iv)
$ES + TS_{unc} \rightarrow E{:}TS_k + S$	$K = (k_{cat})/(k_{unc})$ = rate enhancement	(v)

fore gives the stabilization of the transition state for the uncatalyzed reaction by its binding to the enzyme *diminished by* the stabilization afforded the reactant substrate molecule by its binding to the enzyme. The rate enhancement is thus a measure of the *net transition-state stabilization* by the enzyme.

The two quantities, catalytic proficiency (which has the dimensions M^{-1} in the example above) and rate enhancement (which is dimensionless in the example above), give a valid account of two aspects of enzyme catalysis. The catalytic proficiency, as the equilibrium constant for transition-state binding to the free enzyme, measures quantitatively the affinity of the free enzyme for the transition state. The free-energy equivalent of the catalytic proficiency gives the total transition-state stabilization by the enzyme. The rate enhancement, as the equilibrium constant for the expulsion of a substrate molecule from the active site of the enzyme and its replacement by a transition-state molecule, quantitatively describes the relative affinity of the enzyme for the transition state compared to the reactant-state substrate. The free-energy equivalent of the rate enhancement gives the net transition-state stabilization (the excess of transition-state stabilization over reactant-state stabilization) by the enzyme.

For more complex enzymic reactions, such as those that require more than two rate constants to describe the kinetics, and more complex non-enzymic reactions that also can require complex kinetic expressions, it is possible to define more than two measures of catalytic power. The questions involved are addressed in an appendix (see p. 1071) at the end of this chapter.

4.3.2 Mechanistic Baselines and Enzymic Catalysis

A reasonable ambition for model reactions is that their mechanisms ought to contain some clues about the mechanism of the enzyme-catalyzed reaction also. It has long been realized that it is fruitless simply to build the model-reaction mechanism into an enzyme active site. Such a procedure would entail the view that the factors present and at work in the model system render a complete account of the biological history of the enzyme. There is no reason to expect this to be so, and many reasons to think it would not be so. In the simplest sense, a given enzyme must occupy a niche in a metabolic network that may require its regulation and may influence its structure and mechanistic potentialities in ways that cannot be derived from non-enzymic studies.

What the mechanistic baseline provided by a model reaction does do is to suggest where molecular evolution may have necessarily started, although it must be remembered that substrates and enzymes have co-evolved so that the chemical scope of a primitive substrate may not be that of a modern substrate. Furthermore, a model-reaction mechanism may indicate what processes it was necessary for an enzyme to avoid in order to open the way for a comparatively unfavorable reaction to be catalyzed. One of the most striking examples is provided by the recent studies of Kluger and his coworkers of the unexpected chemical properties of the cofactor thiamin diphosphate [25–29]. Although this work has nothing explicitly to do with hydride transfer, it constitutes a brilliant case of mechanistic deduction that gave rise to new ideas about thiamin-dependent enzymes.

In the action of α-ketoacid decarboxylases, the cofactor thiamin diphosphate adds across the α-keto carbonyl group of the substrate, placing the thiazolium ring so as to delocalize the electron pair liberated by decarboxylation. In the normal course of the enzymic reaction, the delocalized species is protonated on the α-carbon center and the resulting alcohol then undergoes elimination of the thiazolium nucleus to regenerate the cofactor and form the product aldehyde. In the model reaction examined by Kluger and his coworkers, this pathway was dominated by a competing reaction in which the cofactor undergoes fragmentation and is destroyed. Thus the complex evolutionary problem solved in the molecular evolution of the thiamin-dependent decarboxylases involved acceleration of the decarboxylation reaction by a large factor (around 10^{12}) while preventing acceleration of the abortive, indeed self-sacrificial, fragmentation reaction that is chemically preferred. Without establishment of the non-enzymic mechanistic baseline, this fact might never have been known.

4.4 Models for Nicotinamide-mediated Hydrogen Transfer

4.4.1 Events in the Course of Formal Hydride Transfer

Since formal hydride transfer involves the transfer of one proton and two electrons, one can imagine various sequences and combinations by which this end can be achieved. Possible mechanisms can be classified according to the degree to which the transfers of the individual particles occur together in time or separately in time. In the limit that all three particles move precisely at the same time, the transferring entity amounts to a hydride ion and the process may accurately be described as a "true" hydride-transfer reaction. In the opposing limit, when each of three particles moves in a separate chemical reaction, the reaction must involve at least three separate steps and the process is said to be a stepwise transfer of electrons and proton. Sometimes an electron transfer is designated E and a proton transfer P so that the possible sequences are EEP, EPE, and PEE.

Powell and Bruice [30] pointed out that for the simple identity reaction of Eq. (4.3):

H CONH2 ⊕N + H H CONH2 N ⇌ H H CONH2 N + H CONH2 ⊕N (4.3)

in the light of the principle of microscopic reversibility or detailed balance and the necessary symmetry of the reaction, the only possible sequence of transfer steps is EPE: electron, proton, electron. All other sequences are different for forward and reverse reactions and therefore impossible. The two limiting mechanisms are presented in Fig. 4.5.

In the most general sense, the problem is one of proton-coupled electron transfer, as described by Hammes-Schiffer and by Nocera in Volume 1, Chapters 16 and 17, respectively. The two limiting mechanisms described above are the cases of perfect coupling (concerted, one-step hydride transfer) and perfect uncoupling (EPE).

4.4.2
Electron-transfer Reactions and H-atom-transfer Reactions

Electron-transfer reactions and hydrogen-atom-transfer reactions are physically possible with NADH and its analogs. Much evidence indicates such a mechanistic versatility in redox reactions for this class of compounds.

For example, Miller and Valentine [31] in 1988 showed that an extremely simple analog of NADH, 1-benzyl-1,4-dihydronicotinamide, underwent oxidation in propanol solvent by ferricinium ion in a three-step series of transfers (electron-proton-electron) to produce the analog of NAD^+. The absence of an isotope effect for the transferring hydrogen indicates that its transfer does not occur in the rate-limiting step, while the lack of dependence of the rate on electrolyte concentration suggests the rate-limiting step to be electron transfer from the substrate to ferricinium ion with both reactants and transition state bearing a single positive charge. However, the reaction of both this model compound and NADH itself with quinone oxidants in water solution, as opposed to propanol solvent, occurred by one-step hydride transfer, the authors argued, because large isotope effects were observed and the reaction was 10^{5-7}-fold faster than the electron-transfer rate estimated from Marcus theory.

A simple interpretation is that powerful, obligate one-electron oxidants may elicit single-electron donation from NADH, but its reaction with two-electron acceptors is normally a single-step hydride transfer process.

Almarsson et al. [32] found that the NADH-model compound 1-methyl-1,10-dihydroacridan and its deuterated form underwent oxidation by $Fe(CN)_6{}^{3-}$ in aqueous solution by a sequence of:

1. A rapid, reversible one-electron transfer reaction to generate a radical cation.
2. A rate-limiting proton-transfer reaction to a general base with a Brønsted β of 0.2 and deuterium isotope effects of about 5–10, resulting in a neutral free radical.
3. A rapid one-electron transfer to a second molecule of ferricyanide ion.

The sequence is shown in Eq. (4.4), an EPE process as described above.

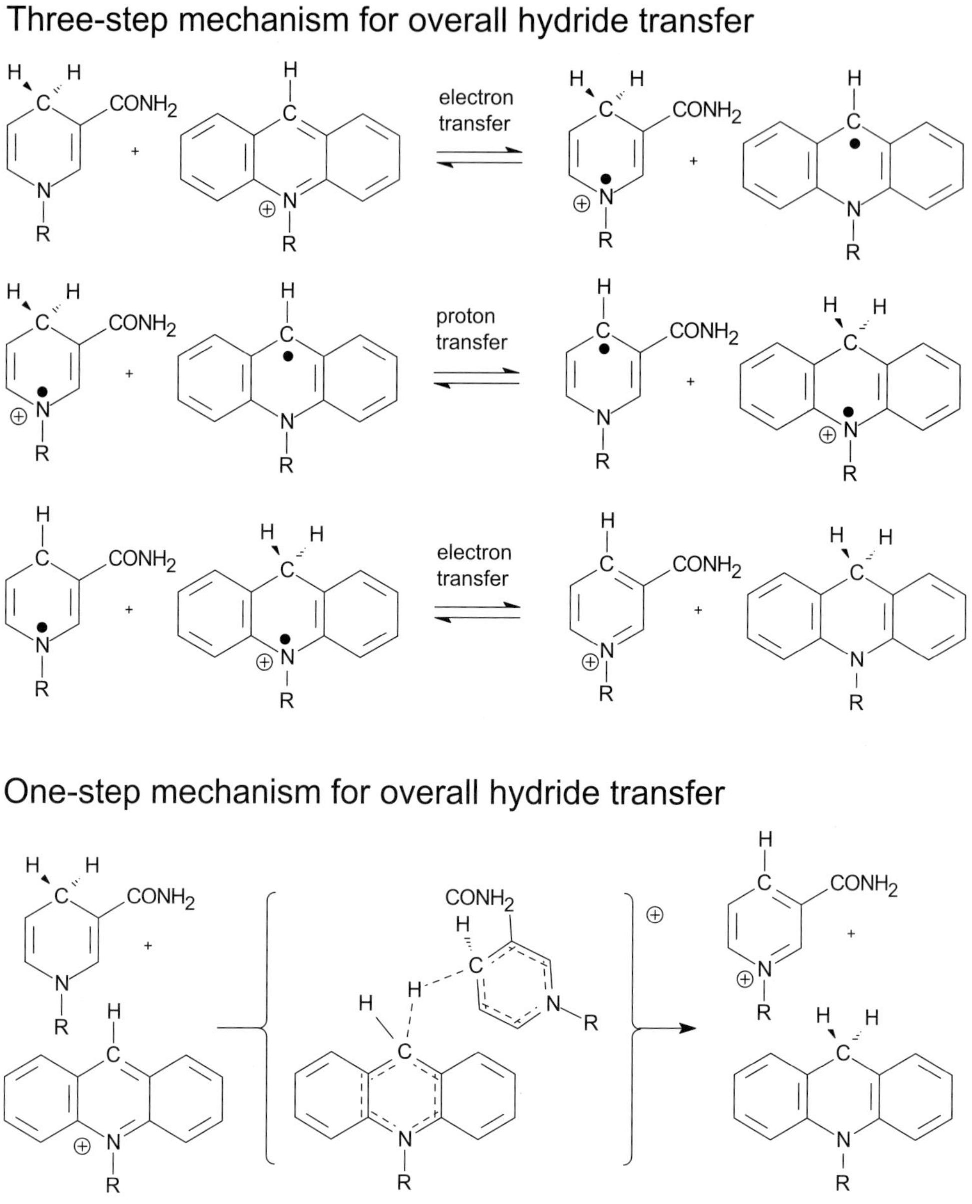

Three-step mechanism for overall hydride transfer

electron transfer

proton transfer

electron transfer

One-step mechanism for overall hydride transfer

Figure 4.5. The limiting mechanisms of three-step hydride transfer and one-step hydride transfer, the former in the canonical order of Powell and Bruice [37], for the overall reduction by NAD(P)H of hydride-acceptor molecules. The operative distinctions are that (a) there are radical/radical-ion intermediates in the multistep mechanism but not of course in the one-step mechanism; (b) the rate-determining step is necessarily the product determining step in a one-step mechanism, but if there are alternative products not shown here, then the two steps may differ in the multistep reaction.

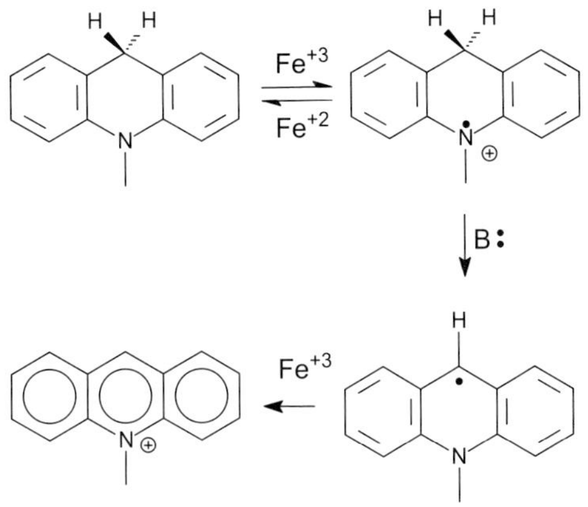

(4.4)

The reaction thus conforms to the rough generalization already given, since ferricyanide ion is an obligate one-electron oxidant.

Matsuo and Mayer [33], on the other hand, found that the same reactant in acetonitrile solution, upon treatment with $Ru^{IV}O^{2+}$, quickly generated the acridinium ion, as might have been anticipated for a simple hydride-transfer reaction, but in only 40–50% yield. Relatively slowly thereafter the acridinium compound and the remaining reactant were converted to the acridinium leuco-base (hydroxide-ion adduct). Matsuo and Mayer concluded that the process shown in Eq. (4.5) was occurring:

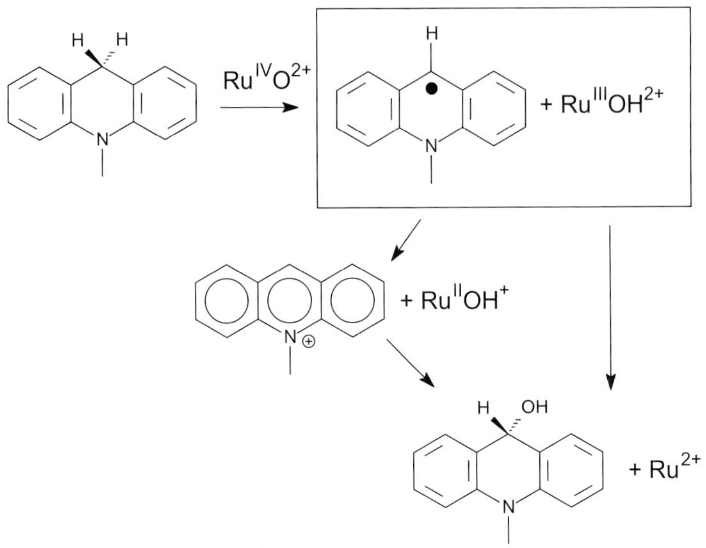

(4.5)

First, hydrogen-atom abstraction occurred to generate the molecular pair shown in the box. This pair then underwent two competing reactions: electron transfer from the acridine radical to ruthenium to produce the observed acridinium ion and $Ru^{II}OH^+$, and hydroxyl-radical donation from $Ru^{III}OH^{2+}$ to produce the leucobase shown at the bottom. Simultaneously the acridinium ion and $Ru^{II}OH^+$ formed along the electron-transfer route react by hydroxide-ion donation to the acridinium partner to produce the same final products. In further support of this scheme, the authors noted that reactant disappearance is accelerated about five-fold under aerobic conditions compared to anaerobic conditions, consistent with the initial hydrogen-atom transfer being to some degree reversible, with dioxygen then trapping away the acridine free radical under aerobic conditions.

Matsuo and Mayer [33] note that the thermodynamic driving forces are not very different for hydrogen-atom and hydride transfers in this system, the faster hydrogen-atom-transfer then suggesting the intrinsic barrier may be lower for the atom-transfer reaction than for the hydride-transfer reaction. A contributing factor to such a difference could be transition-state stabilization in these specific hydrogen-atom-transfer reactions through a donor – acceptor polar effect that operates only in the transition state [34, 35: pp. 77–85]. The oxygen center of an oxo-ruthenium species, such as abstracts the hydrogen atom here, is electronegative and capable of bearing substantial negative charge, while the forming pyridinoid ring of the acridine partner is capable of easily stabilizing positive charge. Such a donor–acceptor pair tends to generate considerable charge dispersion in the favored direction in atom-abstraction transition states and the resulting transition-state stabilization would lead to a reduced intrinsic barrier.

4.4.3 Hydride-transfer Mechanisms in Nicotinamide Models

Particularly in the 1970s, several lines of evidence were taken to suggest a major role for electron-transfer processes in model reactions for the action of nicotinamide cofactors. Bruice and his coworkers [30, 36–38] in 1982–1984 showed that subtle effects rendered these observations deceptive, and that in fact hydride transfer is the only mechanism at work in the aqueous-solution hydrogen-transfer models that had formed the earlier focus. Further relevant references and an extraordinary analysis are given in the review by Westheimer [4]. The main outlines are discussed below.

A key line of evidence for a multistep mechanism, as opposed to the one-step hydride-transfer mechanism, had been derived from isotope effects measured in reduction of various substrates with monodeuterated analogs of NADH. One can compare the observed rate constants k_{HH} and k_{HD}, which in the case of negligible secondary isotope effects should obey the relationship $k_{DH}/k_{HH} = (1 + [k_D/k_H])/2$, allowing the calculation of the primary isotope effect k_H/k_D (if undeuterated, monodeuterated and dideuterated hydride donors are all used, both primary and secondary isotope effects can be obtained). In addition, for an oxidizing agent Acceptor$^+$ one can determine the isotope ratio in the product Acceptor–H/Acceptor–D, called in these studies the product isotope effect Y_H/Y_D.

For a simple one-step hydride-transfer mechanism, these two isotope effects should be identical. For an EPE mechanism, the value of k_H/k_D will reflect which step is more nearly rate-limiting and could vary from unity if one of the electron-transfer steps E is rate-limiting to large values if the proton-transfer step P is rate limiting. If the proton-transfer step is irreversible then, whether it is rate-limiting or not, its isotope effect will fix the value of Y_H/Y_D. Therefore if the two isotope effects are identical, no information about the nature of the mechanism is obtained. But if the two differ, and in particular if Y_H/Y_D (reflecting the true isotope effect on the hydrogen-transfer step) is larger than k_H/k_D (a weighted average of small isotope effects on electron transfer and the true effect for hydrogen transfer), then the one-step hydride-transfer mechanism is excluded.

The reactions of various analogs of NADH with ketones and acridinium cations gave isotope effects on the rate constants for substrate reduction (k_H/k_D) that were different in magnitude from the isotope effects measured by isotope abundances in products compared to reactants (Y_H/Y_D). For example, Y_H/Y_D was found to be constant at around 6 in one series of reactions, while k_H/k_D varied with the structure of the hydride donor from about 3.3 to about 5.7. A hydride-transfer mechanism therefore appeared to be excluded.

Early measurements were quickly shown to be in error as a result of the reversible formation in aqueous solution of hydrates of the dihydronicotinamide analogs and of adducts involving the hydride acceptors, but later studies were conducted in dried aprotic solvents and some made used of acridine-derived analogs of NADH that could not form hydrates or adducts. These later studies continued to exhibit discrepancies in the values of k_H/k_D and Y_H/Y_D.

Reinvestigations of the matter and extended studies were reported by Powell, Wong, and Bruice [36], and by Powell and Bruice [30, 37, 38]. The discrepancies between k_H/k_D and Y_H/Y_D were shown to arise from isotope exchange reactions. For example, when the hydride acceptor *N*-methylacridinium cation $MA(H)^+$ reacted with an NADH analog N(H,D) to form the reduced product MA(H,D), this material could react in a symmetrical hydride-transfer reaction with another molecule of the unreduced reactant $MA(H)^+$, still present in excess. Either of the product hydrogens of MA(H,D) could be transferred. If the H were transferred, as should occur more frequently, a molecule of MA(H,H) was fallaciously added to the apparent product mixture in place of the originally formed MA(H,D) and the deuterium was essentially permanently sequestered as $MA(D)^+$ in a large excess of $MA(H)^+$. This process is facile and quantitatively accounts for the fallaciously large values of Y_H/Y_D that were previously measured. When a full account of all processes at work was constructed by Powell and Bruice [37, 38], there was no isotope-effect discrepancy remaining and all the available evidence favored a hydride-transfer mechanism for nicotinamide reactions. That remains the situation today, one-electron transfers or H-atom transfers arising only in the kinds of limiting circumstances described above.

In a remarkable article published in 1991 [39], Bunting reviewed structure–reactivity studies relevant to the nature of the hydride-transfer process between materials that can be regarded as related to nicotinamide cofactors. Much of the article concerned the large quantity of work published from Bunting's own laboratory, and his interpretative section strikingly presages current thinking, particularly in the area of proton-coupled electron transfer, on the degrees of possible coupling among the motions of proton and electrons in a formal hydride-transfer reaction. Bunting's suggestion was that it is fruitful to think in terms of what he denoted a "merged mechanism" with varying degrees of coupling arising under circumstances that depend on reactant properties and environmental considerations.

4.4.4
Transition-state Structure in Hydride Transfer: The Kreevoy Model

The Marcus formulation of the dependence of the free energy of activation $\Delta G^\ddagger$ on the free energy of reaction ΔG° has been enormously useful as a method of thinking about transition-state structure in solution reactions. Originally developed for electron-transfer reactions, it has been extended to proton-transfer and hydride-transfer reactions and thence to other group-transfer reactions. In an article on methyl-transfer reactions [40], Albery and Kreevoy described the consideration of variations in transition-state structure not just in terms of the *parallel* coordinate connecting reactants to products (along which Marcus theory accounts for the Hammond Postulate, the tendency of exergonic reactions to occur with reactant-like transition states and of endergonic reactions to occur with product-like transition states) but also along the *perpendicular* coordinate (see Fig. 4.6). In such maps as Fig. 4.6, sometimes called "maps of alternate routes," the parallel coordinate represents a trajectory along which bond-order at the transferring H is maintained at unity, while other routes account for the possibility of either total bond orders at H greater than unity (*tight* transition states in the northwestern part of the map) or smaller than unity (*loose* transition states in the southeastern part of the map). Albery and Kreevoy [40] and later Kreevoy, Lee, and their coworkers [12, 41–43] put the concept of both reactant-product and tight-loose characteristics of transition states on a common quantitative basis (Chart 4.2).

In the Albery–Kreevoy–Lee approach, as outlined in Chart 4.2, a hydride-transfer transition state can be described by the structure–reactivity sensitivity factor or Brønsted coefficient $\alpha = d[\ln(k_{i0})]/d[\ln(K_{i0})]$, where k_{i0} is the rate constant for one of a series of hydride-acceptors A_i^+ reacting with a standard hydride donor A_0H, and K_{i0} is the equilibrium constant for the transfer. The Brønsted coefficient in turn is a sum of two terms (Chart 4.2, Eqs. (vii)–(x)).

The first term in Eqs. (x), χ, describes transition-state variation along the reactant–product coordinate and is given by Eq. (viii) of Chart 4.2. This is the normal result of Marcus theory. In effect, $-RT[\ln(K_{i0})] = \Delta G_{i0}{}^\circ$ can vary (for the circumstances we wish to address here) within the limits $-\lambda$ to $+\lambda$. Here λ is the reorganization energy, or the work required to distort the reactant structure to a precise simulacrum of the product (and $\lambda/4$ is the "intrinsic barrier" or reaction barrier in the absence of any thermodynamic driving force). If $\Delta G_{i0}{}^\circ = +\lambda$, then the transition state will itself be a precise simulacrum of the product. If this limit is inserted into Eq. (viii) of Chart 4.2, then $\chi = 1$ as expected for an exactly product-like transition-state structure. At the other limit of $\Delta G_{i0}{}^\circ = -\lambda$, the transition state

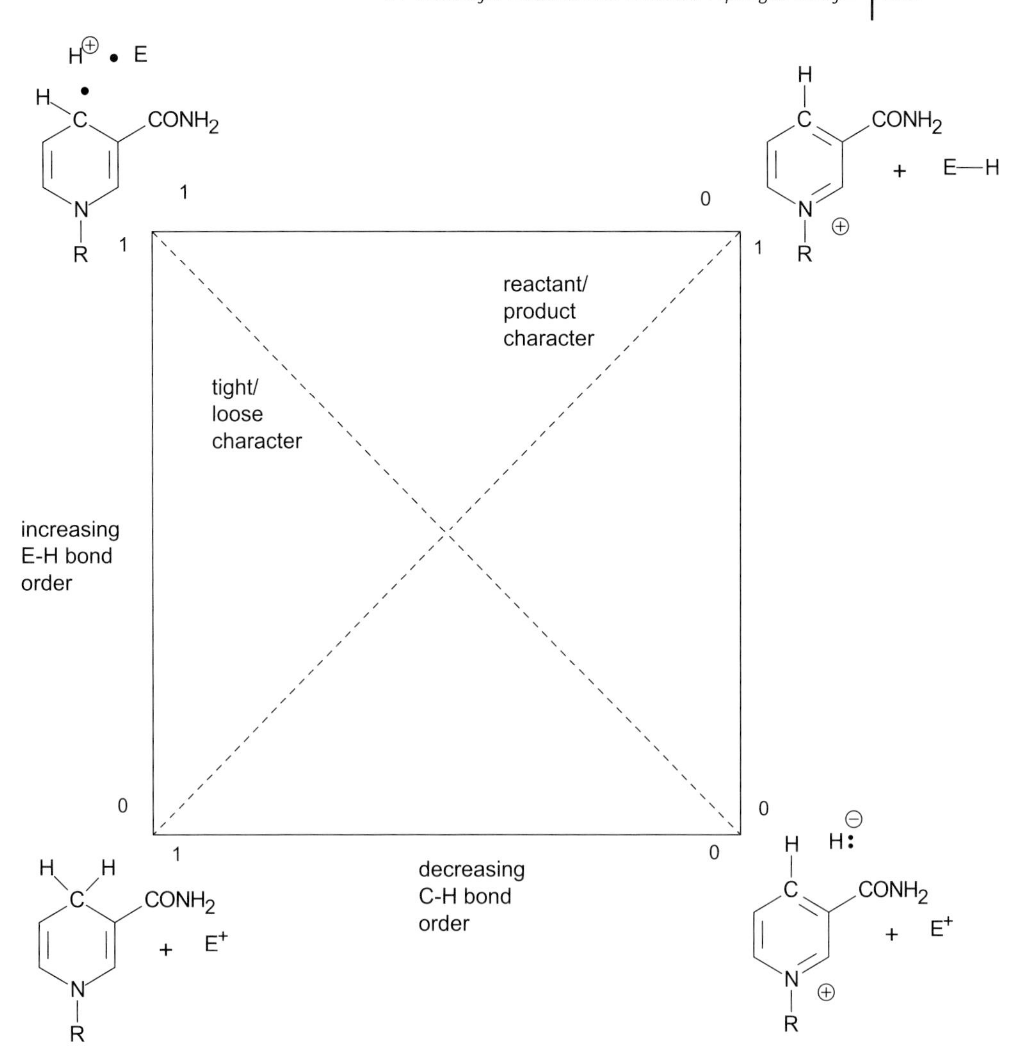

Figure 4.6. A map of alternate routes for overall hydride transfer from NADH or an analog to an electrophile E^+. The reactant structure is shown at bottom left (the "southwest" corner) and the product structure at upper right (the "northeast" corner). The bond order of the C–H decreases from one to zero along the abscissa and the bond order of the E–H bond increases from zero to one along the ordinate. Two hypothetical intermediate structures are shown at the southeast corner (both bond orders about the hydride ion are zero) and at the northwest corner (the E–H bond has fully formed while the C–H bond remains fully intact). Any point in this map represents a possible transition-state structure. Any trajectory that connects the reactant structure, one or more transition-state structure(s), and the product structure represents a possible reaction route.

Chart 4.2. The transition-state descriptors for hydride-transfer reactions as developed by Kreevoy, Han Lee, and their coworkers [12, 40–43][a]

Reaction: $A_i^+ + H\text{–}A_0 = A_i\text{–}H + A_0^+$ (i)

Here the structure of A_i^+ is varied while $H\text{–}A_0$ is a standard compound.

Experimentally accessible: rate constants k_{i0}, k_{ii}; equilibrium constants K_{i0}

Basic relationships:

Thermodynamics: $\ln(K_{i0}) = -\Delta G_{i0}^\circ/RT$ (ii)

Transition-state theory: $\ln(k_{i0}) = -\Delta G_{i0}^\ddagger/RT + \ln(v)$ (iii)

"ultrasimple" version: $v = k_B T/h$

Marcus theory: $\Delta G_{i0}^\ddagger = W^r + [\lambda/4][1 + (\Delta G_{i0}^\circ/\lambda)]^2$ (iv)

$\lambda = (\lambda_{ii} + \lambda_{00})/2$ (v)

From Eqs. (ii), (iii), and (iv):

$\ln(k_{i0}) = a + b[\ln(K_{i0})] + c[\ln(K_{i0})]^2$ (vi)

$a = \ln(v) - (W^r/RT) - (\lambda/4RT)$; $v = \text{Gk nu}$

$b = 1/2$

$c = -RT/4\lambda$

Brønsted coefficient $\alpha = d[\ln(k_{i0})]/d[\ln(K_{i0})]$ *with i varied, 0 constant*

$\alpha = (1/2)\{1 - (RT/\lambda)[\ln(K_{i0})]\} + (1/2)\{d[\ln(k_{ii})]/d[\ln(K_{i0})]\}[1 - \{(RT/\lambda)[\ln(K_{i0})]\}^2]$ (vii)

Define:

Fractional progress from reactant toward product:

$\chi = (1/2)\{1 - (RT/\lambda)[\ln(K_{i0})]\}$ (viii)

(λ as in Eq. (v) above)

Compression from loose toward tight:

(a) K_{i0} varied by changing structure of hydride acceptor:

$\tau = 1 + d[\ln(k_{ii})]/d[\ln(K_{i0})]$ (ix-a)

(b) K_{i0} varied by changing structure of hydride donor:

$\tau = 1 - d[\ln(k_{ii})]/d[\ln(K_{i0})]$ (ix-b)

From putting Eqs. (viii) and (ix) into Eq. (vii):

Acceptor varied: $\alpha = \chi + (1/2)(\tau - 1)(1 - \{(RT/\lambda)[\ln(K_{i0})]\}^2)$ (x-a)

Donor varied: $\alpha = \chi + (1/2)(1 - \tau)(1 - \{(RT/\lambda)[\ln(K_{i0})]\}^2)$ (x-b)

and when $\{(RT/\lambda)][\ln(K_{i0})]\}^2$ *becomes negligible as* $\ln(K_{i0})$ *approaches zero:*

$\alpha = \chi \pm (1/2)(\tau - 1) = 1/2 \pm (1/2)(\tau - 1)$ (xi)

---

[a] Additional assumption: Reactant-state work function W^r and product-state work function W^p are approximately equal.

becomes a precise simulacrum of the reactant, and as expected $\chi = 0$. Between these limits is the point at which reactants and products have equal energy ($\Delta G_{i0}^\circ = 0$) and Eq. (viii) then yields $\chi = 1/2$, signifying a centrally-located or "symmetrical" transition state. The interpretations that have usually been given to the Brønsted coefficient α itself are thus assumed by the χ-term in this formulation.

The second term in Eqs. (x) for the Brønsted coefficient α contains a quantity $(1/2)(\tau - 1)$ or $(1/2)(1 - \tau)$, depending on the site of structural variation, where τ is defined by Eqs. (ix) and purports to describe the transition-state structure along a "tight–loose" coordinate. The quantities $(1/2)(\tau - 1)$ and $(1/2)(1 - \tau)$ are multiplied by a weighting factor $(1 - \{(RT/\lambda)[\ln(K_{i0})]\}^2)$. It should be noted that the weighting factor approaches zero as the equilibrium free-energy change approaches both of its upper and lower limits of $\pm\lambda$, which correspond to reactant-like and product-like structures of the transition state. Thus "tightness" and "looseness" play no role in the value of the Brønsted coefficient α in these limits. Instead the importance of these considerations enters only in the "symmetric" situation, where the weighting factor approaches its maximum value of unity, the term χ approaches a value of 1/2 and, as shown in Chart 4.2 by Eq. (xi), the Brønsted coefficient α itself approaches a value of $\tau/2$ or $[1 - (\tau/2)]$, depending on the site at which structure is varied. A glance at the map of Fig. 4.4 suggests that it is logical that there should be no scope for transition states with reactant-like or product-like structures to develop "tight" or "loose" character, being confined in structure as they are to the southwest and northeast corners of the map. In contrast, there is maximal scope for such influences near the "symmetrical" point.

The quantity τ was originally suggested by Albery and Kreevoy [40] to correspond to the sum of the bond orders about the transferring entity, so that the terminally "loose" transition state ought to have $\tau = 0$, the terminally "tight" transition state ought to have $\tau = 2$, and that for a reaction in which bond-making and bond-breaking are exactly compensatory, the value should be $\tau = 1$. As Eqs. (ix) in Chart 4.2 show, for these values to be achieved puts requirements on the value of the derivative $d[\ln(k_{ii})]/d[\ln(K_{i0})]$, which would need to be equal to -1, for the terminally loose, $+1$ for the terminally tight, and 0 for the fully compensatory transition state. That these requirements agree with expectations for the actual transition states is shown by arguments in Table 4.1, which summarizes the situation with respect to values under various circumstances of the variables χ and τ and of the Brønsted coefficient α.

Kreevoy and his coworkers (and others such as Lee et al. [43] and Würthwein et al. [44]) have applied this formalism to various models for nicotinamide-mediated redox reactions. An immediate result was that the conclusions about transition-state structure based only on the traditional interpretation of α, i.e., ascribing to it the properties of χ, can lead to substantial error.

For example, two systems with different but related structures considered by Lee et al. [42] generated α values of 0.67 (for one system in which structural variation was in the hydride donor) and 0.32 (for a different system in which structural variation was in the hydride acceptor). It is tempting, since the sum of these values is near unity, to imagine that the two systems have identical transition-state structures, with the hydride ion liberated to the extent of about 70% from the donor and attached to the acceptor to the extent of about 30% and the hydride atom itself therefore bearing little or no charge. However, these systems had been thoroughly studied from the viewpoint of Marcus theory and estimates were available to permit the values of χ to be calculated from Eq. (viii) of Chart 4.2. The calculations yielded $\chi = 0.49$ for variation of the hydride donor and $\chi = 0.48$ for variation of the hydride acceptor, suggesting that in fact both transition states were centrally located and of "symmetrical" structure about the hydride moiety. Furthermore

Table 4.1. Values of interest for the reaction-progress variable χ, the compression variable τ, the Brønsted coefficient α, and their significance.

Parameter	Value	Significance
χ	0	Transition state is reactant-like along the reaction-progress coordinate; occurs when $\ln(K_{i0}) = +\lambda/RT$, corresponding to $\Delta G_{i0}{}^{\circ} = -\lambda$, the limiting exergicity.
	1/2	Transition state is central on the reaction-progress coordinate; occurs when $K_{i0} = 1$, corresponding to $\Delta G_{i0}{}^{\circ} = 0$. There is no driving force to influence the structure of the transition state.
	1	Transition state is product-like along the reaction-progress coordinate; occurs when $\ln(K_{i0}) = -\lambda/RT$, corresponding to $\Delta G_{i0}{}^{\circ} = +\lambda$, the limiting endergicity.
τ	0	Transition state has no bonding to H which is therefore in a hydride-like ($H:^-$) circumstance. The electron density available for bonding (2 electrons in a formal hydride-transfer reaction) is thus sequestered on H and shared neither by the donor moiety nor the acceptor moiety in the critical state for the identity reactions (rate constant k_{ii}). For the acceptor moiety, this corresponds to no change in electron density and for the donor moiety a decrease of unit electron density, as in the equilibrium (K_{i0}) reaction. Thus k_{ii} and K_{i0} respond equally to donor/acceptor structure ($d[\ln(k_{ii})]/d[\ln(K_{i0})] = +1$) and $\tau = 1 - 1 = 0$.
	1	Transition state has unit total bond order to H, as is the case in both reactant and product states. The gain in bond order of the forming bond compensates exactly in the transition state for the loss in bond order of the breaking bond. The gain in electron density in the acceptor moiety for the identity reaction is exactly compensated by the loss in electron density from the donor moiety and k_{ii} is independent of donor/acceptor structure ($d[\ln(k_{ii})]/d[\ln(K_{i0})] = 0$). Thus $\tau = 1 - 0 = 1$.
	2	In the transition state, the H is bound by a one-electron bond to each of the donor and acceptor moieties and bears the full positive charge so that the donor and acceptor moieties are electrically neutral. This corresponds in the identity reaction to no change in electron density for the donor moiety upon attaining the transition state, and a gain of unit electron density for the acceptor moiety. The net change is thus equal and opposite to that for the equilibrium (K_{i0}) reaction ($d[\ln(k_{ii})]/d[\ln(K_{i0})] = -1$), so that $\tau = 1 - (-1) = 2$.
α	0	Traditionally taken to signify a reactant-like structure for the transition state, this value could result from such combinations as $\chi = 0$, $\tau = 1$ (transitional conclusion correct) or $\chi = 1/2$, $\tau = 0$ (traditional conclusion incorrect).

Table 4.1 *(continued)*

Parameter	Value	Significance
	1/2	Traditionally taken to signify a central structure for the transition state, this value could result most simply only from $\chi = 1/2$, $\tau = 1$ (traditional conclusion correct). BUT see the text for more complex circumstances.
	1	Traditionally taken to signify a product-like structure for the critical state, this value could result from such combinations as $\chi = 1/2$, $\tau = 2$ (transitional conclusion incorrect) or $\chi = 1$, $\tau = 1$ (traditional conclusion correct).

when Eqs. (x-a) and (x-b) of Chart 4.2 were used to calculate values of τ from the quantity $\alpha - \chi$, values of $\tau = 0.64$ (variation in the donor) and $\tau = 0.68$ (variation in the acceptor) were obtained. Note that because of the difference between Eqs. (x-a) and (x-b), it is possible for these two reactions to have essentially identical values of χ and of τ and yet very different values of α. The suggestion of the complete analysis is then that the two reactions share a common transition-state structure with the hydride moiety about half-transferred (χ around 1/2 in both cases) with less than conserved bond order (a loose transition state with τ about 0.6 to 0.7, while a value near unity would have been expected for conserved bond order and values greater than one for a tight transition state).

Since the question of tunneling arises in the next section, and the essential ubiquity of tunneling in nicotinamide reactions is a current theme in both model studies and enzymic studies, it is worthwhile to consider whether analyses of the sort just described are invalidated if the "ultrasimple" transition-state theory cannot describe the events in question. There seems to be no reason not to continue to use the framework of Chart 4.2 to investigate the nature of the states at or near the maximum of the activation barrier whether or not they require such sophisticated approaches as those described by Hynes, Hammes-Schiffer, Truhlar and Garrett, Warshel, Smedarchina, Klinman, Kohen, Scrutton, and Banerjee in these volumes. Although the results are currently cast in the language of transition-state structure, the measurements are rate-equilibrium comparisons that compare the effect of structure on the work required for barrier crossing (or penetration) with the work required for reaching the equilibrium products. The understanding that is generated by these comparisons, even if phrased in the language of simple transition-state theory as a convenience, should survive translation into the language of new ways of formulating rate-processes whenever that is desired. The treatment of isotope effects and their temperature dependences is more challenging, and such easy interconversions are not to be expected there. Indeed the general subject of how to reconcile findings from structure–reactivity/rate–equilibrium studies with the more complex theoretical treatments currently being introduced and applied needs much more careful study.

4.4.5
Quantum Tunneling in Model Nicotinamide-mediated Hydride Transfer

The articles in these volumes by Hynes, Hammes-Schiffer, Truhlar and Garrett, Warshel, Smedarchina, Klinman, Kohen, Scrutton, and Banerjee provide extensive evidence that enzymic hydride-transfer reactions very frequently, if not universally, proceed by a tunneling mechanism (see also the compilation by Romesberg and Schowen [45]). It is then of substantial interest to know whether tunneling also occurs in the non-enzymic reactions. If it does not, enzymes must in effect be creating *de novo* a tunneling mechanism to replace a slower non-tunneling mechanism, whereas if tunneling occurs also in model reactions, enzymes are simply accelerating an existing tunneling mechanism.

Powell and Bruice ([38]) studied the temperature dependence of the isotope effects for the reaction shown in Eq. (4.6) (L = H, D) from −2 °C to 50 °C in acetonitrile solution.

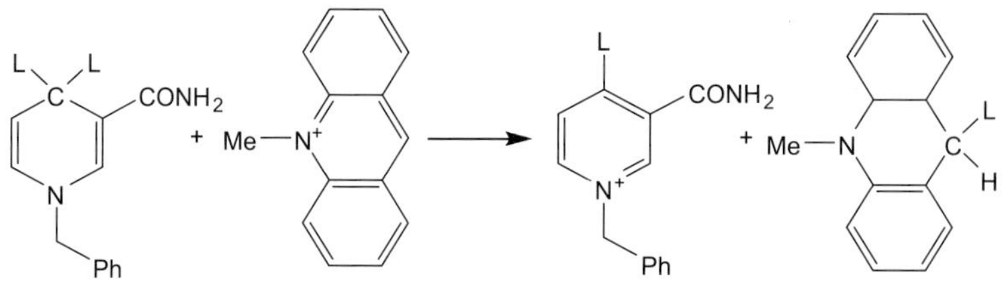

(4.6)

Since two labels are present in the hydride donor, the isotope effects are a product of a primary isotope effect for the transferring hydrogen and a secondary isotope effect for the hydrogen that remains. The two effects were estimated in a related system, the secondary effect emerging as a few percent larger than unity, as expected because the hybridization state is changing from sp^3 to sp^2 and the freedom of motion of the non-transferring center is thus increased in the transition state. Rate constants for both isotopic reactions obey the Arrhenius relationship $k_{LL} = A_{LL} \exp(-E_{LL}/RT)$ very well, the isotope effects k_{HH}/k_{DD} varying from about 6.2 at the lowest temperature to about 4.0 at the highest temperature. Most significantly, the difference in activation energies is far larger than the isotope effects would suggest, with $E_{DD} - E_{HH} = 7.7 \pm 1.0$ kJ mol^{-1}. If this difference alone determined the value of the isotope effects, k_{HH}/k_{DD} would have been around 22 at 25 °C instead of between 4 and 6, as is observed. The ratio A_{DD}/A_{HH}, which should be close to unity if the entire isotope effect arises from zero-point energy differences, was in fact 4.3 ± 1.3.

This combination, of an over-large isotopic activation-energy difference with a ratio of pre-exponential factors that is far from one with A_{DD} greater than A_{HH}, is one of the first-recognized indications of a role for tunneling in solution reactions, having been observed and interpreted by Bell and his collaborators in the 1950s and later [46]. Powell and Bruice made use of the most sophisticated theoretical treatment available at the time (the Bell approach implemented in a computer program of Kaldor and Saunders [47]) and estimated that around 70% to 80% of the reaction might be resulting from tunneling. Current approaches to the treatment of such results would probably lead to a different quantitative conclusion, but it remains clear that at least a considerable fraction of the overall hydride-transfer reaction is a result of tunneling. Indeed, as understanding deepens, it is becoming unclear whether this language of a division between parallel pathways of tunneling and over-the-barrier reaction will survive.

Related studies by Lee et al. [12] produced isotope effects for the reduction of models for NAD^+ by a 1,3-dimethyl-2-phenylbenzimidazoline derivative labeled at the 2-position with either protium or deuterium. The isotope effects k_H/k_D varied from around 4 to about 6.3. It was possible by means of the Marcus formulation described in Chart 4.2 to ascribe the variations with structure to a combination of parallel effects (reactant-like vs. product-like, the parameter χ) and perpendicular effects (tight vs. loose about the hydride center, the parameter τ). The perpendicular effects corresponded very well to an expected linear dependence on the logarithm of the equilibrium constant and were the main source of structure-induced variation in the isotope effect. The smaller parallel effects produced a very good fit to the expected quadratic dependence. The authors point out that the failure to separate these effects is the probable reason for the considerable scatter seen in traditional plots of the phenomenological isotope effects against the free energy of reaction. The isotope effects were consistent with model calculations using variational transition-state theory with inclusion of large-curvature ground-state tunneling (see Chapter 27 by Truhlar and Garrett in Volume 1).

The simplest conclusion is then that tunneling occurs both in enzymic hydride-transfer reactions and in related non-enzymic (model) reactions. It remains to be seen whether enzymic rate acceleration has evolved simply to make the existing tunneling reaction much more efficient or instead to create new tunneling mechanisms distinct from those observed in model systems.

4.4.6 Intramolecular Models for Nicotinamide-mediated Hydride Transfer

An early attempt to construct a close intramolecular model for enzymic hydride transfer was Overman's [48] synthesis of the reactant shown in Eq. (4.7):

NO reaction (4.7)

The pyridinium ring simulates an NAD(P)$^+$ cofactor and the transannular hydroxy function any of a variety of dehydrogenase substrates. Overman noted that space-filling models "show that the hydroxy methine hydrogen and the pyridinium 4-position are held tightly together and that conformations of the 14-membered ring do exist in which hydride addition could occur perpendicular to the plane of the pyridinium ring." However, the transannular redox reaction proved impossible to observe. In aqueous solution at pH 12, a leuco-base was formed by hydroxide-ion addition at the 2-position of the ring. In hexamethylphosphoramide solvent with bases such as sodium hydride, lithium bis-(trimethylsilyl)amide or potassium *tert*-butoxide, no redox reaction occurred over 12 h at 30 °C. The reasonable conclusion reached was that not only approximation of the reactants but other features not accessible to this model compound accounted for the rapid enzymic redox reaction.

Fifteen years later, Meyers and Brown [49], now citing the rapidly accumulating and already large related literature, reported a highly stereospecific intramolecular model reaction, shown in Eq. (4.8).

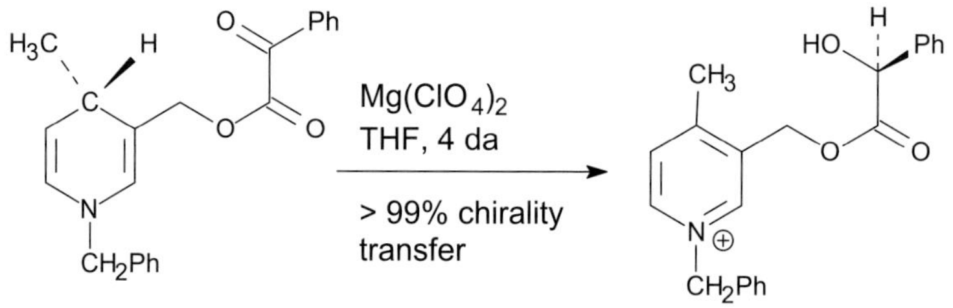

H_3C, H, C, N, CH_2Ph, O, Ph, O, O

$Mg(ClO_4)_2$
THF, 4 da
> 99% chirality transfer

CH_3, $N^{\oplus}$, CH_2Ph, HO, H, Ph, O, O

(4.8)

Here the side-chain is unrestricted but the required magnesium ion is thought to bind to both the carbonyl group to be reduced and the ring nitrogen center to bring the hydride-transfer distance into a range for reaction to occur.

At essentially the same time, Kirby and Walwyn [50, 51] created a closely related model system for lactate dehydrogenase (Eq. (4.9)).

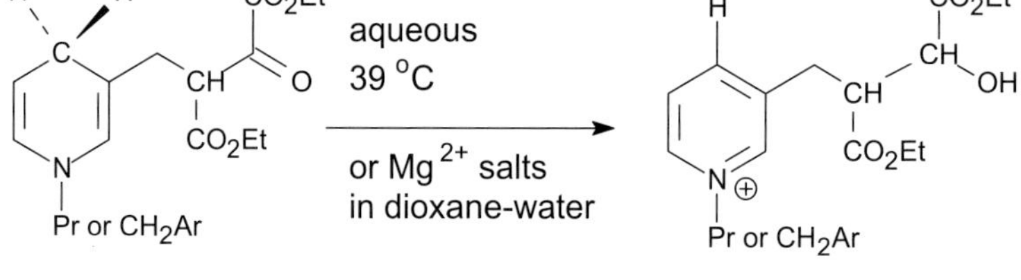

H, H, C, N, Pr or CH_2Ar, CH, CO_2Et, CO_2Et, O

aqueous
39 °C
or Mg^{2+} salts
in dioxane-water

H, $N^{\oplus}$, Pr or CH_2Ar, CH, CO_2Et, CH, CO_2Et, OH

(4.9)

It was then possible for Yang et al. [52] to make use of the results of Kirby and Walwyn to understand a significant mechanistic feature of the NAD$^+$-dependent enzyme *S*-adenosylhomocysteine hydrolase.

This enzyme possesses a molecule of NAD$^+$, bound non-covalently but very tightly to the enzyme, that oxidizes the 2′-hydroxy group of the substrate *S*-adenosylhomocysteine in the first step of the mechanism. The resulting keto-group then activates the adjacent 4′-hydrogen for removal, allowing the elimination of homocysteine from the 5′-position followed by addition of water at the 5′-position. The NADH, formed in the first step of the mechanism, thereafter reduces the product of water addition, 3′-keto-adenosine, to generate adenosine in the last step of the mechanism. Throughout the events between these first and last steps, it is vital that the intermediates are not reduced – if this should happen, the subsequent normal reactions are impossible and the catalytic cycle is aborted.

Yang et al., making use of kinetic studies of Porter and Boyd [53–55], were able to compare the rates of the normal reduction reactions of the enzyme, carried out during the catalytic cycle, with the rate of the abortive reduction reaction, which occurs rarely but for which the rate was carefully measured by Porter and Boyd. The free-energy barriers for the normal reductions were an average of 66 kJ mol^{-1} in height, while that for the model reaction of Eq. (4.8) was 92 kJ mol^{-1} in height (data for aqueous solution), showing that the enzyme during the catalytic cycle was accelerating the reduction by a factor of approximately 40 000 over the acceleration already present in the model reaction. The barrier height for the abortive reaction was 89 kJ mol^{-1}, essentially equal to that for the model reaction (in fact the rate constant for the abortive reaction at 25 °C was 2×10^{-3} s^{-1}, while the rate constant for the model reaction at 39 °C was 3×10^{-3} s^{-1}).

Thus the enzyme prevents the abortive reduction from occurring by suspending, during the central part of the catalytic cycle, its acceleration of the redox reactions from a factor of about 40 000 over the effect of the model reaction to nil, so that the reaction is accelerated by only the approximation effect modeled in the reaction of Eq. (4.8). An examination of several crystal structures of the enzyme suggested that the enzyme may accomplish this suspension of catalytic power by means of a conformation change coupled to the initial and final redox reactions that begin and end the catalytic cycle. The enzyme before oxidation of the ligand appears to have a distance between C-4′ of the cofactor and C-3′ of the substrate (the distance over which the hydride would need to move if its transfer occurred with no change in the positions of cofactor and substrate) of about 3.2 Å. After oxidation, this distance is increased to about 3.6 Å and two histidine residues (H55 and H301) move to buttress the cofactor in this more distant and thus less reactive position. All of these numbers have large errors but the apparent increase in the distances is present in four different structural comparisons, in agreement with the hypothesis that the first redox reaction, by means of a conformation change coupled to it, suspends part of the catalytic power of the enzyme for the redox reaction, and this catalytic power is then only restored when a reverse conformation change, coupled to the final redox reaction, occurs in concert with it.

4.4.7 Summary

Model reactions have been of major importance in the development of our current good understanding of the mechanisms of action of enzymes utilizing nicotinamide cofactors. The cofactors are now generally supposed to effect redox reactions by a hydride-transfer mechanism with structurally variable transition states, and to exhibit one-electron chemistry only in rarely encountered circumstances, e.g., reactions with metal complexes that strongly favor such reactions. Quantum tunneling has been known for over 20 years to be important in nonenzymic redox chemistry of the cofactors. It has been learned over roughly the same period that the enzymic reactions accelerate the tunneling reaction powerfully by means that are the subject of vigorous investigation, much of the work discussed eleswhere in these volumes. In addition, molecular evolution, as reflected in enzymic mechanisms, has been found in some cases to employ chemical principles seen in model reactions, for example an enzyme can (for mechanistic purposes) powerfully reduce the rate of hydride transfer from NADH to an acceptor by lengthening the distance over which the hydride must travel.

4.5 Models for Flavin-mediated Hydride Transfer

Excellent reviews that make a fine starting point for the modern mechanistic history of flavin biochemistry are those of Hemmerich, Nagelschneider, and Veeder [56] and of Walsh [57, 58]. Palfrey and Massey [59] have provided a valuable account up to 1998. On specialized areas, an especially valuable review is that by Ghisla and Thorpe [60]. See also the brief remarks above in Section 4.1.3, including Fig. 4.2 illustrating the structures and numbering system.

A general theme is the stability of the semiquinone form of the cofactors FMN and FAD, the free-radical species that results from one-electron redox reactions, as seen in Eq. (4.10):

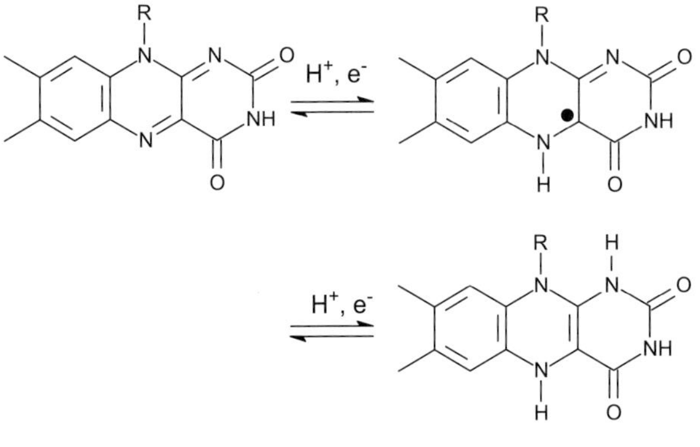

H^+, e^- ⇌ H^+, e^- ⇌

(4.10)

4.5.1 Differences between Flavin Reactions and Nicotinamide Reactions

The stability of the radical species FMNH and FADH is the source of the most dramatic distinctions between the nicotinamide cofactors and the flavins. This stability makes possible a number of reactions for flavins involving single-electron transfers that are rendered essentially impossible by the high energy of the corresponding nicotinamide radical. The source of the relative stability of the semiquinone form would logically be thought to arise from the more extensive delocalization of the unpaired electron, in comparison with the nicotinamide radical. This cannot be the whole story, however, because the mere substitution of N(5) by a CH group to form the 5-deaza derivative renders the cofactor incapable of one-electron reactions (Blankenhorn [61]) or nearly so (Walsh [57]).

The stability of the flavin radical, combined with the versatility of flavins, which also can engage in two-electron reactions, has evolutionarily led to a number of redox enzymes in which flavin-mediated reductions of dioxygen are fed by electrons transmitted from NADH to the flavin coenzyme by means of hydride-transfer events. This arrangement permits the very tight binding of the flavin unit to the enzyme, while its reducing power is restored by the loosely bound, freely circulating NADH. More generally, both FMN and FAD can be thought of as mediators between one-electron chemistry and two-electron chemistry in their interaction with the two-electron nicotinamide cofactors NADH and NADPH, on the one hand, and various one-electron reagents, on the other hand.

4.5.2
The Hydride-transfer Process in Model Systems

Classic work by Powell and Bruice (37) established the hydride-transfer nature of the interaction of flavins with nicotinamide cofactors. Among several lines of evidence, a very impressive experiment (Eq. (4.11)) involved the exposure in *tert*-butyl alcohol solvent at 30 °C of various NADH analogs to the *N*,*N*-bridged flavin analog shown in Eq. (4.11a), which resulted in quick and complete reduction, and exposure to the radical cation shown in Eq. (4.11b), which gave no direct reduction product at all, although as explained below the observations required careful analysis.

H H C R N CH_2Ph N N O NH N O → N N O NH N H O (4.11a)

H H C R N CH_2Ph N N O NH N O → NO DIRECT REACTION (4.11b)

Careful analysis was needed because the acridinium reactant of Eq. (4.11a) was present in the radical cation species of Eq. (4.11b) as a 5% impurity. The observed reaction was then the consumption of one equivalent of the dihydronicotinamide analog and formation of two equivalents of the product of one-electron reduction of the cation radical, just as if the nicotinamide analog were acting as a one-electron reagent in two separate, sequential steps with overall liberation of a proton. The kinetics was more complex than this mechanism suggested, however. In addition, the relative rates of this reaction were essentially those observed for the reduction of the acridinium ion (Eq. (4.11a)) and addition of acridinium ion to the system of Eq. (4.11b) produced a linear increase in the rate. It was thus deduced that the events occurring were (i) direct delivery of a hydride ion from the NADH analog to the acridinium impurity to give the reduced product of Eq. (4.11a); (ii) a sequence of two one-electron transfers, each to a molecule of the radical cation, producing the reduced product of the radical cation, liberating a proton, and regenerating the catalytic acridinium ion.

Finally it was reasoned that if the mechanism of Eq. (4.11a) involved an electron transfer or a hydrogen-atom transfer from the NADH analog to the acridinium species, then such processes should surely occur in the system of Eq. (4.11b). The fact that this did not occur, in spite of the electron transfer being thermodynamically favorable, demonstrated the extreme propensity of the NADH/flavin system for hydride-transfer reaction.

The hydride transfer from reduced nicotinamides to N(5) of the flavin species (Eq. (4.12)) has been studied from other points of view in model systems, for example by Reichenbach-Klinke, Kruppa, and König [62], as shown in Fig. 4.7.

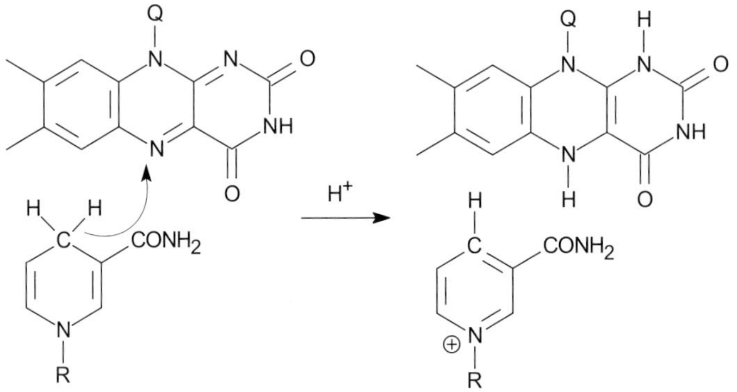

(4.12)

Their aim was to explore the significance of the observation that all known structures of NADH-linked flavoproteins had a very similar, and thus – potentially – evolutionarily conserved, relationship between the NADH binding site and the flavin binding site such that the nicotinamide ring is forced by the enzyme to lie parallel to and about 3–4 Å distant from the central ring of the flavin coenzyme. Possibly this structural relationship has indeed been conserved by evolution because it confers especially favorable properties for the hyride-transfer process.

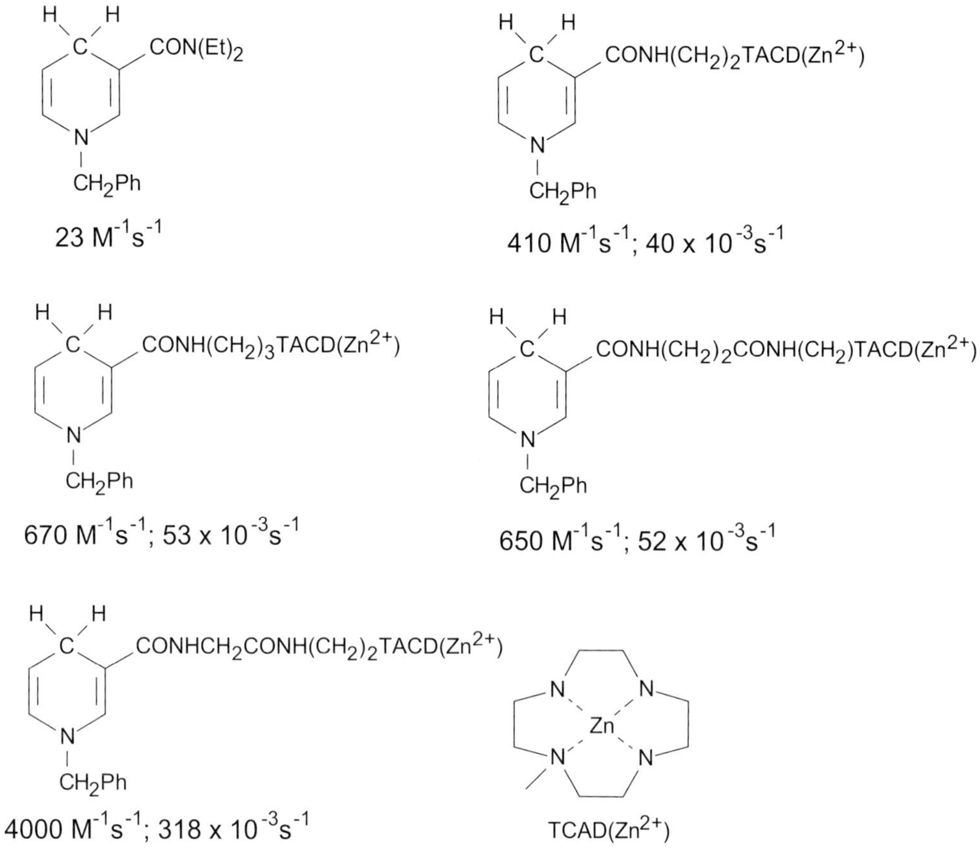

CON(Et)$_2$; CH_2Ph

23 $M^{-1}s^{-1}$

$CONH(CH_2)_2TACD(Zn^{2+})$; CH_2Ph

410 $M^{-1}s^{-1}$; 40 x $10^{-3}s^{-1}$

$CONH(CH_2)_3TACD(Zn^{2+})$; CH_2Ph

670 $M^{-1}s^{-1}$; 53 x $10^{-3}s^{-1}$

$CONH(CH_2)_2CONH(CH_2)TACD(Zn^{2+})$; CH_2Ph

650 $M^{-1}s^{-1}$; 52 x $10^{-3}s^{-1}$

$CONHCH_2CONH(CH_2)_2TACD(Zn^{2+})$; CH_2Ph

4000 $M^{-1}s^{-1}$; 318 x $10^{-3}s^{-1}$

$TCAD(Zn^{2+})$

Figure 4.7. Structures used by Reichenbach-Klinke et al. [62] as agents to reduce riboflavin tetraacetate as a model of the flavin-nicotinamide redox interaction in flavoenzymes. The second-order rate constants shown are approximate values for the redox reaction at 50 μM concentrations of each reactant in aqueous solution at pH 7.4 and 25 °C. The first-order rate constants for compounds with a Zn-center were obtained from variation of the NADH-analog concentration from 50 μM to 0.5 mM, followed by analysis of the initial second-order rate constants on a model that assumed reversible complexation of the reactants followed by unimolecular reaction of the complex. Calculated disssociation constants for complexes ranged from 0.7 to 1.3×10^{-4} M, with an average value of 1.1×10^{-4} M.

To examine the point, Reichenbach-Klinke et al. [62] constructed the NADH-analogs illustrated in Fig. 4.7 and then measured the rate constants for their reduction of riboflavin tetraacetate, the flavin shown in Eq. (4.11) with Q = $CH_2(CHOAc)_3CH_2OAc$). The hypothesis was that the NADH-analogs possessing a zinc binding site could complex the flavin at the Zn-center, most probably through coordination of the ionized imide function. Indeed all analogs equipped with the Zn-center exhibited saturation kinetics. When the NADH-analog concentration was varied in the kinetic studies, the data were consistent with complex dissociation constants that for the various compounds had an average value of about 10^{-4} M and the first-order rate constants for unimolecular reaction within the complex that are shown in Fig. 4.7.

These studies clearly show that a specific approximation of the reactants can lead to an acceleration of the reduction by a factor of some hundreds. That the nature of the approximation is specific is shown by the fact that only one of the analogs had such a large reduction rate, those with the binding site tethered at different distances or with tethers of different rigidity reacting some 10-fold more slowly. A preliminary theoretical exploration showed that the flavin complex of the most rapidly reacting NADH-analog was able to attain – with less strain energy than for any of the other analogs – a structure resembling the enzymic arrangement (nicotinamide parallel to and poised near the central ring of the flavin) with a 3.4 Å ring-to-ring distance. That the hydride-transfer step is under observation is shown by the isotope effect of the monodeuterated version of the same analog. The observed effect on the first-order rate constant is 1.3, which corresponds to $2k_H/(k_H + k_D)$ or, neglecting secondary isotope effects, an isotope effect k_H/k_D of about 2. This is small but too large for a secondary effect and thus indicates the hydrogen transfer to be at least partially rate-limiting. Larger isotope effects consistent with tunneling have been observed in the enzymic equivalents of the reaction [63].

The result is therefore indicative of the importance in the enzymic reaction of an "axial" hydride transfer in which the hydride ion departing from the NADH approaches the flavin N(5) center perpendicular to the ring and thus the plane containing the unshared electron pair of N(5). This study is exemplary of what is becoming a common and vitally important type of model-reaction investigation. In contrast to the historical role of model reactions, which often elucidated the baseline chemistry of biomolecules with very little reliable information about the structure and functional properties of the relevant enzymes, current studies often begin with a sophisticated picture of the facts about the enzymic reaction and are designed to sort out which of the features derive from chemical rules and which from biological factors. The latter often have a basis in metabolic or regulatory imperatives that may not reflect realities in the chemistry of the reactions being effected.

4.6 Models for Quinone-mediated Reactions

There have been some extraordinarily effective contributions of model-reaction studies, particularly by Klinman and Mure [2], to the understanding of quinone-cofactor chemistry, but there seem to have been no uses of this approach with respect to hydride-transfer reactions. Readers who wish to acquaint themselves with the current situation should consult Davidson's volume of 1993 [64], Klinman and

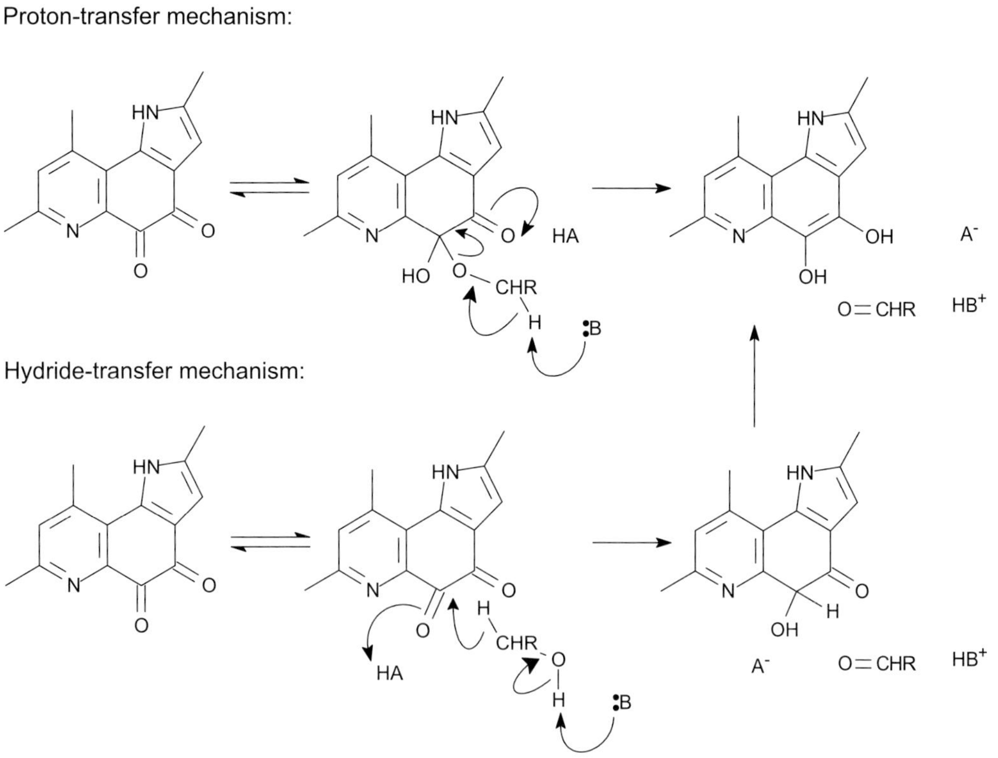

Proton-transfer mechanism:

Hydride-transfer mechanism:

Figure 4.8. Two proposed mechanisms for the reaction in bacterial alcohol dehydrogenases between the substrate alcohol and the cofactor PQQ. In the proton-transfer mechanism at the top, alcohol adds to the C(5) carbonyl group and an enzymic acid–base pair then effects an elimination reaction, leading to the aldehyde oxidation product and to the cofactor in its reduced form. In the hydride-transfer mechanism at the bottom. The acid–base pair acts on the free alcohol to promote hydride transfer from the alcohol to the C(5) center. Again the oxidized alcohol is generated along with, in this case, a ketonic form of the reduced cofactor, which can readily enolize as shown. See the text and Refs. [69, 70].

Mu's review of 1994 [65], Anthony's review of 1998 [66], and the brief reviews of Duine [67], Klinman [68], and Mure [2].

A controversy currently abroad in the field seems a particularly likely candidate for investigations using model reactions. A number of bacterial alcohol dehydrogenases make use of free-standing PQQ (see Fig. 4.3 for the structure) as a cofactor. Figure 4.8 shows two possible mechanisms for a critical step in the mechanism: the question is whether the reaction follows a proton-transfer route or a hydride-transfer route [69, 70]. The question is essentially limited to the alcohol and sugar dehydrogenases, while the enzymes that catalyze amine oxidations tend

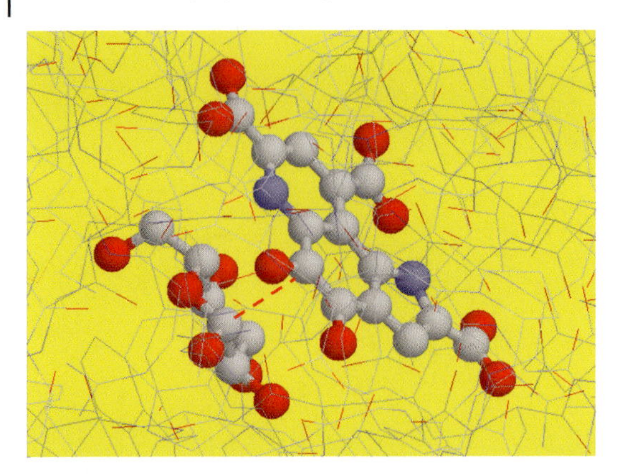

Figure 4.9. Structural evidence favoring a hydride-transfer mechanism for the action of the PQQ cofactor in the soluble glucose dehydrogenase of *Acinetobacter calcoaceticus* (structure and concept of Oubrie et al. [71]; PDB file no. 1CQ1). The dashed red line indicates the 3.2 Å distance from C(1) of glucose to C(5) of PQQ and approximates the hydride-transfer trajectory ($PQQH_2$ was the actual species present in the structure).

to make use of proton-abstraction mechanisms in the Schiff's base intermediates formed from cofactor and amine substrate [71].

In the proton-transfer mechanism, the alcohol is presumed to form a hemiacetal at the C(5) carbonyl group. Then an acid–base pair in the active site performs an elimination reaction, producing the aldehyde product and the reduced cofactor, $PQQH_2$. The hydride-transfer mechanism envisions the approximation of the scissile C–H bond of the alcohol to C(5) of the cofactor, followed by an acid–base catalyzed delivery of hydride ion to C(5), resulting in formation of the aldehyde product and the ketol form of $PQQH_2$, which can readily rearrange to the enediol form.

The proton-transfer mechanism has been favored until recently, in particular up to the publication in 1999 by Oubrie et al. [72] of the structure of a PQQ-dependent bacterial glucose dehydrogenase, in which the active site contained a glucose molecule and a molecule of $PQQH_2$ as a simulacrum of PQQ. As Fig. 4.9 shows, the two ligands are situated relative to each other in the active site, such that a short trajectory for hydride transfer from C(1) of the glucose substrate to C(5) of the cofactor can be identified (dashed red line in Fig. 4.9), the C(1)–C(5) distance being only 3.2 Å. In addition, theoretical work from Zheng et al. [73], along with their re-refinement of an earlier structure of a methanol dehydrogenase (indicating a tetrahedral structure at C(5) in the bound PQQ cofactor) lent further support to this view.

Certainly the situation is ideal, for example, for the design of intramolecular models that would permit the exploration of the hydrogen-transfer systematics under controlled conditions. A further advantage of the model system would be a posssible exploration of both proton-transfer and hydride-transfer mechanisms.

4.7
Summary and Conclusions

1. Studies of model reactions for redox cofactors in general and for hydride-transfer reactions in particular have long formed a major part of the basis for mechanistic knowledge of the enzymic reactions. Model studies are likely to play at least as prominent a role in the future. Particularly as improved enzymological technology permits the dissection of ever finer points of enzyme mechanism, characterization of the scope and limitations of these factors through investigations of highly controlled systems should be more important than ever.

2. The field of model studies of hydride-transfer reactions involving analogs of nicotinamide cofactors is especially well-developed. The powerful preference for one-step hydride-transfer chemistry over multistep processes involving transfers of electrons, protons, and hydrogen atoms under most circumstances has been established very well by model studies. The important role of tunneling in model reactions shows that enzymes are accelerating, rather than originating, tunneling mechanisms in nicotinamide-dependent enzymic reactions. It is now a challenge to students of model reactions to construct analogs capable of evaluating the influence of coupled vibrations in promoting tunneling, as has been argued for various enzymic cases. Model reactions are continuing to expand our understanding of other aspects of nicotinamide-dependent enzyme reactions, including the stereochemistry and regiochemistry of hydride transfer.

3. Hydride-transfer reactions to N(5) of flavin cofactors are indicated by model studies to proceed by single-step hydride-transfer mechanisms, at least with nicotinamide donor/acceptors, and to favor the "axial" donation of hydride along a trajectory perpendicular to the ring plane of the flavin. The required parallel arrangement of the nicotinamide ring and flavin center-ring probably explains the conservation of this arrangement in the active sites of known enzymes that catalyze such hydride-transfer reactions.

4. Quinone cofactors are now thought, on the basis of some enzyme structural information, to prefer a hydride-transfer to a proton-transfer mechanism, at least with alcohol substrates. This distinction could benefit greatly from model studies.

4.8
Appendix: The Use of Model Reactions to Estimate Enzyme Catalytic Power

The examples in Chart 4.1 refer to a unimolecular non-enzymic (standard or model) reaction but there is no difficulty associated with extension of the basic insights of Radzicka and Wolfenden [24] to more general kinds of standard and enzymic reactions. For example, many NAD^+-dependent dehydrogenases have a bimolecular standard reaction ($NAD^+ + S_{red}$ gives $NADH + S_{ox}$) with a second-order rate constant k_{2unc} (dimensions M^{-1} s^{-1}) and a "chemical mechanism" $E + NAD^+$ gives ENAD, which binds S_{red} to give ENAD:S_{red}, which is then transformed to ENADH:S_{ox}, which then releases S_{ox} followed by release of NADH.

The enzymic reaction has four rate constants k_{cat} (s^{-1}), k_{cat}/K_{mNAD} (M^{-1} s^{-1}), k_{cat}/K_{mSred} (M^{-1} s^{-1}), and $k_{cat}/K_{iNAD}K_{mSred}$ (M^{-2} s^{-1}) and thus four measures of catalytic power:

- k_{cat}/k_{2unc}(M), which is the analog of rate enhancement. It is the equilibrium constant for the reaction

  $$\text{ENAD:S}_{red} + \text{T}_{unc} = \text{E:T}_k + \text{NAD}^+ + \text{S}_{red}$$

  and measures the net enzymic stabilization of the transition state for k_{cat} (T_k) in the complex E:T_k over the enzymic stabilization of NAD^+ and S_{red} in the complex ENAD:S_{red};
- $k_{cat}/K_{mNAD}k_{2unc}$ (dimensionless), which is the equilibrium constant for the reaction

  $$\text{E} + \text{T}_{unc} = \text{E:T}_{NADbin} + \text{S}_{red}$$

  and measures the enzymic stabilization of the transition state for binding of NAD^+ (T_{NADbin}) diminished by the cost of liberating the elements of S_{red} from the transition state for the uncatalyzed reaction (T_{unc});
- $k_{cat}/K_{mSred}k_{2unc}$ (dimensionless), which is the equilibrium constant for the reaction

  $$\text{ENAD} + \text{T}_{unc} = \text{E:T}_{k/K} + \text{NAD}^+$$

  and measures the net enzymic stabilization of the transition state for reaction of ENAD with S_{red} ($T_{k/K}$) over the enzymic stabilization of NAD^+ in the complex ENAD.
- $k_{cat}/K_{iNAD}K_{mSred}k_{2unc}$ (M^{-1}), which is the equilibrium constant for the reaction

  $$\text{E} + \text{T}_{unc} = \text{E:T}_{k/K}$$

  and measures the total enzymic stabilization of the transition state $T_{k/K}$ relative to the free transition state for the uncatalyzed (standard) reaction.

The kinetic investigation of enzymic reactions and suitable non-enzymic standard reactions permits, as just described, the numerical calculation of measures of enzyme catalytic power. These measures are ratios of rate constants and they correspond to equilibrium constants for reactions of free enzyme or enzyme complexes with the transition state for the standard reaction to generate complexes of the enzyme with various transition states along the enzymic reaction pathway, sometimes with liberation of other ligands (see the examples above).

The numerical values of the equilibrium constants can of course be converted to standard Gibbs free-energy changes through the relationship $\ln(K) = -\Delta G^\circ/RT$ and it is frequently of interest to attempt interpretations of these free energies in terms of contributions of various individual interactions to the overall value. There are some points that need to be kept in mind when pursuing such attempts. A few of these are explored briefly here. Several of these matters and others have been treated with considerable effect by Miller and Wolfenden [74], Garcia-Viloca et al. [75], Benkovic and Hammes-Schiffer [76], and Sutcliffe and Scrutton [77].

Multiple transition states. Commonly for enzymic reactions there will not be a single rate-determining step with a single transition state for any of the kinetic parameters; instead several steps will contribute to determining the rate in various degrees, and the effective transition state for such a situation has been called a *virtual* transition state. Its properties, including its free energy, will be a weighted average of the properties of the contributing transition states, with those of highest free energy contributing the most (because they most nearly determine the rate). It is rare for sufficient information about any particular enzyme system to be available to permit the situation to be laid out in detail, but it is useful to keep in mind that measures of enzyme catalytic power generally deal with more than a single contributing enzymic transition state. In principle, this certainly may also be true of the standard or model reaction.

Meaning of "transition-state stabilization." When a reaction such as

$$E + T = E{:}T$$

has a large equilibrium constant and a correspondingly negative value of ΔG°, thermodynamicists are accustomed to say, "E stabilizes T in the complex E:T relative to T in the free state." For many of us, this kind of statement may generate a mental picture in which the stabilization is accomplished by the formation of attractive interactions between component structures of E and component structures of T in the complex E:T, the individual (negative) free energies of interaction in a simple case summing up to generate the overall value of ΔG°. Reflection demonstrates of course that such a model is by no means required: indeed every interaction between components of E and T in the complex E:T may be strongly repulsive and make large positive contributions to ΔG°. If E and T in the their free states, however, experience still more strongly repulsive interactions, and these are relieved upon complex formation, then the overall ΔG° may still be quite negative.

Failure to take this complication into account has led to some acrimonious interchanges on occasion, so it is necessary to note that "E stabilizes T in E:T" means only that the combination of E with T to form E:T produces a large equilibrium constant. Nothing is implied about the nature of the interactions that make this true, and indeed the investigation needed to clarify the nature of these interactions may be laborious and difficult.

Novel kinetic formulations. The description given above of the calculation of measures of enzyme catalytic power relies initially on empirically determined rate constants for enzymic and non-enzymic reactions. The numerical results at the initial stage are therefore "theory-free" and may be used for many purposes with perfect confidence.

The further interpretation in terms of equilibrium constants for transition-state binding to enzyme species, however, relies on the "ultrasimple" transition-state theory and there are increasing indications that this formulation may be oversimplified for the interpretation of data for hydride-transfer reactions (see Chapter 10 by Klinman, Chapter 12 by Kohen, and Chapter 19 by Banerjee). It is a not particularly demanding task to modify the language to correspond to the most modern forms of transition-state theory, and for those who use such descriptions to pursue the meaning of the empirical results, there is no great problem.

For the use of other approaches, however, such as have been developed and applied in the context of ideas about vibrationally induced hydrogen tunneling, more careful analysis may be required to assign the origins of the observed measures of catalytic power.

References

1 Bugg, T. D. H. 2004 *Introduction to Enzyme and Coenzyme Chemistry*, 2nd edn., Blackwell Publishing, Oxford UK.

2 Mure, M. 2004 *Acc. Chem. Res. 37*, 131–139: Tyrosine-derived quinone cofactors.

3 Westheimer, F. H. 1985 *Adv. Phys. Org. Chem. 21*, 1–36: The Discovery of the mechanisms of enzyme action, 1947–1963.

4 Westheimer, F. H. 1987 in *Coenzymes and Cofactors, Vol. II Part A, Pyridine Nucleotide Coenzymes*, ed. D. Dolphine, R. Poulsen, and O. Abramović, Wiley, New York, pp. 253–322: Mechanism of action of the pyridine nucleotides.

5 Kosower, E. M. 1962 *Molecular Biochemistry*, McGraw-Hill, New York.

6 Newton, I. 1686 Rules of reasoning in philosophy, in *Philosophiae Naturalis Principia Mathematica*, reprinted in *Newton's Philosophy of Nature*, ed. H. Thayer, Hafner, New York 1953.

7 Jencks, W. P. 1975, *Adv. Enzymol. 43*, 219–410: Binding energy, specificity, and enzymic catalysis: the Circe effect, see particularly pp. 296–305.

8 Bunting, J. W.; Sindhuatmaja, S. 1981 *J. Org. Chem. 46*, 4211–4219: Kinetics and mechanism of the reaction of 5-nitroisoquinolinium cations with 1,4-dihydronicotinamides.

9 Bunting, J. W.; Norris, D. J. 1977 *J. Am. Chem. Soc. 99*, 1189–1196: Rates and equilibria for hydroxide addition to quinolinium and isoquinolinium cations.

10 Ohnishi, Y.; Kagami, M.; Ohno, A. 1975 *J. Am. Chem. Soc. 97*, 4766–4768: Reduction by a model of NAD(P)H. Effect of metal ion and stereochemistry on the reduction of α-keto esters by 1,4-dihydronicotinamide derivatives.

11 Ohno, A.; Ishikawa, Y.; Yamazaki, N.; Okamura, M.; Kawai, Y. 1998 *J. Am. Chem. Soc. 120*, 1186–1192: NAD(P)+-NAD(P)H Models. 88. Stereoselection without steric effect but controlled by electronic effect of a carbonyl group: Syn/Anti reactivity ratio, kinetic isotope effect, and an electron-transfer complex as a reaction intermediate.

12 Lee, I.-S. H.; Jeoung, E. H.; Kreevoy, M. M. 2001 *J. Am. Chem. Soc. 123*, 7492–7496: Primary kinetic isotope effects on hydride transfer from 1,3-dimethyl-2-phenylbenzimidazoline to NAD^+ analogues.

13 Bruice, T. C.; Benkovic, S. J. 1966 *Bioorganic Mechanisms*, W.A. Benjamin, New York, Vol. I, pp. 119–211.

14 Kirby, A. J. 1980 *Adv. Phys. Org. Chem. 17*, 183–278: Effective molarities for intramolecular reactions.

15 Bruice, T. C.; Lightstone, F. C. 1999 *Acc. Chem. Res. 32*, 127–136. Ground state and transition state contributions to the rate of intramolecular and enzymic reactions.

16 Bruice, T. C. 2002 *Acc. Chem. Res. 35*, 139–148: A view at the millennium: the efficiency of enzymatic catalysis.

17 Petsko, G. A.; Ringe, D. 2004 *Protein Structure and Function*, New Science Press, London.

18 Reichardt, C. 2004 *Solvents and Solvent Effects in Organic Chemistry*, 3rd edn., Wiley-VCH, Weinheim.

19 Purse, B. W.; Rebek, J., Jr. 2005 *Proc. Natl. Acad. Sci. USA 102*, 10777–10782: Functional cavitands: Chemical reactivity in structured environments.

20 Rebek, J., Jr. 2005 in NATO Science Series, Series I: Life and Behavioural Sciences, 364 (*Structure, Dynamics and Function of Biological Macromolecules and Assemblies*), pp. 91–105: Recognition, autocatalysis and amplification.

21 Rebek, J., Jr. 2005 *Angew. Chem., Int. Ed. Engl. 44*, 2068–2078: Simultaneous encapsulation: molecules held at close range.

22 Benner, S. A. 1989 *Chem. Rev. 89*, 789–806: Enzyme kinetics and molecular evolution.

23 Wolfenden, R.; Snider, M. J. 2001 *Acc. Chem. Res. 34*(12), 938–945. The depth of chemical time and the power of enzymes as catalysts.

24 Radzicka, A.; Wolfenden, R. V. 1995 *Science 267*, 90–93: A proficient enzyme.

25 Hu, Q.; Kluger, R. 2002 *J. Am. Chem. Soc. 124*, 14858–14859: Reactivity of intermediates in benzoylformate decarboxylase: avoiding the path to destruction.

26 Hu, Q.; Kluger, R. 2004 *J. Am. Chem. Soc. 126*, 68–69: Fragmentation of the conjugate base of 2-(1-hydroxybenzyl)-thiamin: does benzoylformate decarboxylase prevent orbital overlap to avoid it?

27 Hu, Q.; Kluger, R. 2005 *J. Am. Chem. Soc. 127*, 12242–12243: Making thiamin work faster: acid-promoted separation of carbon dioxide.

28 Moore, I. F.; Kluger, R. 2000 *Org. Lett. 2*, 2035–2036.: Decomposition of 2-(1-hydroxybenzyl)thiamin. Ruling out stepwise cationic fragmentation.

29 Moore, I. F.; Kluger, R. 2002 *J. Am. Chem. Soc. 124*, 1669–1673: Substituent effects in carbon-nitrogen cleavage of thiamin derivatives. Fragmentation pathways and enzymic avoidance of cofactor destruction.

30 Powell, M. F.; Bruice, T. C. 1982 *J. Am. Chem. Soc. 104*, 5834–5836: Reinvestigation of NADH analog reactions in acetonitrile: Consequences of isotope scrambling on kinetic and product isotope effects.

31 Miller, L. L.; Valentine, J. R. 1988 *J. Am. Chem. Soc. 110*, 3982–3989: On the electron-proton-electron mechanism for 1-benzyl-1,4-dihydronicotinamide oxidations.

32 Almarsson, Ö.; Sinha, A.; Gopinath, E.; Bruice, T. C. 1993 *J. Am. Chem. Soc. 115*, 7093–7102: Mechanism of one-electron oxidation of NAD(P)H and function of NADPH bound to catalase.

33 Matsuo, T.; Mayer, J. M. 2005 *Inorg. Chem. 44*, 2150–2158: Oxidations of NADH analogues by cis-$[Ru^{IV}(bpys)_2(py)O]^{2+}$ occur by hydrogen-atom transfer rather than by hydride transfer.

34 Roberts, B. R. 1999 *Chem. Soc. Rev. 28*, 25–35: Polarity-reversal catalysis of hydrogen-atom abstraction reactions: concepts and applications in organic chemistry.

35 Huyser, E. S. 1970 *Free-Radical Chain Reactions*, Wiley Interscience, New York.

36 Powell, M. F.; Wong, W. H.; Bruice, T. C. 1982 *Proc. Natl. Acad. Sci. USA 79*, 4604–4608: Concerning $1e^-$ transfer in reduction by dihydronicotinamde: Reaction of oxidized flavin and flavin radical with N-benzyl-1,4-dihydronicotinamide.

37 Powell, M. F.; Bruice, T. C. 1983 *J. Am. Chem. Soc. 105*, 1014–1021: Hydride vs. electron transfer in the redution of flavin and flavin radical by 1,4-dihydropyridines.

38 Powell, M. F.; Bruice, T. C. 1983 *J. Am. Chem. Soc. 105*, 7139–7149: Effect of isotope scrambling and tunneling on the kinetic and product isotope effects for reduced

nicotinamide adenine dinucleotide model hydride transfer reactions.

39 Bunting, J. W. 1991 *Bioorg. Chem.* *19*, 456–91: Merged mechanisms for hydride transfer from 1,4-dihydronicotinamides.

40 Albery, W. J.; Kreevoy, M. M. 1978 *Adv. Phys. Org. Chem.* *16*, 87–157: Methyl transfer reactions.

41 Lee, I.-S. H.; Jeoung, E. H.; Kreevoy, M. M. 1997 *J. Am. Chem. Soc.* *119*, 2722–2728: Marcus theory of a parallel effect on α for hydride transfer reaction between NAD^+ analogs.

42 Lee, I.-S. H.; Chow, K.-H.; Kreevoy, M. M. 2002 *J. Am. Chem. Soc.* *124*, 7755–7761: The Tightness contribution to the Brønsted α for hydride transfer between NAD^+ analogues.

43 Lee, I. S. H.; Ji, Y. R.; Jeoung, E. H. 2006 *J. Phys. Chem. A* *110*, 3875–3881: Reinterpretation of the Bronsted α for redox reactions based on the effect of substituents on hydride transfer reaction rates between NAD^+ analogues.

44 Würthwein, E.-U.; Lang, G.; Schappele, L. H.; Mayr, H. 2002 *J. Am. Chem. Soc.* *124*, 4084–4092: Rate-equilibrium relationships in hydride transfer reactions: The Role of intrinsic barriers.

45 Romesberg, F. E.; Schowen, R. L. 2004, *Adv. Phys. Org. Chem.* *39*, 27–77: Isotope effects and quantum tunneling in enzyme-catalyzed hydrogen transfer. Part I. The experimental basis.

46 Bell, R. P. 1980 *The Tunnel Effect in Chemistry*, Chapman & Hall, London.

47 Kaldor, S. B.; Saunders, W. H., Jr. 1979 *J. Am. Chem. Soc.* *101*, 7594–7599: Mechanisms of eleimination reactions 30. The contributions of tunneling and heavy-atom motions in the reaction coordinate to deuterium kinetic isotope effects in eliminations from 2-phenylethyl derivatives.

48 Overman, L. E. 1972 *J. Org. Chem.* *37*, 4214–4218: Dehydrogenase enzyme models: Approximation of an alcohol and a pyridinium ring.

49 Meyers, A. I.; Brown, J. D. 1987 *J. Am. Chem. Soc.* *109*, 3155–3156: The First nonenzymic stereospecific intramolecular reduction by an NADH mimic containing a covalently bound carbonyl moiety.

50 Kirby, A. J.; Walwyn, D. R. 1987 *Tetrahedron Lett.* *28*, 2421–2424: Intramolecular hydride transfer from a 1,4-dihydropyridine to an α-ketoester in aqueous solution. A model for lactate dehydrogenase.

51 Kirby, A. J.; Walwyn, D. R. 1987 *Gazz. Chim. Ital.* *117*, 667–680: Effective molarities for intramolecular hydride transfer. Reduction by 1,4-dihydropyridines of the neighbouring α-ketoester group.

52 Yang, X.; Hu, Y.; Yin, D.; Turner, M. A.; Wang, M.; Borchardt, R. T.; Howell, P. L.; Kuczera, K.; Schowen, R. L. 2003 *Biochemistry* *42*, 1900–1909: The catalytic strategy of S-adenosyl-L-homocysteine hydrolase: Transition-state stabilization and the avoidance of abortive reactions.

53 Porter, D. J.; Boyd, F. L. 1991 *J. Biol. Chem.* *266*, 21616–21625: Mechanism of bovine liver S-adenosylhomocysteine hydrolase. Steady-state and pre-steadystate kinetic analysis.

54 Porter, D. J.; Boyd, F. L. 1992 *J. Biol. Chem.* *267*, 3205–3213: Reduced S-adenosylhomocysteine hydrolase. Kinetics and thermodynamics for binding of 3′-ketoadenosine, adenosine, and adenine.

55 Porter, D. J. 1993 *J. Biol. Chem.* *268*, 66–73: S-adenosylhomocysteine hydrolase. Stereochemistry and kinetics of hydrogen transfer.

56 Hemmerich, P.; Nagelschneider, G.; Veeger, C. *FEBS Lett.* *8*, 69–83: Chemistry and molecular biology of flavins and flavoproteins.

57 Walsh, C. 1978 *Annu. Rev. Biochem.* *47*, 881–931: Chemical approaches to the study of enzymes catalyzing redox transformations.

58 Walsh, C. 1980 *Acc. Chem. Res.* *13*, 148–155: Flavin coenzymes: At the crossroads of biological redox chemistry.

59 Palfey, B. A.; Massey, V. 1998 in *Comprehensive Biological Catalysis*, ed. M. Sinnott, Vol. III, pp. 83–154: Flavin-dependent enzymes.

60 Ghisla, S.; Thorpe, C. 2004 *Eur. J. Biochem. 271*, 494–508: Acyl-CoA dehydrogenases. A mechanistic overview.

61 Blankenhorn, G. 1976 *Eur. J. Biochem. 67*, 67–80: Nicotinamide-dependent one-electron and two-electron (flavin) oxidoreduction: Thermodynamics, kinetics, and mechanism.

62 Reichenbach-Klinke, R.; Kruppa, M.; König, B. 2002 *J. Am. Chem. Soc. 124*, 12999–13007: NADH Model systems functionalized with Zn(II)-cyclen as flavin binding site – Structure dependence of the redox reaction within reversible aggregates.

63 Basran, J.; Masgrau, L.; Sutcliffe, M. J.; Scrutton, N. S. 2006, in *Isotope Effects in Chemistry and Biology*, ed. Kohen, A.; Limbach, H.-H., pp. 671–689: Solution and computational studies of kinetics isotope effects in flavoprotein and quinoprotein catalyzed substrate oxidations as probes of enzymic hydrogen tunneling and mechanism.

64 Davidson, V. J. (ed.) 1993, *Principles and Applications of Quinoproteins*, Marcel Dekker, New York.

65 Klinman, J. P.; Mu, D. 1994 *Annu. Rev. Biochem. 63*, 299–344: Quinoenzymes in biology.

66 Anthony, C. 1998 in *Comprehensive Biological Catalysis*, ed. M. Sinnott, Vol. III, pp. 155–180: Quinoprotein-catalysed reactions.

67 Duine, J. A. 2000 *Chem. Record*, 74–83: Cofactor diversity in biological oxidations: Implications and applications.

68 Klinman, J. P. 2001 *Proc. Nat. Acad. Sci. USA 98*, 14766–14768: How many ways to craft a cofactor?

69 Kay, C. W. M.; Mennenga, B.; Görisch, H.; Bittl, R. 2006 *Proc. Natl. Acad. Sci. USA 103*, 5267–5272: Substrate binding in quinoprotein ethanol dehydrogenase from *Pseudomonas aeruginosa* studied by electron-nuclear double resonance.

70 Kay, C. W. M.; Mennenga, B.; Görisch, H.; Bittl, R. 2006 *J. Biol. Chem. 281*, 1470–1476: Structure of the pyrroloquinoline quinone radical in quinoprotein ethanol dehydrogenase.

71 Datta, S.; Mori, Y.; Takagi, K.; Kawaguchi, K.; Chen, Z.-W.; Okajima, T.; Kuroda, S.; Ikeda, T.; Kano, K.; Tanizawa, K., Mathews, F. S. 2001 *Proc. Natl. Acad. Sci. USA 98*, 14268–14273: Structure of a quinohemoprotein amine dehydrogenase with an uncommon redox cofactor and highly unusual crosslinking.

72 Oubrie, A.; Rozeboom, H. J.; Kalk, K. H.; Olsthoorn, A. J. J.; Duine, J. A.; Dijkstra, D. W. 1999 *EMBO J. 18*, 5187–5194: Structure and mechanism of soluble quinoprotein glucose dehydrogenase.

73 Zheng, Y.-J.; Xia, Z.-x.; Chen, Z.-w.; Mathews, F. S.; Bruice, T. C. 2001 *Proc. Natl. Acad. Sci. USA 98*, 432–434: Catalytic mechanism of quinoprotein methanol dehydrogenase: A theoretical and x-ray crystallographic investigation.

74 Miller, B. G., Wolfenden, R. 2002 *Annu. Rev. Biochem. 71*, 847–885: Catalytic proficiency: the unusual case of OMP decarboxylase.

75 Garcia-Viloca, M.; Gao, J.; Karplus, M.; Truhlar, D. G. 2004 *Science 303*, 186–195: How enzymes work: analysis by modern rate theory and computer simulations.

76 Benkovic, S. J.; Hammes-Schiffer, S. 2003 *Science 301*, 1196–1202: A perspective on enzyme catalysis.

77 Sutcliffe, M. J.; Scrutton, N. S. 2002 *Eur. J. Biochem. 269*, 3096–3102: A new conceptual framework for enzyme catalysis: Hydrogen tunneling coupled to enzyme dynamics in flavoprotein and quinoprotein enzymes.

5
Acid–Base Catalysis in Designed Peptides

Lars Baltzer

5.1
Designed Polypeptide Catalysts

The purpose of rational design of folded and catalytically active polypeptides is to test critically our understanding of enzyme function and catalysis, and ultimately to provide new catalysts for biotechnical applications [1–3]. The rational design process is based on fundamental principles of organic reactivity implemented in the construction of active sites and any reaction for which the reaction mechanism is understood to a reasonable degree is a target for design. In a scaffold of sufficient complexity, functional groups capable of substrate binding, transition state stabilization, general-acid and general-base catalysis etc. can be introduced and combined to mimic the function of native enzymes or, even better, to catalyze reactions selected from the vast repertoire of chemical transformations developed by organic chemists. In folded polypeptide scaffolds catalytically active residues may be systematically varied and their properties tuned. It is, for example, possible to explore the effect of decreased or increased acidity of a catalytically active residue by introducing charged residues in close proximity to affect its pK_a. It is also possible to explore the effects of charge–charge interactions between substrate and catalyst by variation of the number and position of charged amino acid residues. In the search for optimal active site constellations of amino acids, sequences are easily modified.

Active sites in *de novo* designed polypeptides are, as a rule, built from surface exposed residues, even if the polypeptide is folded. The design from scratch of proteins or polypeptides that fold to form cavities is still in its infancy and systematic variations are, by necessity, difficult in complex structures because the structures may change with amino acid substitutions. Rate enhancements of three to four orders of magnitude have been reported several times in designed catalysts [4–9] but those of typical enzymes are unrealistic in solvent exposed catalytic sites. The number of functional groups that can interact with substrates, intermediates and transition states is limited and the many degrees of freedom of the active site residues reduce the catalytic efficiency for entropic reasons. However, incorporation of an active site developed in a surface catalyst into a constrained hydrophobic pocket

Hydrogen-Transfer Reactions. Edited by J. T. Hynes, J. P. Klinman, H.-H. Limbach, and R. L. Schowen

ISBN: 978-3-527-30777-7

will affect ionization constants, the strengths of charge–charge interactions and the degrees of freedom of rotable bonds and it is possible, and even likely, that catalytic efficiencies will increase significantly for these reasons alone [10]. In the design process the access to an easily modified scaffold in which principles can be tested is more important than the actual rate enhancement. Surface catalysts serve the purpose of defining the roles of individual residues and what catalytic functions are needed.

Following this ambition, polypeptide catalysts have been designed for e.g. ester hydrolysis [6, 7, 11, 12], transesterification [6], amidation [13], transamination [8], chemical ligation [5] and decarboxylation reactions [4, 14]. The reactions have been studied and mechanisms elucidated to various degrees and as a result considerable mechanistic understanding has been generated. Recently, computational approaches have been applied to the re-engineering of native proteins to introduce catalytic sites for monooxygenase [9] and triosephosphateisomerase [15] activities. The results are impressive and demonstrate that enzymatic activity may be introduced in folded proteins with no prior catalytic functions. Computational methods are now emerging as the most promising approach to *de novo* protein design and, in combination with mechanistically driven *de novo* design, may prove to be an efficient road to new enzymes.

5.1.1 Protein Design

The design of folded polypeptides and proteins has now reached a level where sequences of 100 residues or more can be designed from scratch and many well-defined proteins have been reported as well as folded but structurally not uniquely defined polypeptides [1–3, 16, 17]. For catalyst design a number of sequences may be taken from the literature and adapted to a catalytic problem since the synthesis of 40–50 residue sequences are, by todays standards, almost routine in a peptide chemistry laboratory. The choice of sequence may seem an enigmatic problem but most of the many helical bundle motifs that were reported ten years ago are likely to be adequate scaffolds for catalyst design, even though they do not have the properties of native proteins. They fold into several, but similar, conformers that are in rapid equilibrium and the advantage of using sequences that do not fold cooperatively is that they are very tolerant to modifications and that the introduction or deletion of charged residues will not significantly alter the balance between conformations. In contrast, a protein that folds cooperatively may undergo substantial conformational changes and may be difficult to redesign if residues that are critical for folding are replaced. The advantage of using folded proteins with well-defined tertiary structures is that crystal and NMR structures are available making it considerably easier to design active sites and to determine the relationships between structure and function.

Helical folds have dominated in the pioneering development of *de novo* protein design [18] and have also been the most common in the design of new catalysts [4–9, 11–14], Fig. 5.1. The robustness of the motif has been a contributing factor

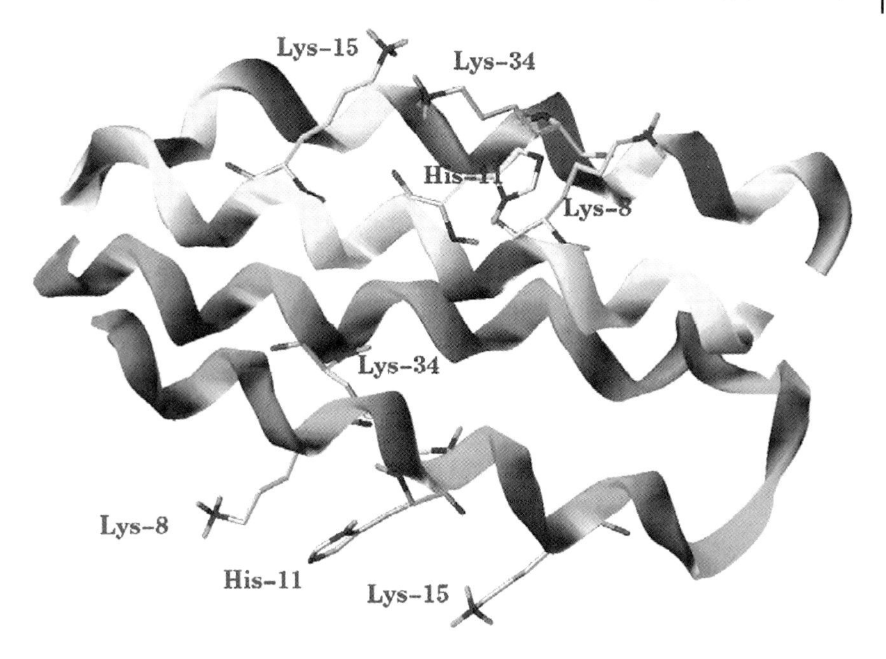

Figure 5.1. Modelled structure of a 42-residue peptide folded into a helix–loop–helix motif and dimerized to form a four-helix bundle protein. Helices are amphiphilic with a hydrophobic and a polar face. Due to the robustness and ease of synthesis this has become a popular motif in *de novo* protein design.

to their popularity and so has the regularity of the folded structure that makes it highly suitable for active site engineering. The design principles have been elucidated in great detail [1–3]. A peptide with helix propensity is in an equilibrium between the folded and unfolded forms and the formation of a helix is favored by interactions that preferentially stabilize the folded state. Amino acids like alanine favor helix formation for steric reasons and every amino acid residue has a well-defined propensity for helix formation, high or low. Salt bridges between side chains of opposite charge may be introduced between residues four positions apart in the sequence to stabilize the helical conformations and charged residues are introduced at the helical ends to stabilize the macroscopic dipole resulting from helix formation. In addition, the amide protons and carbonyl oxygens in the first and last turn of the helix, for which no acceptors and donors of hydrogen bonding are available in the peptide backbone, may be stabilized by suitably selected side chain functional groups. Tertiary interactions are, however, the most important for structure formation of helical proteins in aqueous solution.

Helical bundles are conveniently described in terms of the heptad repeat pattern $(a, b, c, d, e, f, g)_n$, Fig. 5.2, according to which the a and d positions form one side of the folded helix, as do the b and e positions and the g and c positions. Helical bundles are formed from amphiphilic helices and the a and the d positions are

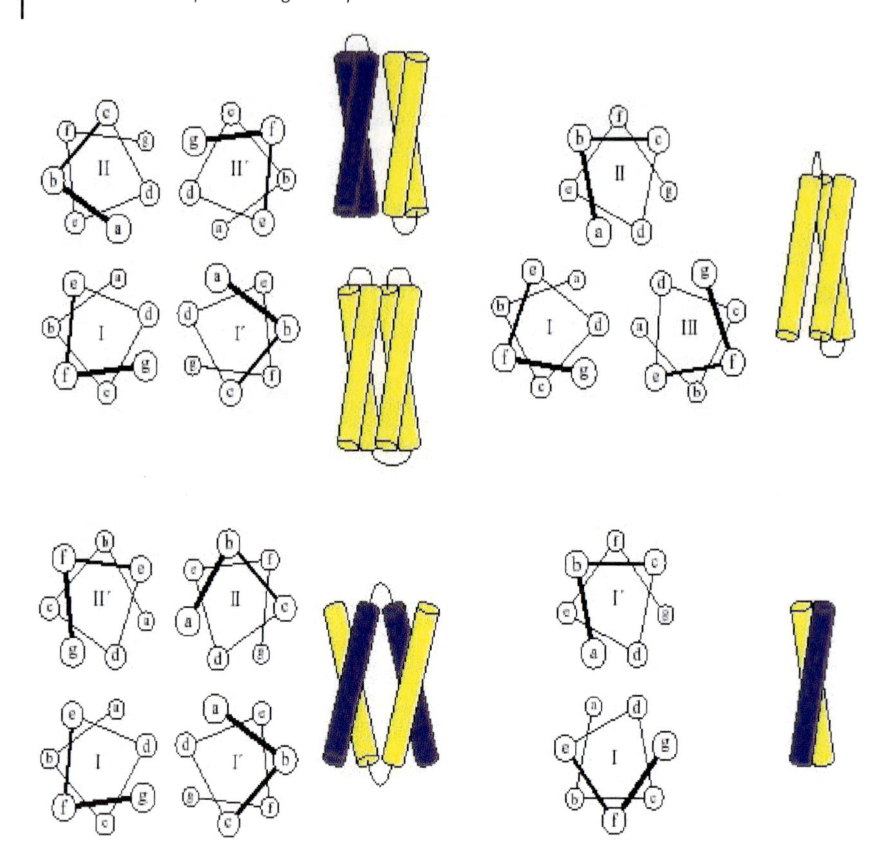

Figure 5.2. Helical wheel representation of helical bundle proteins to illustrate principles of design. Helix–loop–helix dimers are predominantly antiparallel to neutralize helical dipole moments and can fold in two ways with consequences for which residues can form active sites. Hydrophobic residues in *a* and *d* positions form cores in the folded state and drive folding. Residues in *b* and *e* positions control aggregation and organization of helical subunits. The three helices in a three-helix bundle may form a clockwise or an anticlockwise structure depending on the complementarity of the charged residues in *b* and *e* positions.

normally occupied by hydrophobic residues so that in the folded four helix bundle, for example, a core is formed from hydrophobic residues from four helices. The binding energy from these interactions is the main driving force for bundle formation. If the hydrophobic faces of the helices are shape complementary the polypeptides will fold to form well-defined tertiary structures, comparable to those of evolved proteins. If the shapes are not perfectly complementary the polypeptides may well fold anyway but not into a uniquely defined three-dimensional structure but into groups of similar conformations in rapid exchange. These structures are usually referred to as molten globules and are very useful as scaffolds for many purposes because, although the hydrophobic cores are slightly disordered, the overall fold is that of a helical bundle.

In the antiparallel helix–loop–helix motif that has been used in several model catalysts, the *b* and *e* positions are occupied by charged residues and charge–charge interactions control the mode of dimerization. While charge–charge interactions are not sufficient to drive folding by themselves, charge repulsion between the residues in the *b* and *e* positions of each monomer subunit in a helix–loop–helix dimer is a powerful determinant of structure formation and can be used to completely inhibit dimerization.

5.1.2 Catalyst Design

The g and *c* positions that form one face of a four-helix bundle protein are preferentially used for active site engineering. The surface area of a folded helix–loop–helix hairpin is approximately 20×25 Å and roughly a dozen residues may be given catalytic functions. The distance between the α carbon atoms four residues apart in the sequence of a folded helix is 6.3 Å, and the corresponding distance between residues three positions apart is 5.2 Å. These distances appear to be well suited for e.g. acyl transfer between two residues, Fig. 5.3. While these distances are suitable for placing functional groups in positions to interact with different residues of substrates, intermediates and transition states, they cannot be fine tuned, as the helix is a structurally well-defined entity. The distance between the g and *c* positions in neighboring helices in a helix–loop–helix hairpin is roughly 10 Å. Residues in neighboring helices may be positioned at the helix–helix interface in positions that makes the side chains of residues from one helix come into proximity with side chains of amino acids from the other helix and bind substrates, intermediates and transition states cooperatively. In contrast to distances between residues within a single helix the distances between residues in neighboring helices may be modulated.

The available residues in g and *c* positions may not only be used to participate in bond making and bond breaking, but also to tune the properties of residues that

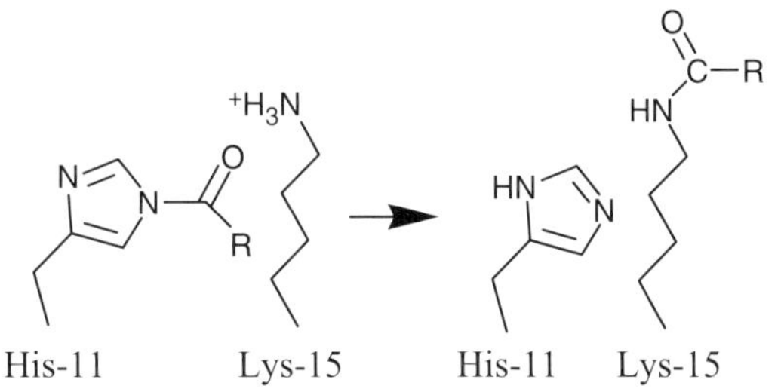

His-11 Lys-15 His-11 Lys-15

Figure 5.3. Intramolecular acyl transfer between residues four positions apart in the sequence in a helical conformation, the key step in the site-selective functionalization reaction [13].

are part of the actual catalytic machinery. Positively charged arginine residues were shown to depress the pK_a values of His residues by 0.5 units when one turn apart in the *g* and *c* positions of a helix [19]. The depressive effect was found to be approximately additive and two Arg residues in close proximity decreased the pK_a of a His by roughly one pK_a unit. Glu residues had the opposite effect. The ability to tune residue properties is an important asset in mechanistic investigations and it was shown to be possible to vary by rational design the pK_a of a His in a solvent exposed position from 5.2 to 7.2 [6]. The random coil value of a His is 6.4 [20]. The residues in the *f* positions are also used to tune the properties of residues in the *g* and *c* positions. It has been suggested that buttressing effects of side chains in *f* positions are capable of forcing side chains in *g* and *c* positions into more reactive conformations [4].

The use of helical bundles in catalyst design expands the possibilites for fine tuning further. The binding energy of a helical bundle is predominantly provided by the hydrophobic residues in the *a* and *d* positions. The hydrophobic interactions are short range and proportional to the contact surface area of the hydrophobic side chains [21]. In 40 to 50 residue sequences with approximately 20 residues assigned to each helix, at least four *a* and *d* positions in each helix are occupied by hydrophobic amino acids, more than enough to make the sequence fold. The replacement of one or two of these residues by charged ones has large effects on the pK_a values of the charged residues and thus on their reactivities but does not disrupt the fold of the bundle. The pK_a of a lysine residue in a *d* position was measured to be 9.2, in comparison with 10.4, which is the random coil value [22]. The pK_a of a His residue in a *d* position was 5.6, in comparison with 6.4 which is the random coil value [23]. Depending on the pH and type of reaction these changes can give rise to substantial effects on reactivity. Although residues formally in *a* and *d* positions according to the heptad repeat pattern are, in principle, buried in the hydrophobic core, they are very reactive and thus capable of participating in bond making and breaking in active sites. The modulation of pK_a values has proven to be an important tool in mechanistic investigations of catalytic activity.

The techniques commonly used for structural characterization of folded polypeptides are NMR and CD spectroscopy and analytical ultracentrifugation. NMR spectroscopy is informative at many levels, and simple one-dimensional ^{1}H NMR spectra provide very useful, qualitative information about substrates, intermediates and products under reaction conditions and about whether they bind to the macromolecular catalyst [12]. The sharp resonances of small molecules are easily observed in the presence of the broad peaks of biomacromolecules and the binding of a small molecule by a macromolecule is reflected in the increased line width of the small molecules upon binding. The chemical shift dispersion and linewidths in the ^{1}H NMR spectrum of the polypeptide catalyst provide qualitative information about whether it is well defined or unordered. The chemical shift dispersion and temperature dependence will reveal whether it is close to being well defined (well dispersed, slow on the NMR time scale) or poorly defined (poor dispersion, fast on the NMR time scale) [24]. High resolution solution structures may also be obtained, but only after considerably greater effort and in specialist laboratories.

Highly repetitive peptide sequences, and sequences dominated by only a few amino acids are not easily studied by NMR spectroscopy since assignment of the NMR spectrum is difficult. A forward-looking aspect of the design process is therefore to vary the amino acid sequence as far as possible to enable one to carry out NMR spectroscopic analyses.

The CD signature of a helical peptide, with minima at 208 and 222 nm, is a powerful source of information about solution structure and molecular interactions with molecules that affect the helical content. The dissociation behavior of a helix–loop–helix dimer is conveniently monitored by CD spectroscopy and provides critical information about what species dominates in solution under reaction conditions [6]. Information about the state of aggregation is obtained by analytical ultracentrifugation, which is especially interesting with regards to higher order aggregation. CD spectroscopy is, as a rule, not informative in this respect since helicity does not seem to increase when helical bundles aggregate further.

5.1.3 Designed Catalysts

Several designed polypeptide catalysts have been reported to date, together with reasonably complete reaction mechanistic analyses. These may serve as good introductions to rational catalyst design and are listed here for reference purposes.

A helical 14-residue peptide, rich in lysine residues, was reported by Benner and coworkers in 1993 to catalyze the decarboxylation of oxaloacetate [4], Scheme 5.1. The catalytic efficiency was at least partially due to the depression of lysine pK_a values caused by the presence of neighboring protonated lysines in the folded helix, and the resulting increased propensity for imine formation. Although the peptide was partially disordered there appeared to be a correlation between helical content and catalysis and specific acid catalysis was an important feature of the reaction mechanism. Follow up publications by Allemann in ordered polypeptide scaffolds showed enhanced activity [25].

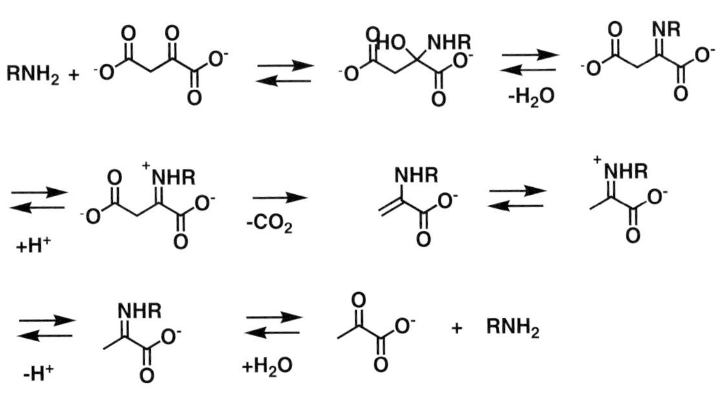

Scheme 5.1

Native chemical ligation, Scheme 5.2, is a two-step peptide ligation reaction in which a C-terminal thiol ester is reacted with the side chain of a N-terminal cysteine under release of the thiol leaving group, to form an intermediate that reacts further to form a peptide bond. The amino group replaces the cysteine side chain in an intramolecular rearrangement and forms the thermodynamically more stable product. Ghadiri et al. designed peptide templates on which the peptide fragments were assembled prior to reaction and catalysis was due mainly to proximity effects [5]. Selectivity was introduced by Chmielewski et al. and controlled by electrostatic interactions between catalyst and peptide [26]. Product inhibition was a problem.

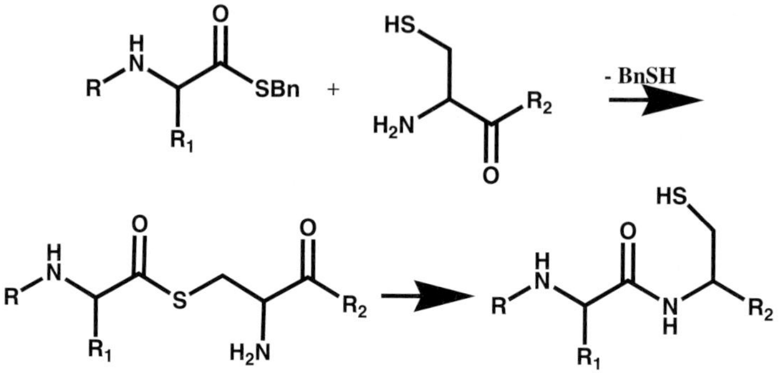

Scheme 5.2

A 42-residue polypeptide that folded into a helix–loop–helix motif and dimerized to form a four-helix bundle was shown to catalyze the hydrolysis and transesterification reactions of active esters, Scheme 5.3 [6, 11, 12]. The solution structure and reaction mechanism were extensively studied, see Section 5.2, and the rate en-

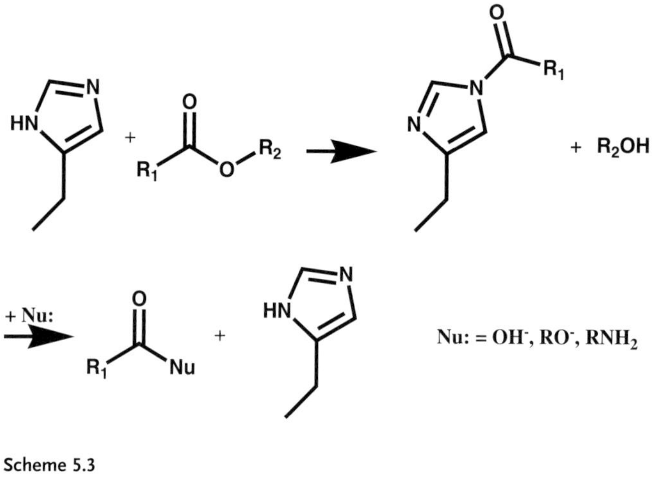

Scheme 5.3

hancement was shown to depend on cooperative nucleophilic and general-acid catalysis. A $HisH^+$-His pair was found to be the basic catalytic unit and supplementing charged residues in the catalyst enhanced the reactivity by factors that corresponded well to what was expected from transition state binding by salt bridge formation. Similar constructs were used to catalyze transformations of phosphate mono- and diesters, for example the cyclization of the RNA mimic uridine 3′-2,2,2-trichloroethylphosphate with a leaving group pK_a of 12.3, although the reaction mechanistic analyses were not as complete as for catalysis of ester hydrolysis (unpublished).

The same scaffold was used to design catalysts for pyridoxal phosphate-dependent deamination of aspartic acid to form oxaloacetate, one half of the transamination reaction [8], and oxaloacetate decarboxylation [14]. Catalysis was due to binding of pyridoxal phosphate in close proximity to His residues capable of rate limiting 1,3 proton transfer. A two-residue catalytic site containing one Arg and one Lys residue was found to be the most efficient decarboxylation agent, more efficient per residue than the Benner catalyst, most likely due to a combination of efficient imine formation, pK_a depression and transition state stabilization.

In order to bypass the problem of designing a pocket from scratch, Bolon and Mayo [27] introduced a catalytically active His residue in thioredoxin, a well-defined 108-residue protein for which much structural and functional information was available. The design was based on the well-known reaction mechanism of *p*-nitrophenyl acetate hydrolysis and thioredoxin was redesigned by computation to accommodate a histidine with an acylated side chain to mimic transition state stabilization. The thioredoxin mutant was catalytically active and the reaction followed saturation kinetics with a k_{cat} of 4.6×10^{-4} s^{-1} and a K_M of 170 μM. The catalytic efficiency, after correction for differential protonation and nucleophilicity, can be estimated to be a factor of 50 greater than that of 4-methylimidazole, due to nucleophilic catalysis and proximity effects, see Section 5.2.3.

A four-helix bundle protein, S-824, reported by Wei and Hecht [7], selected from a focused binary patterned library showed catalytic activity towards *p*-nitrophenyl acetate. S-824 exhibited a larger catalytic efficiency than that of KO-42 at pH 5.1 (>2) and a larger k_{cat} than that of PZD2 at pH 7 (>10) whereas k_{cat}/K_M at pH 7 was slightly smaller than that of PZD2. The pH profile was bell-shaped with a maximum at pH 8.5 and the reaction followed saturation kinetics. Since the catalyst has not yet been sufficiently characterized with regards to the identity of the catalytic machinery and the reaction mechanism it is not possible to make detailed comparisons with other catalysts, especially in the light of the fact that there are 12 histidines in the sequence. The pH profile suggests that acid–base catalysis is likely to play a role and the observation of the maximum rate at pH 8.5 suggests that His residues might function as nucleophiles and residues with pK_a values of around 10 provide general-acid catalysis or transition state stabilization. A more detailed analysis should provide a better platform for further development of an enzyme-like catalyst.

A rationally designed four-helix bundle diiron metalloprotein was shown by

Kaplan and DeGrado [9] to catalyze the oxidation of aminophenol by a two-electron transfer pathway, Scheme 5.4. The catalyst has been extensively characterized with regards to structure and the designed protein provided not only an active site cavity but also a channel through which the active site could be reached. The reaction mechanism and reactive site residues were determined.

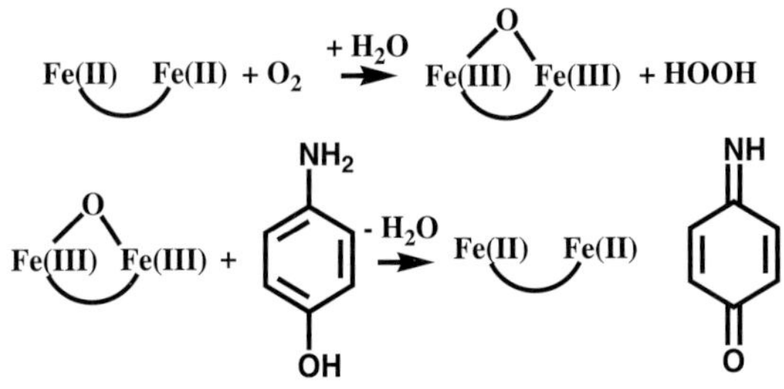

Scheme 5.4

The re-engineering of the ribose binding protein to introduce triose phosphate isomerase activity represents a major advance in enzyme design. Triose phosphate isomerase catalyzes the interconversion of dihydroxyacetone phosphate (DHAP) and glyceraldehyde 3-phosphate (GAP), Scheme 5.5. Computational design by Dwyer et al. [15] enabled not only the binding of substrate and product with micromolar affinities but also binding of the enediolate intermediate and the introduction of catalytically crucial Glu, His and Lys residues to catalyze the rate-limiting proton transfer reaction and to stabilize the intermediate. The catalyst NovoTim 1.2.4 catalyzed the DHAP to GAP transformation with a k_{cat}/K_M of 1×10^3 M^{-1} s^{-1} and the reverse reaction with a k_{cat}/K_M of 2.1×10^4 M^{-1} s^{-1}, only two and three orders of magnitude, respectively, lower than the rate constants measured for the wildtype enzyme. This achievement is remarkable, especially in view of the fact that triose phosphate isomerase is an enzyme operating at the diffusion controlled limit.

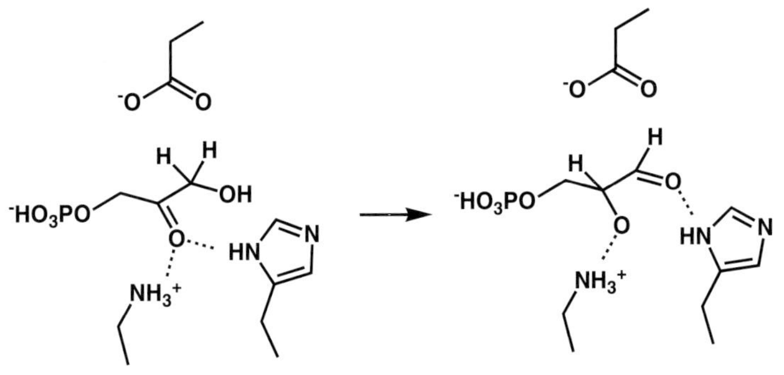

Scheme 5.5

5.2
Catalysis of Ester Hydrolysis

One of the most studied chemical reactions is that of ester hydrolysis, Scheme 5.6. The well-defined tetrahedral intermediate in which partial negative charge develops and the well understood dependence of reactivity on leaving group pK_a make it an obvious target for catalyst design. The acyl group and the leaving group may be optimized for binding and reactivity and the chromogenic properties may be adapted to simplify kinetic measurements. Furthermore, hydrolytic reactions are less prone to product inhibition than addition reactions and it is an important reaction in biology which makes it a target for drug development. From a mechanistic perspective, general-acid, general-base and nucleophilic catalysis, as well as transition state stabilization, may be considered for implementation in design and the success by which two or more of these principles can be incorporated cooperatively in a catalyst is a good measure of our understanding of the rational design principles. The hydrolysis of mono-*p*-nitrophenyl fumarate catalyzed by a designed 42-residue helix–loop–helix motif is one of the mechanistically best characterized polypeptide catalyzed systems. Since it exhibits cooperativity in catalysis it will be described here in some detail.

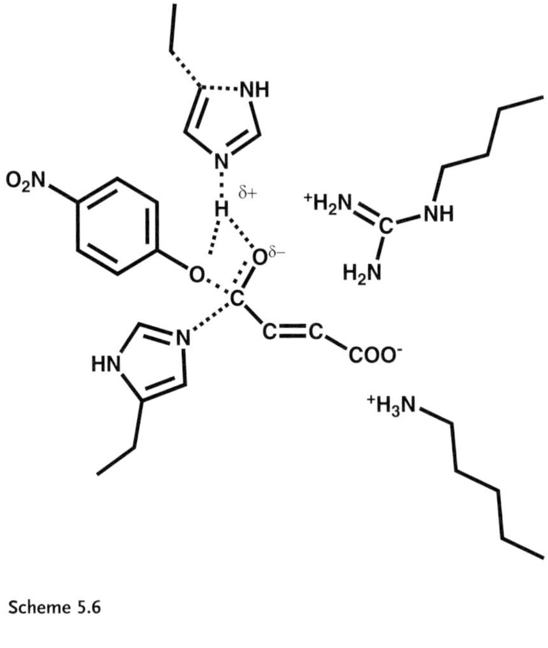

Scheme 5.6

5.2.1
Design of a Folded Polypeptide Catalyst for Ester Hydrolysis

The 42-residue sequence KO-42 was designed to fold into a helix–loop–helix motif and dimerize to form a four-helix bundle [6]. The design principles followed those of the previously described four-helix bundle SA-42 [28], and only five residues differed between the sequence of KO-42 and that of SA-42. These five residues were all in *g* and *c* positions on the surface of the folded motif and they were modified to introduce catalytic activity. Choosing from the common amino acids, histidine was the primary choice for catalysis of hydrolysis, because the imidazole side chain has a pK_a of 6.4 in a random coil peptide and is capable of general-acid, general-base and nucleophilic catalysis at around neutral pH. Imidazole catalysis was elucidated in great detail previously by Bruice and Jencks [29, 30] and its reactions with active esters follow a two-step mechanism. In the first and rate-limiting step an acyl intermediate is formed under the release of the leaving group, and in the second step the acyl group reacts with the most potent nucleophile in solution to form the reaction product. If the hydroxide ion is the most efficient nucleophile the overall reaction is hydrolysis. If alcohols are the most efficient the overall reaction is transesterification, Scheme 5.3.

Cooperativity in catalysis requires more than one catalytically active residue, but even more demandingly it requires the simultaneous catalytic activity by more than one residue in the rate-limiting step. The difficulty in predicting with high precision how to introduce two or more residues in positions and conformations where the probability is high for cooperative bond making and bond breaking in the transition state of the reaction suggests that a first generation catalyst should be designed with more than one possible combination of active site residues. KO-42 was designed with six His residues in *g* and *c* positions on the surface of the helix–loop–helix motif in order for the catalyst to provide several alternative configurations for catalysis, Fig. 5.4, [6]. His-11, His-15, His-19, His-26, His-30 and His-34 formed the catalytic surface of the folded polypeptide. Within each helix they represented a fixed i, $i+4$ pattern in which the structural relationship between

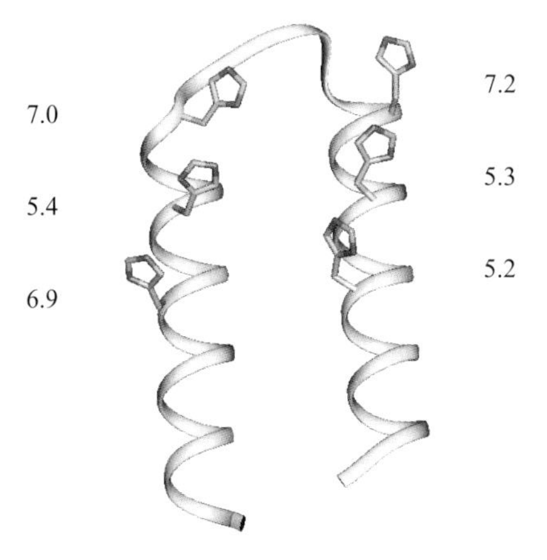

Figure 5.4. Schematic representation of the design of KO-42 with measured pK_a values indicated next to the His residues. In solution under reaction conditions KO-42 is a dimer.

each residue was well defined. A larger variation in inter-residue distances and orientations is represented by combinations of His residues in different helices. Every conceivable catalytically competent combination of His residues in a helix–loop–helix motif was represented in a single design using this approach. Not only were a large number of geometrical combinations represented but also a large number of pK_a combinations. A protonated His will suppress the pK_a of a neighboring His residue by electrostatic repulsion of the protonated and positively charged form of the flanking His. All in all, six ionizable residues represents $2^6 = 64$ states of protonation in a single molecule and each one will, in principle, have a different microscopic pK_a value for each His residue. The combination of six His residues in a single polypeptide therefore represented a large chemical library with inter-residue distances, rotamer populations and pK^a values as the variables. If no catalytic activity was found in this catalyst the likelihood of finding one based on the reactivity of His residues would be very small.

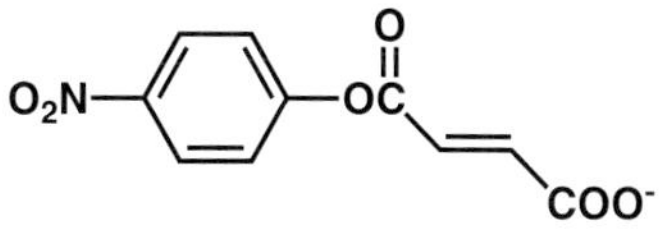

Scheme 5.7

The substrate designed and synthesized for initial kinetic investigations of catalysis was mono-*p*-nitrophenyl fumarate, Scheme 5.7, a rigid, negatively charged ester with an even more negatively charged transition state. This substrate was considered optimal for catalysis by a positively charged catalyst and due to the relatively low pK_a of the *p*-nitrophenol leaving group it was expected to be susceptible to nucleophilic, general-acid or general-base catalysis, all of which could be executed by His residues. The negative charge of the acyl group of the ester should be able to bind to positively charged residues and, at a later stage, the fumarate group could be replaced by hydrophobic residues to probe whether substrate binding by hydrophobic forces could provide more efficient catalysis due to proximity effects. The use of activated esters in catalyst development has been questioned because the reaction mechanisms may not be applicable to less reactive substrates [31]. Nucleophilic catalysis depends critically on the relative magnitudes of the pK_a values of the nucleophile and the leaving group and although a nucleophile with a pK_a of 6.4 may provide efficient catalysis with *p*-nitrophenyl esters, it will not with alkyl esters. Nevertheless, the use of active esters in the early stages of catalyst design is necessary because, for very primitive catalysts, the reaction rates would otherwise be intolerably slow.

5.2.2
The $HisH^+$-His Pair

The polypeptide KO-42 catalyzed the hydrolysis of mono-*p*-nitrophenyl fumarate with a second-order rate constant k_2 of 0.1 M^{-1} s^{-1}, a rate enhancement of more than three orders of magnitude over that of the 4-methylimidazole catalyzed reaction (k_2 of 8.8×10^{-5} M^{-1} s^{-1}) in aqueous solution at pH 4.1 and 290 K [11]. In comparison with the uncatalyzed background reaction the rate enhancement was a factor of 43000 at pH 5.1. The pH profile was recorded and showed that catalysis depended on the unprotonated form of a residue with a pK_a of around 5, suggesting histidine catalysis. The ^{1}H NMR spectrum of KO-42 was assigned and the apparent pK_a value of each His residue was determined by titration of the non-exchangeable 2- and 4-protons of the imidazole ring. The apparent pK_a values were found to be in the range 5.2 to 7.2, with His-34 exhibiting the lowest value. The kinetic solvent isotope effect, k_{H2O}/k_{D2O} was determined at pH 4.7 and found to be 2.0, a value that showed strong hydrogen bonding in the transition state, suggesting general-acid catalysis. The pH dependence and the kinetic solvent isotope effect taken together showed that there were two residues involved, one of which should be unprotonated and one of which provided strong hydrogen bonding in the transition state but not in the ground state. The unprotonated form of a His residue would be capable of nucleophilic as well as general-base catalysis, but the kinetic solvent isotope effect did not lend support to an interpretation of general-base catalysis. Based on previous studies suggesting that imidazole catalyzes the hydrolysis of *p*-nitrophenyl acetate by a nucleophilic mechanism it was concluded that the unprotonated residue acted as a nucleophilic catalyst, Scheme 5.6.

The number and the closely similar pK_a values of His residues made it difficult to assign the catalytic activity to specific residues. In a series of polypeptides histidines were partially replaced by the residues used in the sequence of SA-42 to form a library of catalysts derived from KO-42 but with less complexity [12, 23]. Essentially, the catalytic site of KO-42 was divided into its components and analyzed. It was assumed that the sequence modifications had only minor effects on structure and that the rate constants of the resulting peptides could be directly compared. The peptide MN, closely related to KO-42 but with His-26, His-30 and His-34, reverted to the SA-42 residues Gln, Gln and Ala, catalyzed the reaction at pH 5.1 and 290 K with an efficiency that was less than 10% of that of the KO-42 catalyzed reaction. The peptide JN, in which His-11, His-15 and His-19 were reverted to Ala, Gln and Lys, exhibited a second-order rate constant that was 20% of that of KO-42. The sum of the second-order rate constants of the MN and JN catalyzed reactions was therefore not equal to the second-order rate constant of KO-42, and there appeared to be cooperativity between residues in the two helices of the helix–loop–helix motif. However, the pK_a values were affected and increased as a result of the modifications, which may account for at least part of the observed discrepancy.

The complexity of the histidine site was further reduced and all combinations of two His residues four positions apart in the sequence were synthesized and analyzed. All were catalytically active with second-order rate constants in the range 0.008–0.055 M^{-1} s^{-1}, values that were higher than that of 4 methylimidazole, 0.00074 M^{-1} s^{-1} by a factor of 10–75. The largest second-order rate constant was found for the peptide JNII, Figure 5.5, in which His-30 and His-34 were the only His residues and, in general, the lower the pK_a the larger the rate constant. The sum of the second-order rate constants for the hydrolysis reactions catalyzed by

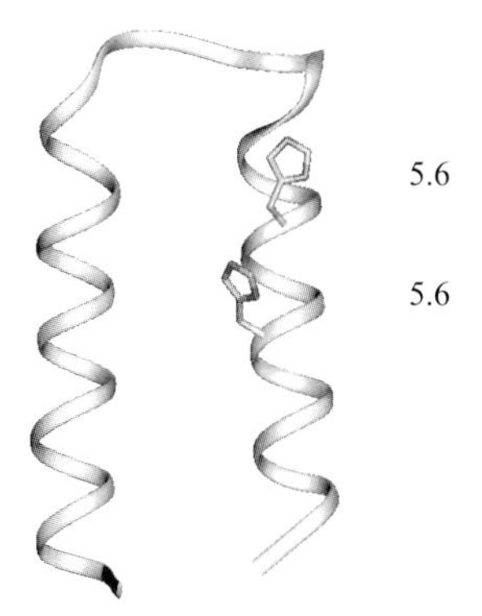

Figure 5.5. The 42-residue catalyst JNII has two His residues that catalyze the hydrolysis of mono-*p*-nitrophenyl fumarate by a combination of nucleophilic and general-acid catalysis. The pK_a values are indicated and the relative rate enhancement over that of the 4-methylimidazole catalyzed reaction is the largest at a pH below that of both His residues.

peptides with two His residues corresponded well to the measured rate constants for the reactions catalyzed by peptides containing three His residues. Consequently, within a single helix, there was no cooperativity between three histidine groups, but clearly between two. It was concluded that the sites with two His residues were the basic catalytic units in the observed catalysis of ester hydrolysis. The pH dependence revealed that one of them should be unprotonated and the kinetic solvent isotope effect showed that one of them should be protonated. The basic catalytic unit was therefore the $HisH^+$-His pair [11].

5.2.3
Reactivity According to the Brönsted Equation

The reactivity of a nucleophile is described by the Brönsted equation according to which

$$\log k_2 = A + \beta \mathrm{p}K_a \tag{5.1}$$

The Brönsted coefficient β for imidazole catalysis of *p*-nitrophenyl acetate hydrolysis is 0.8 [29], and the second-order rate constants of all His residues can therefore be related to that of 4-methylimidazole to determine whether there are effects on reactivity beyond those of differential nucleophilicity and levels of protonation. The reactivity of His residues in the pH independent region may be estimated from rate constants, pH and pK_a values. The second-order rate constant of the 4-methylimidazole catalyzed hydrolysis of mono-*p*-nitrophenyl fumarate at pH 5.85 and 290 K is 1.02×10^{-2} M^{-1} s^{-1}. From this value and the pK_a of 7.9, the second-order rate constant of the unprotonated form of 4-methylimidazole was readily calculated to be 1.15 M^{-1} s^{-1}. From this value the second-order rate constants of each unprotonated His residue of KO-42 could be estimated using the Brönsted equation, and that of His-34 (pK_a 5.2) would be 0.008 M^{-1} s^{-1}, whereas that of His-26 (pK_a 7.2) would be 0.32. The sum of the rate constants of all His residues in KO-42 is 0.75, very close to the second-order rate constant of the KO-42 catalyzed reaction at high pH, which is 0.74 M^{-1} s^{-1}. At a pH where all His residues are unprotonated KO-42 behaves catalytically merely as the sum of a number of independent imidazole groups.

At a pH below the pK_a of each histidine the situation is more complex. Here the more acidic His is the most efficient catalyst because log k_2 is proportional only to $0.8 \times$ pK_a, whereas the fraction of unprotonated and catalytically active residues is directly proportional to pK_a. The more acidic catalyst is more efficient than one with a higher pK_a by a factor of $10^{0.2\Delta pKa}$ because the decrease in nucleophilicity of a residue with a lower pK_a is outweighed by the increased availability of the catalytically active form. If nucleophilicity and the degree of protonation were the only factors involved, His-34 would be a better catalyst than 4-methylimidazole at pH 4.1 by a factor of $10^{0.2\times2.7}$, or a factor of 3.5. The sum of second-order rate constants estimated for each His residue in KO-42 under these conditions would be 1×10^{-3} M^{-1} s^{-1}, a factor of 100 less than the experimentally determined value of 0.1 M^{-1} s^{-1}. The pH dependence of imidazole catalysis reveals that general-acid catalysed hydrolysis of ester hydrolysis by individual imidazolium ions at pH 4.1 is a very inefficient reaction and can be disregarded. The catalytic efficiency of KO-42 at low pH is therefore larger than the sum of that of each individual residue by a factor of 100, a factor that is most likely due to cooperativity between nucleophilic catalysis and general-acid catalysis.

5.2.4 Cooperative Nucleophilic and General-acid Catalysis in Ester Hydrolysis

The nature of the cooperativity was further characterized based on results from kinetic measurements. The two $HisH^+$-His pairs in helix II catalyzed the hydrolysis of mono-*p*-nitrophenyl fumarate at pH 5.1 and 290 K with second-order rate constants of 0.01 M^{-1} s^{-1} (JNI, His-26, His-30) and 0.055 M^{-1} s^{-1} (JNII, His-30, His-34), respectively, and a rate constant ratio JNII/JNI of 5.5. The pK_a values of both His residues in JNII are the same, so for the analysis it does not matter which residue is the nucleophile and which one is the acid. In JNI, however, the pK_a values are 6.9 for His-26 and 5.6 for His-30. The rate constant ratio of 5.5 should therefore arise due to the difference in nucleophilicity or due to the difference in acidity, or if both residues in the pair can be both nucleophile and acid, from a mixture of the two. If His-30 functions as a general acid in JNI, then the rate constant ratio should arise from the difference in nucleophilicity between two nucleophiles with the pK_a values 5.6 and 6.9. We can, however, calculate the reactivity difference as in Section 5.2.3 to find that $10^{0.2\times1.3} = 1.8$, one third of the observed ratio of 5.5. If, on the other hand, the rate constant ratio is due to a difference in general-acid catalysis by two residues with pK_a values of 5.6 and 6.9, then the the Brönsted equation for general-acid catalysis can be applied

$$\log k_2 = A - \alpha \mathrm{p}K_\mathrm{a} \qquad (5.2)$$

A value of the Brönsted coefficient α for general-acid catalysis of 0.56 gives rise to a rate constant ratio of 5.5 for two acids with a difference in $\mathrm{p}K_\mathrm{a}$ of 1.3. If it is taken into consideration that an acid with a $\mathrm{p}K_\mathrm{a}$ of 5.6 is only 75% protonated at pH 5.1 then the second-order rate constant of the JNII catalyzed reaction is only 75% of its true value, and k_2 for the HisH$^+$-His pair should be corrected to 0.073 $\mathrm{M}^{-1}\ \mathrm{s}^{-1}$. Even so a Brönsted coefficient of 0.66 would account for the rate constant ratio. Both of these values fall within the range of Brönsted coefficients typically observed for general-acid catalysis. An acid with a $\mathrm{p}K_\mathrm{a}$ of 6.9 is largely protonated at pH 5.1 and no corrections are required. From these considerations it is likely that the stereochemistry of the helix favors nucleophilic catalysis by the His residue with the highest number in the sequence and general-acid catalysis by the residue with the lowest number.

5.2.5
Why General-acid Catalysis?

It has been claimed that highly activated esters do not require catalysis for their hydrolysis and that *p*-nitrophenyl esters in general are degraded so rapidly that it is very difficult to find a catalyst efficient enough to make a difference [32]. This statement, however, is contradictory to the statement made by others that the hydrolysis of *p*-nitrophenyl esters is so easy to catalyze that catalysts for active esters are irrelevant for biologically significant esters [31]. The KO-42 catalyzed hydrolysis of mono-*p*-nitrophenyl fumarate, described in detail in Sections 5.2.2–5.2.4, was as efficient as that provided by six imidazoles with comparable $\mathrm{p}K_\mathrm{a}$ values, at a pH that in a broad sense is higher than or equal to the $\mathrm{p}K_\mathrm{a}$ of the corresponding acid of the *p*-nitrophenolate leaving group. At a pH lower than the $\mathrm{p}K_\mathrm{a}$ of *p*-nitrophenol the rate enhancement was two orders of magnitude larger than that of six imidazoles with comparable $\mathrm{p}K_\mathrm{a}$ values, and suggested by the experimental evidence to be due to general-acid catalysis. In comparing the high and low pH reactions it may be noted that they were measured under conditions where the released leaving group exists predominantly in its unprotonated and protonated forms, respectively. At low pH the leaving group therefore requires a proton, normally provided by the solvent water. A Hammet ρ value of 1.4 was determined by hydrolyzing a set of phenyl esters covering $\mathrm{p}K_\mathrm{a}$ values from 3.96 to 8.28 [33], and it was found that when comparing the ρ value for hydrolysis with that of ionization, which is 2.2, approximately half a negative charge was found to reside on the phenolate oxygen in the transition state of the reaction. A plausible explanation is therefore that general-acid catalysis may operate when the leaving group, after expulsion, exists predominantly in its protonated form. A complication with this interpretation is that at high pH the His residues of the catalyst are unprotonated and therefore incapable of proton donation.

In order to determine whether the lack of general-acid catalysis at high pH was due to a lack of proton donors or whether it was due to the fact that the leaving group was not in need of a proton an even more activated ester was synthesized, 2,4-dinitrophenyl acetate, with a pK_a of 3.96 [33]. The second-order rate constant of 7.2 M^{-1} s^{-1} for the JNII catalyzed reaction was determined at pH 5.1 and compared to that of 4-methylimidazole, 0.0081 M^{-1} s^{-1}. The rate constant ratio was almost three orders of magnitude. Estimating the catalytic efficiency of a hypothetical imidazole derivative with a pK_a of 5.6, in the absence of differential protonation effects, as in Section 5.2.3, showed JNII to be a more efficient catalyst by a factor of 68, apparently due to very efficient nucleophilic catalysis. At pH 3.1 which is a value that is lower than that of the pK_a of 2,4-dinitrophenol, the rate enhancement was 1800, the intrinsic catalytic efficiency was a factor of 140 and the kinetic solvent isotope effect 2.1. No kinetic solvent isotope effect was determined for JNII at pH 5.1, but a helix–loop–helix motif JNIIOR, that contained two His residues in the same positions as in JNII and differed by only two residues, gave rise to a kinetic solvent effect at pH 3.1 but not at pH 5.1. In these scaffolds at pH 5.1 protonated His residues were present to provide general-acid catalysis but none occurred. It was concluded that the leaving group state of ionization determines the need for general-acid catalysis in ester hydrolysis. Consequently, general-acid catalysis is efficient for the hydrolysis of active as well as inactive esters.

5.3 Limits of Activity in Surface Catalysis

While folded polypeptide catalysts are excellent vehicles for catalyst design and for mechanistic investigations of catalysis, they are limited with regards to the catalytic efficiency that can be expected. Experimentally, rate enhancements of the order of 10^3–10^4 have been reported, but rate enhancements of the order of magnitude of even the slowest of enzymes have not been observed, with enzymatic efficiencies considered to be rate enhancements of not less than 10^6 over background [34]. The reasons for this are only understood at the hypothetical level since it is not easy to determine conclusively why a catalyst is not as fast as expected. Some aspects of catalysis may, however, be tested in detail in a folded polypeptide scaffold. The $HisH^+$-His pair has, for example, been systematically varied with regards to nucleophilicity and acidity, as described in Sections 5.2.3–5.2.5, to probe the intrinsic reactivity of the catalytic machinery. Tuning pK_a values to optimal catalytic efficiencies at a given pH is an important aspect of enzyme catalysis. The use of active sites and pockets is also characteristic of enzymes, perhaps due to the reduced degrees of freedom and optimized positions of amino acid side chains involved in the making and breaking of bonds and to the strength of electrostatic interactions in a low dielectric medium. In a helix–loop–helix motif the relative positions of the $HisH^+$ and the His can be varied but residues on the surface of folded polypeptides and proteins have many degrees of freedom and cannot be locked in fixed positions. Charge–charge interactions are weak but measurable. These aspects of catalysis were tested within the context of HisH$^+$-His-based catalysis of ester hydrolysis.

5.3.1
Optimal Organization of His Residues for Catalysis of Ester Hydrolysis

In the transition state of HisH$^+$-His-based catalysis of ester hydrolysis there is a partial bond between one of the nitrogen atoms of the imidazole ring and the carbonyl carbon of the ester and a hydrogen bond from the protonated imidazolium ring to one or both of the oxygen atoms of the carbonyl group of the substrate. The preferred conformation of each histidine side chain is determined by principles that are well understood and controlled by steric factors with anti, staggered conformations being energetically the most favorable, although gauche conformations are populated as well. In a folded helix the bonds between the α and β the carbons of amino acid side chains is not orthogonal to the helix axis but points slightly towards the N-terminal. Depending on the relative positions of the HisH$^+$ and the His groups in the polypeptide scaffold bond formation between scaffold residues and substrate would be expected to be more or less energetically favorable, and rate enhancements correspondingly different. Several polypeptide sequences were designed and their activity towards model substrates determined [35]. Within each helix of the helix–loop–helix motif several combinations in which the two residues were four residues, or one helical turn, apart. In addition, a number of catalysts were designed in which the two His residues were in separate helices. Against this background of geometrical diversity it was expected that the kinetic results would provide some guidance as to what would be an optimal catalyst structure. No such guidance was found and all the catalysts studied provided rate constants that were in agreement with a model in which pK_a values, and thus nucleophilicity and general-acid catalysis, were the dominant factors. It is likely that covalent bond formation between nucleophile and carbonyl group was near optimal and that proton donation from the imidazolium group was tolerant to a range of hydrogen bond distances and angles.

Although residue side chains in polypeptides have preferred conformations they are free to rotate in solvent exposed sites since there are no contraints posed by neighboring groups. If there is only one catalytically active conformation then there is a cost in entropy associated with reaching the transition state of the reaction due to a reduction in the degrees of freedom. The side chain of a His residue has two rotatable carbon–carbon bonds and in an aliphatic substituent the entropy loss due to inhibition of rotation has been measured and estimated to be 0.9 kcal mol^{-1} per bond [36]. Based on this estimate it is possible that as much as 3.6 kcal mol^{-1} of the free energy of activation is due to the mobility of the His side chains in a surface exposed catalytic site, corresponding to almost three orders of magnitude in rate decrease, in comparison with that of a preorganized site. Although a further rate enhancement of three orders of magnitude in addition to the three due to cooperative nucleophilic general-acid catalysis would make an impressive catalyst, it does not bring it within range of native enzymes. Dynamics in the scaffold structure could contribute even more to the free entropy of activation and the relative positions of the catalytically active residues may not be optimal.

5.3.2 Substrate and Transition State Binding

Substrate binding is the hallmark of native enzymes and makes the chemical transformations in effect intramolecular. The hydrolysis of mono-*p*-nitrophenyl fumarate was investigated with regards to substrate binding and found to follow saturation kinetics with a k_{cat} of 0.00017 s^{-1} and a K_M of 1 mM [12]. The interactions between substrate and catalyst were further probed by NMR spectroscopy and upon addition of the substrate under reaction conditions to a solution containing the catalyst the resonances of the hydrophobic residues of the folded polypeptide were shifted. Due to the spectral differences between substrate and product it could be established that the substrate and not the products interacted with the polypeptide. The reasons for the observed binding were probably that the hydrophobic *p*-nitrophenyl residue interacted with the hydrophobic core of the four-helix bundle while the fumarate group bound to the positively charged surface residues of the scaffold. The binding of the substrate by the catalyst does not prove that the binding is productive but only that a complex is formed at concentrations that are compatible with the observed dissociation constant. The binding site was not designed and the interactions of the substrate with hydrophobic core and positively charged residues may not have been specific. Further attempts to initiate specific binding of the substrate included the incorporation of arginine and lysine substituents in the neighbouring helix to interact with the negatively charged fumarate group and with the developing negative charge in the tetrahedral transition state, Figure 5.6, [12]. The number of methylene groups in the side chains of the flanking residues were varied in order to investigate whether the interactions could be optimized, again for the purpose of probing specificity. Rate enhancements were obtained upon introduction of flanking, positively charged residues close to the $HisH^+$-His pair to show that increased transition state binding could be introduced by rational design. The effects were significant but not larger than factors of 2–3, corresponding to ≈ 0.5 kcal mol^{-1} of binding energy, in agreement with what has been measured for a salt bridge in a helix [37]. Charge–charge interactions are inversely dependent on the dielectric of the solvent and would be stronger by a factor of ten or more in the dielectric of a hydrophobic pocket. Differential transition state binding by 5 kcal mol^{-1} would result in an increase in rate constant by almost four orders of magnitude at room temperature. The transfer of the reactive site designed for ester hydrolysis into a hydrophobic environment would therefore be expected to enhance the catalytic efficiency considerably. On the surface of a folded polypeptide, practically in aqueous solution, the weakness of forces between residues is a major reason for the poor efficiency of the catalyst. Substrate and transition state binding is weak due to the high dielectric constant of the solvent water

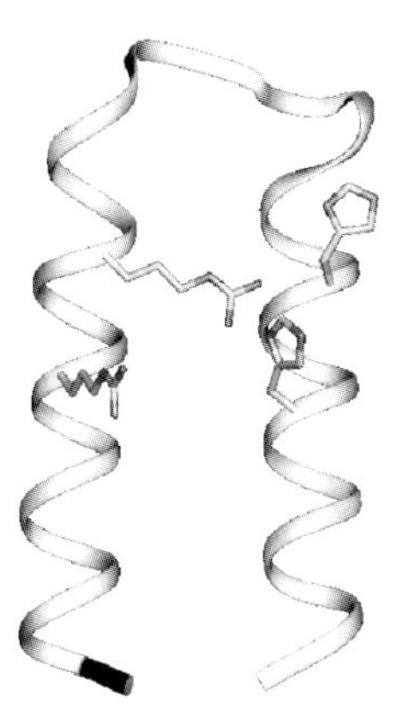

Figure 5.6. The incorporation of two His residues in one helix as well as one Arg and one Lys residue in the neighboring helix led to a catalyst capable of cooperative catalysis and transition state stabilization. The catalyst has enzyme-like properties but lacks catalytic efficiency in comparison with native enzymes.

and due to the complex solvation equilibria. Nevertheless, the active site can be expected to function better in the pocket of a protein and thus serves as a good model system for new biocatalysts.

5.3.3 His Catalysis in Re-engineered Proteins

On a naturally occurring scaffold with better defined structure it may not be straightforward to graft new catalytic sites because the effect on structure is difficult to predict. The introduction of a general acid or a general base is, however, a minor invasion that may be tolerated by the protein, and can be achieved by a single residue. From a mechanistic point of view such a modification may open new reaction pathways and allow us to test in a protein scaffold the structural requirements for general-acid and general-base catalysis. Human glutathione transferase is a detoxification enzyme that rids our bodies of hydrophobic compounds by catalyzing the conjugation of the non-endegenous molecule to the tripeptide glutathione, and secreting it. The active site is covered by a helix and two His mutations four residues apart in the helix, A216H and F220H, were selected to introduce the $HisH^+$-His pair [38]. *S*-benzoylglutathione was selected as substrate since its position in the active site was well defined and determined by crystallography. Although at a predictive level both His residues were within bond forming distance from the thiol ester, the reaction mechanism followed a different pathway and in the first step of the reaction the acyl group was transferred to a tyrosine residue to form an ester, and in the second step of the reaction His 216 catalyzed the hydrolysis of the tyrosyl ester, most likely by general-base catalysis. His-220 was too far away from the tyrosine side chain to be able to contribute by general-acid catalysis.

The fact that His-216 was observed in the crystal structure of the mutant showed that its position was well defined with a distance that did not accommodate covalent bond formation to the carbonyl carbon of the ester. A water molecule would, however, nicely bridge the distance of 7 Å between the nitrogen and the carbon. The pH dependence showed that catalysis was due to a residue in its unprotonated form and by elimination it was concluded that general-base catalysis was the key to hydrolysis. The hydrolysis of *S*-benzoylglutathione did not take place in the wild-type enzyme where general-base catalysis did not operate, and the introduction of general-base catalysis opened a new reaction pathway. The rate enhancement in comparison with the wildtype enzyme cannot be calculated since the reaction does not take place in the absence of a catalytically active residue, it can only be concluded that it is very efficient. It may be that the distance of 7 Å between the imidazole nitrogen and carbonyl carbon is not optimal, but its well-defined orientation towards the ester group may be the reason for the catalytic efficiency.

5.4
Computational Catalyst Design

Early strategies in *de novo* protein design were influenced by respect for the difficulties in predicting structure from sequence and the idea that well behaved proteins could be built from combinations of stable secondary structure motifs. Much work was directed towards understanding the factors that controlled helix formation and helix stability, and subsequently also β-sheet formation and stability, although uncontrolled aggregation remained a long-standing problem in β-sheet design. Shape and charge complementarity were engineered into helices to control docking and drive folding, and an understanding of how to design well-behaved compact proteins emerged. The work by DeGrado pioneered the field of *de novo* protein design [3]. The design of pockets and cavities needed for sophisticated functions such as enzyme-like catalysis was, however, beyond this approach, although metal ion complexation was used in an effort to enable partial separation of secondary structure elements and form at least first generation clefts and hollows [39]. In this process the power of computation was appreciated but calculations of free energies of the possible conformers of a polypeptide in the search for global minima proved to be too demanding in terms of computer capacity.

The redesign of proteins known to fold has been a considerably more successful approach, by reducing the computational problem using a simple but powerful assumption. Proteins known to fold are expected to fold in the same way, even if several amino acid side chains are replaced by others and by compounds that are not linked to the polypeptide scaffold. In simple terms the backbone of a selected protein is locked in its native conformation and a "hole" carved out by removing amino acid side chains in a part of the protein structure. After introducing a small organic molecule or transition state model into the hole by computation, the rest of the cavity is filled with amino acid side chains to form a compact structure. If a small molecule is introduced the result is a receptor for this molecule and if a transition state analog is introduced the result is a catalyst capable of transition state stabilization. More sophisticated versions allow also the introduction of residues that enable acid–base catalysis etc. Computational methods following this strategy show great promise in the engineering of new enzymes. Some examples are described below.

5.4.1
Ester Hydrolysis

A site for His-dependent nucleophilic catalysis of *p*-nitrophenyl acetate hydrolysis was introduced into the thioredoxin protein scaffold by computational design. The acyl intermediate described in Section 5.2.1 formed at the side chain of a His residue was used as the starting point for design and after side chain rotamer library generation and analysis a His residue with an acylated side chain was introduced in the most favorable side chain conformation [27]. The surrounding protein residues were selected to stabilize the acyl intermediate and two resulting thioredoxin mutants were expressed and analyzed with regards to catalytic power. Wildtype thioredoxin is capable of His-mediated ester hydrolysis due to the presence of a surface exposed residue, but with a low efficiency, whereas catalysis by the mutant PZD2 followed saturation kinetics in aqueous solution at pH 6.95 with a k_{cat} of 4.6×10^{-4} s^{-1} and a K_M of 170 μM. A comparison with the background reaction showed that k_{cat}/k_{uncat} for PZD2 was 180, and k_{cat}/K_M was 25 times larger than the second-order rate constant of the 4-methylimidazole catalyzed reaction. An analysis of the catalytic efficiency as conducted in Section 5.2.3 is difficult since the pK_a of the catalytically active His residue of PZD2 was not reported. Under the assumption that it is 6.4, as in a random coil peptide, 4-methylimidazole, with a pK_a of 7.9 is an intrinsically better catalyst than the His of PZD2 by a factor of two because the concentration of unprotonated and active His is larger than that of 4-methylimidazole by a factor of eight but the histidine is a weaker nucleophile by a factor of sixteen. The rate constant ratio $(k_{cat}/K_M)/k_2$ was reported as 25 but the ratio of catalytic efficiencies is better described as 50. The comparison with KO-42 at pH 6.95 is not interesting because at pH 6.95 KO-42 behaves as six unprotonated His residues and there is little general-acid catalysis in operation. At pH 4.1 the catalytic efficiency of the polypeptide catalyst, in comparison with that of 4-methylimidazole, is slightly better than that of PZD2 at pH 6.95, due to the cooperativity between nucleophilic and general-acid catalysis. The catalytic activity of PZD2 was due to a combination of proximity effects and covalent catalysis by His but no attempt to incorporate a second catalytic mechanism was described. When this can be achieved considerably larger rate enhancements would be expected.

5.4.2
Triose Phosphate Isomerase Activity by Design

Although the ribose binding protein is not a peptide, the topic of this chapter, but a mature folded protein, it is discussed here because the simultaneous introduction of residues capable of acid–base catalysis in addition to substrate, intermediate and transition state binding represents a major advance in enzyme design. The key step in the dihydroxyacetone phosphate to glyceraldehyde 3-phosphate transformation is a 1,2 proton transfer between two carbon atoms. Efficient proton abstraction from a carbon acid with a pK_a of ≈ 18 and its subsequent delivery to a neighboring carbon atom requires a general base in a precisely organized position and would be expected to be difficult, for example, on the surface of a helical bundle. A Glu, a His and a Lys residue were introduced into the protein scaffold to abstract a proton, to provide a proton to the enediolate intermediate and to stabilize the negative charges in the active site. The precise organization of catalytically active residues demonstrated by Dwyer et al. [15] suggests that computational design may be used to engineer several new enzymes, primarily those that have simple reaction mechanisms. The rate enhancements were reported to be within two and three orders of magnitude, respectively, of the forward and reverse reactions of the native enzyme. The pH profiles for the forward and reverse reactions were bell-shaped with maxima between 7 and 8 and similar to those of the wild-type TIM. Single- double- and triple-alanine mutations of the three putatively catalytic residues resulted in loss of enzymatic activity. While the individual role of each one of these residues has not been unequivocally established the bulk of the evidence is compatible with the design of the catalyst where Glu is the base and His and Lys are involved in hydrogen bonding and electrostatic stabilization of the developing charges in the transition state.

5.5 Enzyme Design

The description in this chapter of several designed catalysts with the capacity to enhance reaction rates of selected reactions by several orders of magnitude is intended to impress upon the reader that the understanding of how to implement catalytic sites into polypeptides and proteins is slowly emerging. Computational methods in particular have reached a level where the precise positioning of amino acid residues in protein and polypeptide scaffolds has become possible. In combination with an increased understanding of reaction mechanistic principles it may well prove to be the strategy for the future. The introduction of residues capable of general acid, general base and covalent catalysis has been demonstrated in several designs, as has the introduction of residues capable of transition state stabilisation and substrate binding. The rational design of new enzymes for practical purposes is slowly becoming reality.

References

1 L. Baltzer, H. Nilsson, J. Nilsson, *Chem. Rev.* 101 (2001) 3153–3163.

2 L. Baltzer, J. Nilsson, *Curr. Opin. Biotechnol.* 12 (2001) 355–360.

3 W. F. DeGrado, C. M. Summa, V. Pavone, F. Nastri, A. Lombardi, *Annu. Rev. Biochem.* 68 (1999) 779–819.
4 K. Johnsson, R. K. Allemann, H. Widmer, S. A. Benner, *Nature.* 365 (1993) 530–535.
5 K. Severin, D. H. Lee, A. J. Kennan, R. M. Ghadiri, *Nature.* 389 (1997) 706–709.
6 K. S. Broo, L. Brive, P. Ahlberg, L. Baltzer, *J. Am. Chem. Soc.* 119 (1997) 11362–11372.
7 Y. N. Wei, M. H. Hecht, *Protein Eng. Des. Sel.* 17 (2004) 67–75.
8 M. Allert, L. Baltzer, *ChemBioChem.* 4 (2003) 306–318.
9 J. Kaplan, W. F. DeGrado, *Proc. Natl. Acad. Sci. USA* 101 (2004) 11566–11570.
10 F. Hollfelder, A. J. Kirby, D. S. Tawfik, *J. Org. Chem.* 66 (2001) 5866–5874.
11 K. S. Broo, H. Nilsson, J. Nilsson, A. Flodberg, L. Baltzer, *J. Am. Chem. Soc.* 120 (1998) 4063–4068.
12 K. S. Broo, H. Nilsson, J. Nilsson, L. Baltzer, *J. Am. Chem. Soc.* 120 (1998) 10287–10295.
13 K. Broo, L. Brive, A.-C. Lundh, P. Ahlberg, L. Baltzer, *J. Am. Chem. Soc.* 118 (1996) 8172–8173.
14 M. Allert, L. Baltzer, *Chem. Eur. J.* 8 (2002) 2549–2560.
15 M. A. Dwyer, L. L. Looger, H. W. Hellinga, *Science*, 304 (2004) 1967–1972.
16 R. B. Hill, W. F. DeGrado, *J. Am. Chem. Soc.* 120 (1998) 1138–1145.
17 S. Olofsson, L. Baltzer, *Folding Des.* 1 (1996) 347–356.
18 S. P. Ho, W. F. DeGrado, *J. Am. Chem. Soc.* 109 (1987) 6751–6758.
19 K. S. Broo, L. Brive, R. S. Sott, L. Baltzer, *Folding Des.* 3 (1998) 303–312.
20 C. Tanford, *Adv. Protein Chem.* 17 (1962) 69–165.
21 A. Fersht, *Structure and Mechanism in Protein Science: A Guide to Enzyme Catalysis and Protein Folding*, W. H. Freeman, New York, 1999, Ch. 11.
22 L. K. Andersson, G. T. Dolphin, L. Baltzer, *ChemBioChem* 3 (2002) 741–751.
23 L. Baltzer, K. S. Broo, H. Nilsson, J. Nilsson, *Bioorg. Med. Chem.* 7 (1999) 83–91.
24 G. T. Dolphin, L. Brive, G. Johansson, L. Baltzer, *J. Am. Chem. Soc.* 118 (1996) 11297–11298.
25 C. J. Weston, C. H. Cureton, M. J. Calvert, O. S. Smart, R. K. Allemann, *ChemBioChem* 5 (2004) 1075–1080.
26 S. Yao, I. Ghosh, R. Zutshi, J. Chmielewski, *Angew. Chem. Int. Ed. Engl.* 37 (1998) 478–479.
27 N. D. Bolon, S. L. Mayo, *Proc. Natl. Acad. Sci. USA* 98 (2001) 14274–14279.
28 S. Olofsson, G. Johansson, L. Baltzer, *J. Chem. Soc., Perkin Trans 2.* (1995) 2047–2056.
29 T. C. Bruice, R. Lapinski, *J. Am. Chem. Soc.* 80 (1958) 2265–2272.
30 D. G. Oakenfull, K. Salvesen, W. P. Jencks, *J. Am. Chem. Soc.* 93 (1971) 188–194.
31 M. J. Corey, E. Corey, *Proc. Natl. Acad. Sci. USA* 93 (1996) 11428–11434.
32 A. Fersht, *Structure and Mechanism in Protein Science: A Guide to Enzyme Catalysis and Protein Folding*, W. H. Freeman, New York, 1999, Ch. 2.
33 J. Nilsson, L. Baltzer, *Chem. Eur. J.* 6 (2000) 2214–2220.
34 A. Fersht, *Structure and Mechanism in Protein Science: A Guide to Enzyme Catalysis and Protein Folding*, W. H. Freeman, New York, 1999, Ch. 1.
35 J. Nilsson, K. S. Broo, R. S. Sott, L. Baltzer, *Can. J. Chem.* 77 (1999) 990–996.
36 M. I. Page, W. P. Jencks, *Proc. Natl. Acad. Sci. USA* 68 (1971) 1678–83.
37 Z. S. Shi, C. A. Olson, A. J. Bell, N. R. Kallenbach, *Biopolymers* 60 (2001) 366–380.
38 S. Hederos, K. S. Broo, E. Jakobsson, G. J. Kleywegt, B. Mannervik, L. Baltzer, *Proc. Natl. Acad. Sci. USA* 101 (2004) 13161–13167.
39 G. R. Dieckmann, D. K. McRorie, D. L. Tierney, L. M. Utschig, C. P. Singer, T. V. O'Halloran, J. E. Penner-Hahn, W. F. DeGrado, V. L. Pecoraro, *J. Am. Chem. Soc.* 119 (1997) 6195–6196.

Part II
General Aspects of Biological Hydrogen Transfer

Proton abstraction from carbon occupies a very substantial niche among enzyme catalyzed reactions, occurring for example in glucose oxidation and in the citric acid cycle. Initially, studies of the proteins within these pathways were focused on individual mechanistic features, whereas in recent years the focus has moved toward the establishment of general principles. Gerlt lays out the problems of rate acceleration by enzymes catalyzing proton loss from carbon: how to remove a proton from a site with an inherent pK_a substantially higher than for any catalytic functional group within the active site? Marcus theory is introduced as a useful tool, in particular through its separation of the reaction barrier into driving force, ΔG°, and reorganization energy, λ. Gerlt argues that even in the event of no contribution of λ to the reaction barrier, the inherently uphill process of proton loss from carbon requires the stabilization of intermediates to obtain the observed enzymatic rate accelerations. The role of the enzyme in decreasing ΔG° could be approached by measuring the concentration of enzyme bound carbanion intermediates, but these species can be very difficult to detect and quantify. Additionally, few experimental data are available to compare the value of λ in enzymatic deprotonations to their solution counterparts – a clear challenge for the future. Readers will want to compare Gerlt's thesis of electrostatic/H-bonding stabilization of carbanion intermediates as a dominant factor in enzymes, with that of Herschlag and co-workers (Kraut DA, Sigala PA, Pybus B, Liu CW, Ringe D, Petsko GA, Herschlag D., PLoS Biol. 2006 Apr;4(4):e99. Epub 2006 Mar 28.), according to which the rate acceleration from electrostatic stabilization is at most 300-fold in the paradigmatic proton abstracting enzyme, ketosteroid isomerase. The chapter by Spies and Toney is focused on the enzymes that catalyze racemization and epimerization, largely by proton abstraction. Their discussion of alanine racemase is an elegant demonstration of experimental approaches that can demonstrate the formation of a carbanion intermediate when none can be observed directly. They show how kinetic isotope effects distinguish a step-wise from a concerted reaction, thereby implicating the elusive carbanionic intermediate. They suggest that maintaining a very low concentration for a carbanion may be beneficial, to minimize or prevent undesirable chemical side reactions. The chapter by Warshel and coworkers presents an excellent account of the use of the empirical valence bond approach (EVB) to calculate rates and their attendant properties for proton abstraction reactions. They empha-

Hydrogen-Transfer Reactions. Edited by J. T. Hynes, J. P. Klinman, H.-H. Limbach, and R. L. Schowen

ISBN: 978-3-527-30777-7

size the importance of fluctuations of the environment and the quantum nature of the hydrogen transfer, while pointing out the complexities that can arise when there is substantial mixing between reactant and product states as occurs in the transfer of the charged proton nucleus. In contrast to a focus on the role of ΔG°, Warshel has concluded (for many years) that a reduction in λ is the dominant mechanism whereby enzymes catalyze proton transfer. In the concluding remarks, he and his coauthors address two issues that reappear in later chapters: the role of dynamical effects in enzyme reactions, and the extent to which tunneling effects may be different between enzymes and their solution counterparts. The reader should be aware that Warshel's definition of dynamics is confined to the rate of barrier re-crossing, quite different from the use by some of the term dynamics to refer to protein motions and their possible coupling to the C–H activation step.

6 Enzymatic Catalysis of Proton Transfer at Carbon Atoms

John A. Gerlt

6.1 Introduction

Reactions in which protons are abstracted from carbon are ubiquitous in biochemistry. The proton that is abstracted is rendered "acidic" by its location either adjacent (α-proton) or vinylogously conjugated to a carbonyl or carboxylate group. With the ability to stabilize the resulting negative charge on the adjacent or conjugated carbonyl/carboxyl oxygen, the values of the pK_as of carbon acids range from 18 to 20 for aldehydes, ketones, and thioesters, 22 to 25 for carboxylic *acids* (presumably never encountered in enzymatic reactions at neutral pH), and 29 to 32 for carboxylate *anions* [1, 2]. Although depressed relative to the pK_a of an alkane hydrogen, these values significantly exceed those of the pK_as of the conjugate acids of the active site bases to which the protons are transferred (<7), thereby explaining the intellectual interest in understanding the mechanisms of these reactions and the strategies that enzymes use to accelerate their rates relative to their nonenzymatic counterparts.

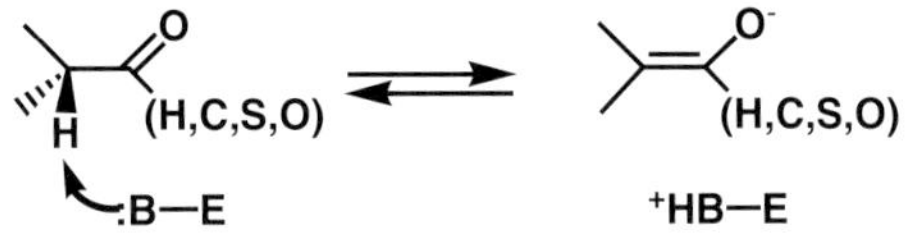

O H (H,C,S,O) :B—E ⇌ O⁻ (H,C,S,O) ⁺HB—E

The glycolytic pathway includes three such reactions: glucose 6-phosphate isomerase (1,2-proton transfer), triose phosphate isomerase (1,2-proton transfer), and enolase (β-elimination/dehydration). The tricarboxylic acid cycle includes four: citrate synthase (Claisen condensation), aconitase (β-elimination/dehydration followed by β-addition/hydration), succinate dehydrogenase (hydride transfer initiated by α-proton abstraction), and fumarase (β-elimination/dehydration). Many more reactions are found in diverse catabolic and anabolic pathways. Some enzyme-catalyzed proton abstraction reactions are facilitated by organic cofactors, e.g., pyridoxal phosphate-dependent enzymes such as amino acid racemases and transaminases and flavin cofactor-dependent enzymes such as acyl-C-A dehydrogenases; others,

Hydrogen-Transfer Reactions. Edited by J. T. Hynes, J. P. Klinman, H.-H. Limbach, and R. L. Schowen

ISBN: 978-3-527-30777-7

those that are the focus of this chapter, are either divalent metal-ion assisted or require no cofactor. Yet, despite the wide-spread occurrence of these reactions, the enzymological community has only recently recognized the structural and mechanistic strategies that enzymes employ to allow these reactions to occur at biologically acceptable rates.

This chapter summarizes (i) the intellectual problem associated with understanding the rates of these reactions, i.e., the required reduction in the kinetic barrier for proton transfer from carbon to an active site general base so that k_{cat} will not be limited by the rate of this overall reaction; (ii) the active site structural features that allow the necessary reduction in kinetic barrier; and (iii) specific enzymatic examples of how these strategies are employed.

6.2 The Kinetic Problems Associated with Proton Abstraction from Carbon

The high values of the pK_as of carbon acid substrates and the associated instability of enolate anion intermediates in nonenzymatic reactions first led to the expectation that these intermediates could not be rendered kinetically competent in enzymatic reactions [3]. As a result, the expectation was that these reactions must be concerted, thereby avoiding the problem of how an active site might provide sufficient, significant stabilization of the intermediates. However, the weight of the experimental evidence now is that enzymes that abstract protons from carbon acids are able to sufficiently stabilize enolate anion intermediates so that they can be kinetically competent.

A convincing demonstration of a step-wise reaction involving a stabilized enediolate anion intermediate was provided by studies of mandelate racemase (MR). MR catalyzes a 1,1-proton transfer reaction in which the enantiomers of mandelate are equilibrated by a two-base mechanism [4]: Lys 166 is the (*S*)-specific base that mediates proton transfers to/from (*S*)-mandelate [5]; His 297, hydrogen bonded to Asp 270 in a His-Asp dyad, is the (*R*)-specific base [6]. The Asn mutant of His 297 (H297N) catalyzes exchange of the α-proton of (*S*)-mandelate with D_2O solvent, in the absence of racemization, at nearly the same rate the wild-type enzyme catalyzes racemization [6]. The only reasonable explanation is that Lys 166 retains its ability to abstract the α-proton of (*S*)-mandelate to generate a stabilized intermediate; following rotation of the C_ε–N_ε bond, a solvent-derived deuteron can be delivered to the intermediate, resulting in exchange without racemization. In the reaction catalyzed by the wild-type enzyme, the enediolate intermediate would be competitively protonated on either face by Lys 166 *and* His 297, with the latter resulting in both incorporation of solvent hydrogen and racemization. Given the nearly equivalent rates of exchange catalyzed by H297N and racemization catalyzed by the wild-type enzyme, the inescapable conclusion is that an enolate anion intermediate is present on the reaction coordinate for the reactions catalyzed by both the mutant and wild-type enzymes. Now, enolate anions in which negative charge is localized on and, therefore, stabilized by the carbonyl/carboxyl oxygen are as-

Table 6.1. Examples of the rate accelerations for enzyme-catalyzed proton abstraction from carbon acids.

Enzyme	Substrate	pK_a	k_{non} (s^{-1})	k_{cat} (s^{-1})	Rate acceleration
triose phosphate isomerase	glyceraldehyde 3-phosphate	18[a]	8×10^{-4}[b]	8300[c]	1.1×10^7
	dihydroxyacetone phosphate	20[a]	6×10^{-7}[b]	600[c]	1×10^9
ketosteroid isomerase	5-androstene-3,17-dione	12.7[d]	1.7×10^{-7}[e]	3.8×10^4[f]	2.2×10^{11}
	4-androstene-3,17-dione	16.1[d]	2.5×10^{-12}[e]	–[g]	
enoyl-CoA hydratase	3-hydroxybutyryl-CoA	21[a]	3×10^{-7}[h]	600[i]	2×10^9
mandelate racemase	mandelate	29[j]	3×10^{-13}[k]	650[l]	2.2×10^{15}
enolase	2-phosphoglycerate	32[m]	1.1×10^{-14}[l]	80[n]	7.3×10^{15}

[a] Ref. [2]. [b] Ref. [78]. [c] Ref. [8]. [d] Ref. [79]. [e] Ref. [53]. [f] Ref. [9].
[g] $K_{eq} = 2400$, so this value is not available; Ref. [53]. [h] Ref. [65]. [i] Ref. [80]. [j] Ref. [1]. [k] Ref. [81]. [l] Ref. [10]. [m] Ref. [82]; assuming that the value of the pK_a of 2-phosphoglycerate is the same as that of malate.
[n] Ref. [83].

sumed to be intermediates in all reactions involving abstraction of a proton from a carbon acid. Of course, this requires that those that study these reactions be able to provide a structural explanation for how the active sites stabilize the enolate anion intermediates so that they can be kinetically competent.

As summarized in Table 6.1, the rate constants for nonenzymatic (k_{non}) and enzymatic (k_{cat}) proton abstraction have been measured directly for a number of carbon acids substrates, thereby allowing the rate accelerations to be quantitated. Despite the considerable range for the values of the pK_as and, therefore, of k_{non}, the values of the k_{cat}s fall in a narrow range. As a result, the rate accelerations vary over a large range and, in some cases, are very significant.

As first espoused by Knowles and Albery, the limiting selective pressure on enzymatic function is the diffusion-controlled limit by which substrates bind and products dissociate [7]. In the case of triose phosphate isomerase [8], ketosteroid isomerase [9], mandelate racemase [10], and proline racemase [11], the energies of various transition states on the reactions coordinates have been quantitated, with the result that the free energies of the transition states for the proton transfer reactions to and from carbon are competitive with those for substrate association/product dissociation. However, as discussed in later sections, the energies of the enolate anion intermediates usually have not been quantitated but in some cases have been estimated by calculations. Irrespective of the identity of the substrate, the value of the pK_a of its α-proton, and the rate acceleration, the necessary conclusion is that the active sites provide significant stabilization of the enolate anion intermediates as well as the transition states for proton transfer.

6.2.1
Marcus Formalism for Proton Transfer

Marcus formalism has been applied to understanding the strategies by which active sites stabilize the transition states for proton transfer to and from carbon [12, 13]. This is based on the usual assumption that the equation for an inverted parabola can be used to describe the dependence of the free energy, G, on the reaction coordinate, x, where $x = 0$ for carbon acid substrate/general base and $x = 1$ for the enolate anion intermediate/conjugate acid of the general base.

$$G = -4\Delta G^{\ddagger}{}_{\text{int}}(x - 0.5)^2 + \Delta G^{\circ}(x - 0.5) \tag{6.1}$$

The equation employs two parameters, the thermodynamic barrier, ΔG°, and the intrinsic kinetic barrier, $\Delta G^{\ddagger}{}_{\text{int}}$, to describe the dependence of G on the reaction coordinate. The value of ΔG° is determined by the difference in the values of the pK_as of the proton that is abstracted and of the conjugated acid of the active site general base to which it is transferred. The value of $\Delta G^{\ddagger}{}_{\text{int}}$ is an "extra" activation energy barrier associated with proton abstraction and is defined as the barrier that would remain if the proton transfer reaction were isoenergetic, i.e., the value of ΔG° were reduced to zero.

According to Marcus formalism, the value of the activation energy barrier, $\Delta G^{\ddagger}$, is specified by the following equation:

$$\Delta G^{\ddagger} = \Delta G^{\ddagger}{}_{\text{int}}(1 + \Delta G^{\circ}/4\Delta G^{\ddagger}{}_{\text{int}})^2 \tag{6.2}$$

Inspection of this equation reveals that a reduction in $\Delta G^{\ddagger}$ from that associated with the nonenzymatic reaction (k_{non}) to that associated with the enzymatic reaction (k_{cat}) can be accomplished by reducing the value of ΔG°, $\Delta G^{\ddagger}{}_{\text{int}}$, or both.

If the structural strategies for achieving the rate accelerations associated with enzymatic proton abstraction from carbon are to be completely understood, the reduction in $\Delta G^{\ddagger}$ must be partitioned into the individual contributions from ΔG° and $\Delta G^{\ddagger}{}_{\text{int}}$ so that these can be separately interpreted. From the previous equation, the amount of the contribution of $\Delta G^{\ddagger}{}_{\text{int}}$ to $\Delta G^{\ddagger}$ depends on the value of ΔG°: if ΔG° is small, the contribution from $\Delta G^{\ddagger}{}_{\text{int}}$ will approach the value of $\Delta G^{\ddagger}{}_{\text{int}}$; if the value of ΔG° is large, the contribution from $\Delta G^{\ddagger}{}_{\text{int}}$ will approach zero. For example, if $\Delta G^{\ddagger}{}_{\text{int}} = 12$ kcal mol^{-1} (a typical value for nonenzymatic proton abstraction from a carbon acid) and $\Delta G^{\circ} = 0$ kcal mol^{-1} (the substrate and intermediate states are isoenergetic), $\Delta G^{\ddagger} = 12$ kcal mol^{-1}, i.e., the "extra" activation energy associated with $\Delta G^{\ddagger}_{int}$ is the "full" 12 kcal mol^{-1}. However, if the value of ΔG^{o} were increased to 10 kcal mol^{-1}, $\Delta G^{\ddagger} = 17.5$ kcal mol^{-1}, i.e., the "extra" activation energy associated with $\Delta G^{\ddagger}_{int}$ is "only" 5.5 kcal mol^{-1}. This dissection requires that the value of ΔG^{o} be known, i.e., the concentration of the enolate anion intermediate must be quantitated. With this value and the measured value of $\Delta G^{\ddagger}$, the value of $\Delta G^{\ddagger}_{int}$ can be calculated. However, the values of ΔG^{o} are usually unknown for enzyme-catalyzed reactions because the concentrations of the intermediates are too small to detect and quantitate, so the values of $\Delta G^{\ddagger}_{int}$ are also unknown.

6.2.2 ΔG^{o}, the Thermodynamic Barrier

The factors underlying reductions in ΔG^{o} are relatively easy to understand. As noted previously, the values of the pK_as of biochemically relevant carbon acids range from 18 to 20 for aldehydes, ketones, and thioesters to 29 to 32 for carboxylate *anions* [1, 2]. The values of the pK_as of the conjugate acids of the amino acid functional groups that participate as general bases are necessarily no larger than ~7; otherwise, they would be protonated and unable to be kinetically competent as bases. Thus, the value of ΔG^{o} is given by the following equation:

$$\Delta G^{o} = 2.303RT\Delta pK_a \tag{6.3}$$

where ΔpK_a is the difference between the pK_a of the carbon acid and the conjugate acid of the active site base.

If the value of $\Delta G^{\ddagger}$ were determined only by ΔG^{o}, the value of k_{cat} would be specified by the equation

$$k_{cat} = 6.2 \times 10^{12} 10^{(-\Delta Go/2.3RT)}\ s^{-1} \tag{6.4}$$

So, for example, in the case of the mandelate racemase-catalyzed reaction, for which the value of the pK_a of mandelate anion is 29 [1] and the value of the pK_a of Lys 166, the (S)-specific base, is 6 [6], the value of k_{cat} would no larger than 6.2×10^{-11} if the enolate anion intermediate were not stabilized in the active site; this value is ~10^{13}-fold less than the observed value for the k_{cat}, 500 s^{-1}. Recall that an enolate anion is necessarily on the reaction coordinate, so the value of ΔG^{o} must be reduced for the enolate anion to be kinetically competent irrespective of whether $\Delta G^{\ddagger}_{int}$ can be reduced. Thus, the active site of mandelate racemase *must* decrease ΔG^{o} from the value predicted from the values of the substrate carbon acid and the active site base in solution. The obvious strategy to accomplish this reduction is preferential stabilization of the enolate anion intermediate: relative to the carbon acid substrate, the increased negative charge on (or proton affinity of) the carbonyl/carboxylate oxygen of the enolate anion intermediate provides a convenient handle for enhanced electrostatic or hydrogen bonding interactions with the active site.

6.2.3 $\Delta G^{\ddagger}_{int}$, the Intrinsic Kinetic Barrier

The interpretation of $\Delta G^{\ddagger}_{int}$ is not so straightforward, even disregarding the work term associated with bimolecular reactions that describes the energetic costs of bringing the reactants together. For convenience, the work terms for the nonenzymatic and enzymatic reactions are ignored if (i) the value of k_{non} refers to that of an encounter complex of the carbon acid and base in the nonenzymatic reaction; and (ii) the value of k_{cat}, the reactivity of the enzyme–substrate complex, is used to quantitate the rate of the enzymatic reaction. That the values of $\Delta G^{\ddagger}_{int}$ for nonenzymatic proton transfer from carbon acids (13 kcal mol^{-1}) are considerably greater than those for proton transfer between heteroatoms ($\leq$ 3 kcal mol^{-1}) has been attributed to both the temporal requirements for reorganization of solvent that occurs as charge is localized on the carbonyl/carboxyl oxygen as well as for rehybridization of the tetrahedral carbon from which the proton is abstracted. Presumably, within the enzyme–substrate complex, the carbonyl/carboxyl oxygen will be prearranged relative to active site charges, dipoles, and hydrogen-bond donors and the α-carbon can be proximal to positive charge (associated with either an active site function group or resulting from protein structure, e.g., the N-terminal end of an α-helix). The author and others have proposed that this preorganization may provide a structural basis for a reduction in $\Delta G^{\ddagger}_{int}$ by enzyme active sites relative to the nonenzymatic reaction [12, 14]. However, this argument and the possible effects of active site structure on the value of $\Delta G^{\ddagger}_{int}$ remain controversial.

Despite efforts, only limited success has been made in measuring the values of ΔG° and $\Delta G^{\ddagger}_{int}$ for enzyme-catalyzed proton abstraction reactions. Indeed, despite the importance of the reaction catalyzed by triose phosphate isomerase in our understanding of the conceptual strategies by which catalytic efficiency may evolve (uniform binding of substrate/intermediate/product, differential binding of bound species so that the conversion of bound substrate to an intermediate can be isoenergetic, and stabilization of transitions states for chemical steps so that diffusion-controlled and chemical steps can have competitive rates [7]), the concentration of the enediolate intermediate in this reaction has not been measured. Indeed, the intermediate has no useful UV/visible spectroscopic properties nor can it be chemically prepared and used as an alternate "substrate" for the enzyme-catalyzed reaction. So, although the derived theory about evolution of enzymatic activity assumes that ΔG° for formation of the enolate intermediate is significantly reduced, thereby making the bound substrate, intermediate, and product nearly isoenergetic, the effects of the active site of triose phosphate isomerase on both ΔG° and $\Delta G^{\ddagger}_{int}$ are unknown. Similar situations exist for the "quantitative" reaction coordinates for both the Mg^{2+}-dependent mandelate racemase- and metal-independent proline racemase-catalyzed reactions that also have provided considerable insight into understanding enzyme-catalyzed proton abstraction from carbon. Bearne and coworkers quantitated the reaction coordinate for mandelate racemase [10], and Knowles and Albery and coworkers also studied proline racemase [11], but in neither case could the concentration of the enolate anion intermediate be determined.

Although not a subject of this chapter, Toney and coworkers have quantitated the reaction coordinate of a PLP-dependent L-alanine racemase [15]. Despite the expectation that the cofactor provides resonance stabilization of the carbanion/enolate anion (quinonoid) intermediate derived by abstraction of the α-proton, the spectroscopic and kinetic analyses for the wild type racemase at steady-state provided no evidence for the intermediate in the reaction catalyzed by the wild type enzyme. Indeed, Toney had previously demonstrated that a kinetically competent quinonoid intermediate accumulates in the impaired R219E mutant [16]; Arg 219 is hydrogen-bonded to the pyridine nitrogen of the cofactor. For the wild type racemase, the derived transition state energies for conversion of the bound enantiomers of alanine, ~12 kcal mol^{-1}, could be explained by a value of $\Delta G^{\ddagger}{}_{int}$ that need not differ from that associated with nonenzymatic reactions if the quinonoid intermediate is present but at a concentration that is too low to be detected spectrophotometrically. However, in the case of the proton abstraction reactions catalyzed by ketosteroid isomerase, the concentration of the dienolate anion intermediate is known with some certainty [9]. The conjugated and unconjugated enone substrate/product are among the most acidic carbon acids found in enzymology (Table 6.1). The dienolate anion intermediate can be prepared chemically by treatment with NaOH and rapid neutralization with buffer (exploiting the slow protonation of the intermediate on carbon due to the large value of $\Delta G^{\ddagger}{}_{int}$), and the rates at which the neutralized intermediate partitions to substrate and product could be quantitated [17]. Analyses of the data revealed that, at pH 7, the value of ΔG° for formation of acetate-catalyzed formation of the intermediate is reduced from 10 kcal mol^{-1} in solution to ~0 kcal mol^{-1} in the active site in the direction in proton abstraction from the more acidic nonconjugated enone substrate ($pK_a = 12.7$); in the reverse direction with the conjugated enone product ($pK_a = 16.1$), the intermediate is ~5 kcal mol^{-1} higher in energy (at equilibrium, the conjugated enone is favored). With these values and the measured rates of proton abstraction, the values of $\Delta G^{\ddagger}{}_{int}$ could be calculated and were found to be decreased by 3 kcal mol^{-1} from the nonenzymatic reaction (from 13 kcal mol^{-1} to 10 kcal mol^{-1}), so the kinetic barriers for the relatively slow proton transfer reactions are determined almost entirely by the value of $\Delta G^{\ddagger}{}_{int}$. This provides persuasive evidence that, despite a preorganized active site structure in which stabilizing hydrogen-bonding Tyr and Asp residues are located proximal to the substrate carbonyl group, the value of $\Delta G^{\ddagger}{}_{int}$ is effectively unchanged.

A similar conclusion can be reached for proton abstraction catalyzed by (*S*)-mandelate dehydrogenase, a FMN-dependent reaction. In this active site, an active site His abstracts the α-proton, and the enolate anion is located proximal to oxidized isoalloxazine ring. A long wavelength charge-transfer complex is transiently formed, suggesting the accumulation of a "significant" amount of the intermediate [18]. Although the value of the extinction coefficient for the complex is uncertain, the rate of formation of the intermediate, 400 s^{-1}, is consistent with a large, i.e., unperturbed, value for $\Delta G^{\ddagger}{}_{int}$. This conclusion is consistent with observations on acyl-CoA dehydrogenases that demonstrate the accumulation of charge-transfer complexes from substrate analogs that are enolizable but unable to reduce the cofactor by hydride transfer, e.g., 3-thiaoctanoyl-CoA [19].

In summary, insufficient data are available to decide whether enzymes that catalyze proton abstraction from carbon acids either need or are able to reduce the values of $G^{\ddagger}{}_{int}$ from those measured for nonenzymatic reactions. But the conclusion is inescapable: these enzymes must significantly stabilize the enolate anion intermediate if the observed values of k_{cat} and the associated rate acceleration are to be understood.

6.3 Structural Strategies for Reduction of ΔG°

As noted previously, the vast majority of enzymes that catalyze proton abstraction from carbon acids *must* be able to reduce the value of ΔG° from that used to describe the nonenzymatic reaction. Focusing on reactions that do not involve organic cofactors, stabilization of the enolate anion intermediate is most reasonably accomplished either by hydrogen-bonding or electrostatic interactions with active site components.

6.3.1 Proposals for Understanding the Rates of Proton Transfer

In the early 1990s, considerable controversy emerged about the possible importance of hydrogen bonding in stabilizing anionic intermediates. The author and the late Gassman [12, 20] and, also, Cleland and Kreevoy [21] suggested that not all hydrogen bonds are "created equal" and that the strengths of these could increase sufficiently as a proton is abstracted from a carbon acid substrate to allow significant stabilization of an enolate anion intermediate. In an early application of site-directed mutagenesis to elucidate the structural bases of catalysis, Fersht and coworkers systematically examined the interactions of the acyladenylate intermediate with the active site of tyrosyl-tRNA synthetase [22]. From their results, that demonstrated that charged hydrogen bonds are somewhat stronger than neutral hydrogen bonds, the differential strengths of hydrogen bonds apparently could contribute only modestly to catalysis. Thus, suggestions to the contrary by the Gerlt–Gassman/Cleland–Kreevoy proposals were immediately met with resistance [23].

The Gerlt–Gassman/Cleland–Kreevoy proposals were virtually the same, although the semantics differed in "bonding" detail and, therefore, perceived emphasis. The structures of enzymes catalyzing proton abstraction from carbon acids without the participation of a divalent metal ion suggested the required presence of a hydrogen bond donor spatially proximal to the carbonyl/carboxylate acceptor oxygen of the substrate [1]. In an enzyme–substrate complex, the proton affinity (as assessed by value of the pK_a) of the carbonyl/carboxylate oxygen is low; in contrast, neutral His and Tyr are often the hydrogen bond donors, so their proton affinities are much greater. In fact, Knowles and coworkers had demonstrated that the neutral form of His 95 is the electrophile in the active site of triose phosphate isomerase and determined that the pK_a of the imidazole proton is >14 [24, 25]. Both proposals also noted that proton abstraction from the carbon acid substrate results in an increased proton affinity for the carbonyl/carboxylate oxygen so that it can approach that of the hydrogen bond donor when the enolate anion was formed. In the extreme, the proton affinities of the enolate anion and active site donor could be the same, depending on the identities of the substrate and active site hydrogen bond donor.

6.3.2
Short Strong Hydrogen Bonds

A large body of experimental work in the physical chemistry community has addressed the properties of hydrogen bonds in which the proton affinities of the donor and acceptor are equal [26]. These exist in the crystalline state and in non-aqueous, but not aqueous, solutions. Their enthalpies of formation are measured and/or calculated to be 25 to 30 kcal mol^{-1}, their lengths are as short as 2.29 Å in the $[HO-H-OH]^-$ ion, and the hydrogen-bonded proton is located equidistant between the donor and acceptor. In contrast to asymmetric hydrogen bonds in which the hydrogen is located proximal to one heteroatom and must traverse an energy barrier to be transferred to another, these "low barrier" hydrogen bonds have covalent character with no energy barrier impeding the transfer of the proton from one heteroatom to another. Cleland–Kreevoy used the term "low barrier" to describe these hydrogen bonds; Gerlt–Gassman used the term "short, strong" to describe the properties of the hydrogen bond that would result as the developing negative charge resulting from abstraction of the α-proton is localized on the more electronegative oxygen. The existence and properties of a low-barrier hydrogen bond sometimes can be studied by measuring ^1H NMR chemical shift ($\delta > 16$ ppm) and isotope fractionation factor ($\Phi < 0.5$) of the hydrogen-bonded proton [27, 28].

Although Cleland had previously pointed out the potential importance of "low barrier" hydrogen bonds in the interpretation of low deuterium fraction factors, the Gerlt–Gassman/Cleland–Kreevoy proposals focused on the largely unappreciated strengths (at least among biochemists) of "pK_a-matched" hydrogen bonds in stabilizing enolate and other anionic intermediates in enzymatic reactions. Indeed, the proposals suggested that these hydrogen bonds could provide much, if not all, of the previously elusive energetic contribution required for sufficient stabilization of enolate anions so that they could be kinetically competent.

6.3.3
Electrostatic Stabilization of Enolate Anion Intermediates

The proposed importance of hydrogen bonds in providing significant differential stabilization of oxyanion intermediates was quickly challenged. Kluger and Guthrie pointed out that electrostatic effects are likely to be significant in the apolar environment of enzyme active sites [23]: in media of low dielectric constant, the Coulombic stabilization provided by an ion-pair interaction between a dianionic intermediate and a divalent cation (as in the active site of mandelate racemase) could be as large as 18 kcal mol^{-1}. So, for those enzymes in which a divalent metal is coordinated to a carbonyl/carboxylate oxygen of the substrate, e.g., mandelate racemase and enolase, enolate anion intermediates undoubtedly are significantly stabilized by "simple" electrostatic effect. But, not all enzymes that stabilize anionic intermediates require divalent metal ions, e.g., triose phosphate isomerase, ketosteroid isomerase, and enoyl-CoA hydratase, so another strategy *must* be used to stabilize the intermediates in such reactions. Perhaps, in some cases, this strategy may be electrostatic effects that are not associated with divalent metal ions. However as detailed in later sections of this chapter, the more prevalent strategy, as deduced from X-ray structures and supported by mutagenesis of the hydrogen bonding residues, is hydrogen bonding interactions with active site hydrogen bond donors. As discussed in the next sections, the magnitudes of these interactions are, in fact, sufficient to provide the necessary stabilization of the enolate anion intermediate so that it can be kinetically competent.

6.3.4
Experimental Measure of Differential Hydrogen Bond Strengths

Herschlag interpreted the Gerlt–Gassman/Cleland–Kreevoy proposals as invoking a "special stabilization" of an enolate anion intermediate when the proton affinities of the hydrogen bond donor and acceptor are matched ($\Delta pK_a = 0$) [29]; however, the proposals did not specify a function that related hydrogen bond strength to ΔpK_a but simply pointed out the enhanced strengths of hydrogen bonds involving pK_a matched donors and acceptors relative to hydrogen bonds characterized by large values of ΔpK_a.

Herschlag determined this dependence by studying the dependence of intramolecular hydrogen bond strength (expressed as K^{HB}) on the ΔpK_a between the neutral phenolic OH group (hydrogen bond donor) and the anionic carboxylate group (hydrogen bond acceptor) for a series of substituted salicylates [30]. The experiments were conducted in both water and dimethylsulfoxide, the latter assumed to mimic the environment of an active site. The difference, if any, in the values of the slopes of plots of log K^{HB} versus ΔpK_a (Brønsted β) in these solvents would describe the influence of the active site on the strengths of hydrogen bonds. Any enhanced effect (greater negative value of the Brønsted β) would produce a larger rate acceleration for the enzyme-catalyzed reaction if stabilization of an enolate anion intermediate by hydrogen bonding is an important feature of the reaction coordinate. The strengths of the intramolecular hydrogen bonds showed little dependence on the value of ΔpK_a in aqueous solution: the Brønsted slope of a plot of log K^{HB} versus ΔpK_a is -0.05. However, in DMSO solution, the Brønsted slope is -0.73, describing a significant increase in hydrogen bond strength as the value of

ΔpK_a decreases. The difference in the values of the Brønsted βs, 0.68, predicts an enhancement of hydrogen bond strengths in active sites.

In reactions involving the the enolization of carbon acid substrates, the pK_a of the active site hydrogen bond donor is always much larger than the pK_a of the conjugate acid of the substrate carbonyl group acceptor, i.e., $\Delta pK_a \gg 0$; however, the pK_a of the hydrogen bond donor is usually similar to that of the conjugate acid of the enolate anion acceptor, i.e., $\Delta pK_a \sim 0$. Thus, as the α-proton is abstracted, the ΔpK_a between the hydrogen bond acceptor and donor decreases significantly. From Herschlag's studies, the enhanced importance of hydrogen bonding in stabilizing an enolate anion intermediate in an active site, $\Delta\Delta G^\circ$, is quantitated by the following equation

$$\begin{aligned}\Delta\Delta G^\circ &= -1.36 \text{ kcal mol}^{-1} \times (0.73 - 0.05)\Delta pK_a \\ &= -1.36 \text{ kcal mol}^{-1}(0.68)\Delta pK_a \end{aligned} \tag{6.5}$$

where ΔpK_a quantitates the effect of the increase in pK_a of the conjugate acid of the carbonyl/enolate anion oxygen as the α-proton is abstracted.

From this equation, the increase in the hydrogen bond strength available to stabilize an enolate anion intermediate is predicted to be substantial. For example, in the active site of enoyl-CoA hydratase, the thioester enolate anion intermediate is assumed to be stabilized by hydrogen bonding interactions with two peptide backbone NH groups in an oxyanion hole [31]. The pK_a of a peptidic NH, the hydrogen bond donor, is ~18; the pK_a of the conjugate acid of the thioester oxygen of crotonyl-CoA, the hydrogen bond acceptor in the enzyme–substrate complex, is ~−2; and the pK_a of the conjugate acid of the enolate anion resulting from abstraction of the α-proton by Glu 164, the acid site general base, is ~10. Thus, assuming that the pK_a of the peptidic NH is unchanged in the enzyme–substrate and enzyme–intermediate complexes, the 12 unit change in pK_a of the acceptor is predicted to result in an 11 kcal mol^{-1} increase in the strength of the hydrogen bond, *even though the pK_as of the conjugate acid of the enolate anion intermediate and the active site hydrogen bond donor are not matched.* In experimental support of this prediction, the value of k_{cat} for the hydration reaction catalyzed by the G141P mutant of enoyl-CoA hydratase is decreased by 10^6-fold, corresponding to destabilization of the transition state for proton abstraction 8.4 kcal mol^{-1} [32]. Gly 141, located at the N-terminal end of an α-helix, provides one of two NH groups in an oxyanion hole occupied by the thioester carbonyl oxygen, so the G141P mutant lacks one of the NH groups. The second peptidic NH group is associated with a residue that is not appropriately located for a similar analysis. However, assuming that the peptidic NH groups provide independent stabilization of the enolate anion intermediate, hydrogen bonding interactions could provide as much as 17 kcal mol^{-1} of stabilization, more than needed to account for the value of k_{cat}.

The important message in this analysis, which follows directly from Herschlag's experimental studies, is that there is no need for any "special" stabilization associated with "matching" of the pK_as of the hydrogen donor and acceptor as the enolate anion is formed. A substantial increase in hydrogen bond strength is obtained simply from a marked increase in the affinity of the oxygen as the α-proton is abstracted and the enolate anion intermediate is formed. To realize this stabilization, the enzyme "simply" needs to bind the substrate so that the substrate carbonyl/carboxyl group is hydrogen bonded to a weakly acidic donor, e.g., neutral His, neutral Tyr, or a peptide NH group. In fact, in those enzymes that do not require divalent cations, these hydrogen bond donors are (almost) always found in the active sites!

So, the substance of the Gerlt–Gassman/Cleland–Kreevoy proposals appears correct, i.e., hydrogen bonding can provide much more stabilization of enolate anion intermediates under conditions of appropriate local dielectric environment than had been previously recognized by the bioorganic and biochemical communities.

6.4 Experimental Paradigms for Enzyme-catalyzed Proton Abstraction from Carbon

In recent years, several enzymes have been subjected to sufficient structural, mechanistic, and, in some cases, computational scrutiny so that the underlying principles by which these catalyze facile proton abstraction from carbon are reasonably well understood. This section highlights the current state of knowledge for four of these.

6.4.1 Triose Phosphate Isomerase

HO–CH₂–CO–CH₂–OPO_3^- (DHAP) $\overset{TIM}{\rightleftharpoons}$ OHC–CH(OH)–CH₂–OPO_3^- (G3P)

DHAP G3P

Triose phosphate isomerase (TIM), that catalyzes the cofactor-independent interconversion of dihydroxyacetone phosphate (DHAP) and D-glyceraldehyde 3-phosphate (G3P) in the glycolytic pathway, continues as a paradigm for understanding enzymatic strategies for proton abstraction from carbon as a result of the classical work of Knowles and workers who dissected the catalytic roles of active site residues [24, 33, 34] and, also, of Knowles and Albery and coworkers who quantitated the free energy profile for the reaction (with the exception of the energy/stability of the enediolate anion intermediate) [8]. However, as summarized in this section, some notable deficiencies remain in our understanding of the energetics of this seemingly simple reaction.

The values of the pK_as of the dihydroxyacetone phosphate (DHAP) substrate and the glyceraldehyde 3-phosphate (G3P) product for the TIM-catalyzed reaction are estimated as 18 and 20, respectively [2]. Knowles established the importance of

Glu 165 as the active site general base that abstracts a proton from carbon-1 of DHAP and delivers a proton to carbon-2 of G3P; the value of its pK_a has been estimated as between 3.9 and 6.5 [35, 36]. The value of k_{cat} using the less reactive DHAP as substrate is 750 s^{-1} [8]. Given these values for the pK_as of Glu 165 and the DHAP substrate, the ΔG_o for formation of the enediolate anion intermediate from DHAP must be reduced by at least 10 kcal mol^{-1} in the active site of TIM for it to be kinetically competent (making the unlikely assumption that $\Delta G^{\ddagger}_{int}$ is reduced to zero). The value of the rate acceleration associated with the TIM-catalyzed reaction is $\sim 10^9$, confirming the importance of $\Delta G^{\ddagger}_{int}$ in understanding the enzyme-catalyzed reaction (Table 6.1).

The equilibrium constant for the TIM-catalyzed reaction is 22 in the direction of DHAP for the reactive, nonhydrated forms of DHAP and G3P [8]; as a result, steady-state kinetics as well as the fates of substrate- and solvent-derived protons can be studied using either DHAP or G3P as substrate. In a series of landmark papers, Knowles and Albery quantitated the free energy profile for the reaction (Fig. 6.1) [37–42]. These studies utilized protiated, deuteriated, and tritiated forms of both DHAP and G3P in both unlabeled and tritiated water to follow the course of the transferred proton(s) during the course of the reaction. Several important conclusions were reached: (i) for protiated substrates, the transition state for binding/release of G3P is the highest point on the energy diagram; (ii) for tritiated

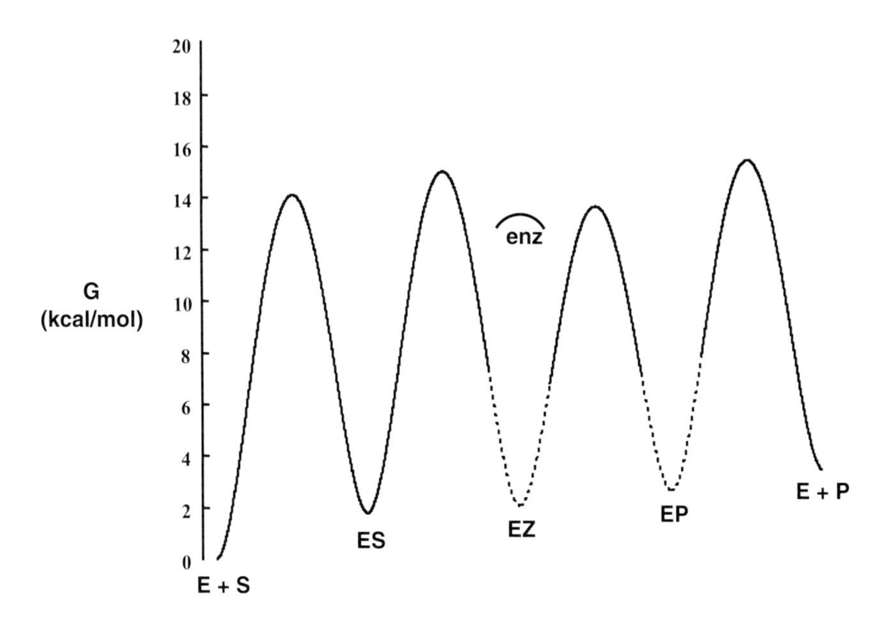

Figure 6.1. Reaction coordinate for the reaction catalyzed by TIM. The DHAP substrate is S, the enediolate intermediate is Z, and the G3P product is P. The dashed lines indicate uncertainties in the concentrations of bound intermediate (EZ) and product (EP). The barrier labeled "enz" is that for the exchange of the conjugate acid of Glu 165 with solvent hydrogen.

substrates, the transition state for proton abstraction from DHAP is the highest point (the details of the reaction coordinates depend upon the identities of the substrate and solvent isotopes); (iii) the energies of bound DHAP and G3P are similar; and (iv) exchange of the substrate-derived proton with solvent hydrogen occurs in competition with intramolecular proton transfer from carbon-1 of DHAP and to carbon-2 of G3P. Unfortunately, the free energy(ies) of the enediolate/enediol intermediates could not be quantitated.

The intramolecular 1,2-proton transfer catalyzed by TIM occurs in competition with exchange of the solvent-derived proton with solvent protons. Knowles and Albery reported that with [1R-^{3}H]-DHAP as substrate, the formation of G3P is accompanied by ~6% intramolecular transfer of the isotopic label [41]; realizing that the exchange of a triton from the conjugate acid of Glu 165 will be suppressed relative to exchange of a proton, the reaction of protiated DHAP is expected to occur with a lower level of intramolecular transfer. More recent experiments by Amyes and Richard using protiated DHAP and G3P in D_2O revealed 18% intramolecular proton transfer starting with DHAP and 49% intramolecular transfer starting with G3P [43, 44]. Again, assuming discrimination against solvent deuterium in the exchange reaction, lower levels of intramolecular proton transfer are expected in H_2O. Although none of these experiments allow quantitation of the exchange that would occur with protiated substrates in H_2O, they clearly demonstrate that a substrate-derived proton can be transferred, proving the formation of an enediolate intermediate rather than a hydride ion migration as in some other aldo–keto isomerization reactions. However, the challenge has been to provide a structural description of the mechanisms by which substrate-derived protons exchange with solvent-derived protons.

On the basis of the early structures of TIM determined with bound analogs of the enediolate intermediate, e.g., phosphoglycolohydroxamate, His 95 was proposed to be an electrophilic catalyst, polarizing the carbonyl groups of DHAP/G3P, thereby rendering the α-proton more acidic [45]. Lys 12 was also known to be proximal to the ketone oxygen of DHAP as well as the phosphate group. In accord with these structures, site-directed mutants of Glu 165, His 95, and Lys 12 were constructed and found to be markedly defective in catalysis [24, 33, 34].

A very high resolution X-ray structure of the complex with substrate DHAP is now available (Fig. 6.2) [46]. One carboxylate oxygen of Glu 165 is positioned in close proximity to the proton on carbon-1 as well as to carbon-2; the other carboxylate oxygen is proximal to the 1-OH group. The $N_\varepsilon H$ group of neutral His 95 participates in a bifurcated hydrogen bond to both the 2-ketone oxygen and the 1-hydroxyl oxygen of the bound DHAP. With this geometry, the carbonyl group is polarized but intramolecular hydrogen transfer from O1 to O2 is impeded; the potential for this transfer is important in understanding the energetics of the reaction coordinate. The electrophilic ε-ammonium group of Lys 12 is hydrogen bonded to the 2-ketone oxygen, the bridging oxygen of the phosphate ester, and via an intervening water molecule to one of the nonesterified oxygens of the phosphate ester.

On the basis of theses structures, three mechanisms have been proposed for the TIM-catalyzed reaction (Fig. 6.3):

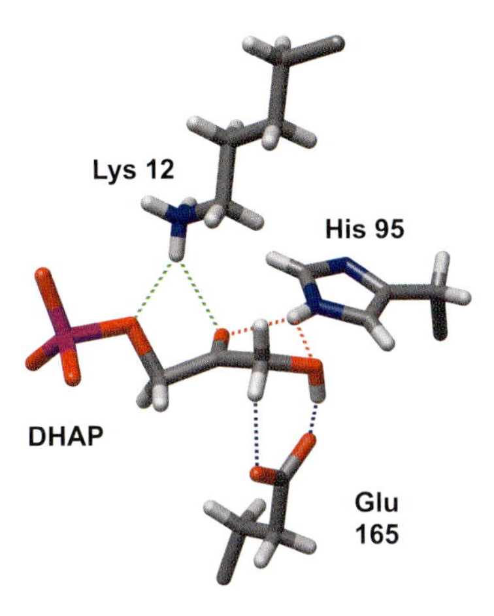

Figure 6.2. The active site of TIM.

1. In the "classical" mechanism (path A in Fig. 6.3), Glu 165 abstracts the 1-pro*R* proton from DHAP to generate an initial enediolate intermediate in which negative charge is localized on oxygen-2 and a proton is located on oxygen-1. The oxyanion is transiently stabilized by the $N_{\varepsilon}H$ group of neutral His 95, and a proton is transferred from His 95 to the enediolate intermediate to generate a neutral enediol and the conjugate base of His 95. The imidazolate anion so generated then accepts the proton from the 1-OH group to yield a second enediolate intermediate in which negative charge is localized on the 1-oxygen. (In the original formulation of the reaction coordinate, Knowles and Albery did not differentiate the tautomeric enediolate intermediates [8].) In competition with the intramolecular 1,2-proton transfer, the substrate-derived proton initially located on the carboxylate group of Glu 165 can exchange (by an unspecified mechanism) with a solvent-derived proton so that Glu 165 can deliver a fractional mixture of a substrate- and a solvent-derived proton to carbon-2 to generate G3P. The rate and extent of the exchange of the substrate-derived proton with a solvent-derived proton is of considerable interest.
2. In a variant of this mechanism (path B), the $N_{\varepsilon}H$ group of neutral His 95 stabilizes the enediolate anion in which negative charge is localized on oxygen-2 [24, 25]. Intramolecular proton transfer from the oxygen-1 to oxygen-2 then occurs without the involvement of His 95 as an acid/base catalyst, and the resulting tautomeric enediolate intermediate in which negative charge is localized on oxygen-2 is stabilized by the $N_{\varepsilon}H$ group of neutral His 95. A proton is delivered from Glu 165 to carbon-2 to generate G3P. During the course of this reaction, the substrate-derived proton initially located on Glu 165 can exchange with a solvent-derived proton, as is assumed in the classical mechanism. The difference between this mechanism and the previous mechanism is whether His 95 participates as an acid–base catalyst.

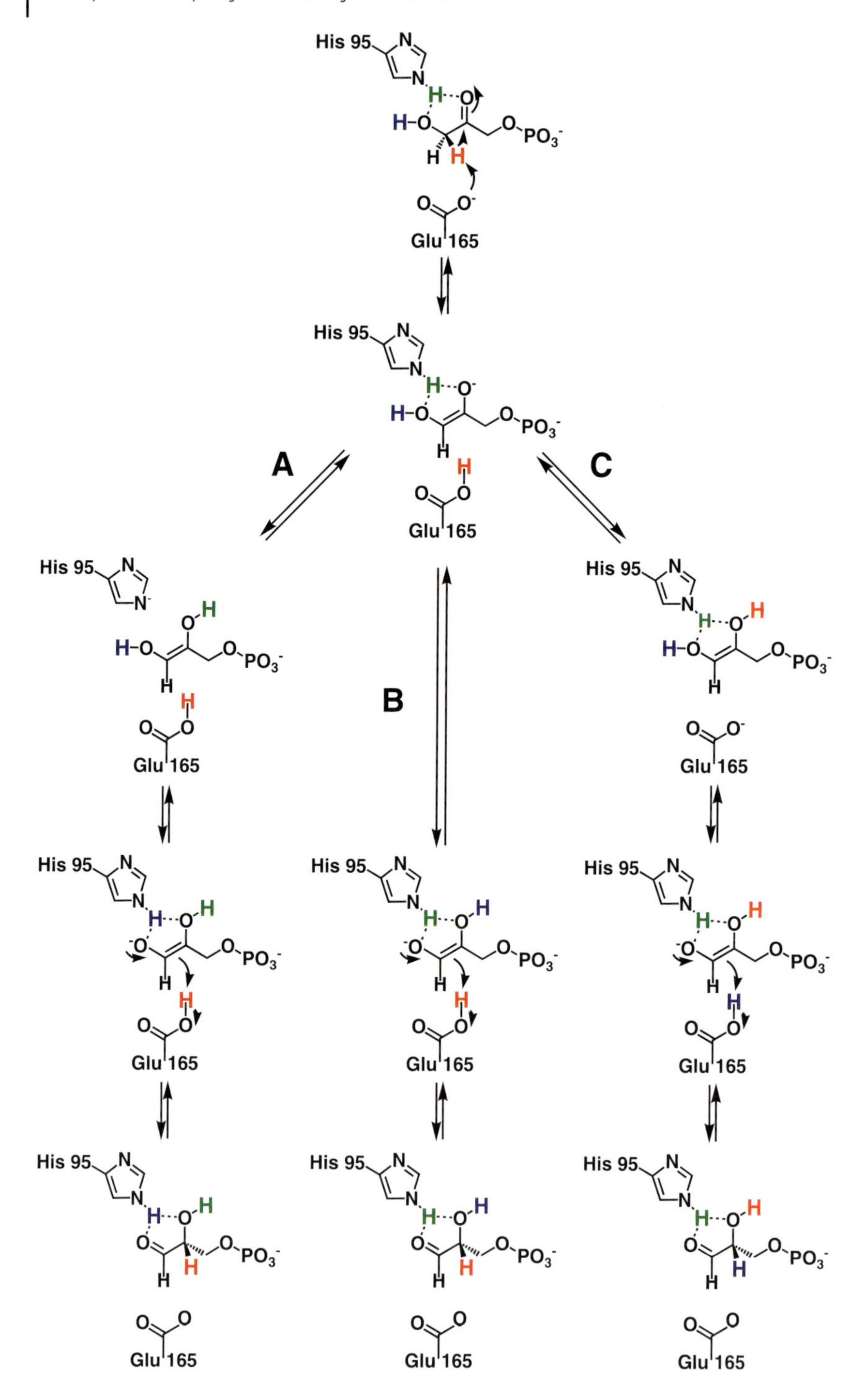

Figure 6.3. Possible mechanisms for the reaction catalyzed by TIM.

3. In the "criss-cross" mechanism (path C), Glu 165 abstracts a proton from carbon-1 of DHAP to generate an enediolate intermediate in which negative charge is localized on oxygen-2; this intermediate is stabilized by hydrogen-bonding to the $N_\varepsilon H$ group of neutral His 95 [47–49]. The conjugate acid of Glu 165 then delivers the DHAP-derived proton to oxygen-2 to generate a neutral enediol intermediate. The deprotonated Glu 165 then abstracts the proton from the 1-OH group of the intermediate and delivers it to carbon-2, thereby generating G3P. In this mechanism, the proton delivered to carbon-2 of the G3P product is predicted to be derived exclusively from solvent; the proton abstracted from DHAP is predicted to be necessarily "lost" to solvent via its transfer to oxygen-2. Although this mechanism has been proposed to explain the behavior of the H95Q mutant [47, 48], the fact that the reaction catalyzed by wild-type TIM is accompanied by partial retention of the DHAP-derived proton in the G3P product requires the competing participation of one of the first two mechanisms in the reaction it catalyzes. In addition, the exchange of the solvent-derived proton with solvent required in the first two mechanisms may be explained by this mechanism.

The energies of the enediolate/enediol intermediates relative to the bound DHAP/G3P have not been evaluated experimentally: they do not accumulate sufficiently to allow spectroscopic detection. However, the proposals put forth by Albery and Knowles regarding the evolution of catalytic efficiency are based, in part, on the assumption that the various bound species, substrate, intermediates, and products, are isoenergetic on the reaction coordinate ("differential binding" to achieve a reduction in ΔG_o) and that the transition states for the proton transfer reactions can be selectively stabilized ("catalysis of elementary steps" to achieve a reduction in $\Delta G^\ddagger_{int}$) [7]. Without a measure of the stabilities of the enediol/enediolate intermediates relative to DHAP and G3P, the importance of reductions in ΔG_o and/or $\Delta G^\ddagger_{int}$ cannot be dissected.

Without the ability to measure experimentally the energies of the enediol/enediolate intermediates produced by the various mechanisms, computational approaches may provide otherwise inaccessible insights. In recent studies reported by Friesner and coworkers [50], the starting coordinates were those obtained from the high resolution structure of the DHAP substrate complex (1.2 Å) [46]. In earlier studies reported by Cui and Karplus [51, 52], the starting coordinates were those of the lower resolution structure of the complex with phosphoglycolohydroxamate, an analogue of the enediol/enediolate intermediates (1.9 Å) [45]. Although these structures are similar, they differ in some potentially significant details, e.g., the location of the ε-ammonium group of Lys 12 relative to the enediol/enediolate intermediate and its phosphate group. Both studies were in agreement that the mechanism in which intramolecular proton transfer from oxygen-1 to oxygen-2 occurs without the participation of either His 95 or Glu 165 is accompanied by a significant energy barrier that is incompatible with the measured kinetic parameters: the bifurcated hydrogen bond involving the $N_\varepsilon H$ group provides a steric and electrostatic barrier to this proton transfer.

Unfortunately, the computational studies differ in quantitative detail regarding the importance of the mechanisms that involve either Glu 165 or His 95 as the acid–base catalysts to catalyze interconversion of the tautomeric enediolate intermediates. Friesner and coworkers concluded that the transition state for proton abstraction from DHAP is the highest point on the energy diagram; after formation of the enediolate anion intermediate, the calculations predict that the barrier for the "criss-cross" mechanism catalyzed by Glu 165 is $\sim$3 kcal mol^{-1} lower than that for "classical" mechanism involving catalysis of tautomerization of the enediolate intermediates by His 95, so the "criss-cross" mechanism is predicted to be the favored mechanism. In contrast, Cui and Karplus concluded that transition state energies for tautomerization of the enediolate anion intermediates via an enediol intermediate are isoenergetic for both the classical and criss-cross mechanisms.

The extent of intramolecular proton transfer in the reaction catalyzed by wild-type TIM is low but measurable; however, quantitation of the potential competition between the classical and criss-cross mechanisms can be experimentally achieved only by "extrapolation" of the effect of isotopic substitutions to the all protium situation, i.e., the behavior of protiated substrates in H_2O. The observed low level of intramolecular transfer from DHAP to G3P can be explained by the Friesner calculations assuming that the disfavored intramolecular proton transfer from carbon-1 of DHAP to carbon-2 of G3P involves no, or inefficient, exchange of the DHAP-derived proton with solvent hydrogen. The Karplus calculations better accommodate the intramolecular transfer of a proton via the classical mechanism, but a more extensive significant exchange of the conjugate acid of Glu 165 with solvent hydrogen would be required to explain the observed low levels of intramolecular proton transfer. Because the details of the process by which the conjugate acid of Glu 165 exchanges with solvent are unknown (presumably with a small "pool" of water molecules in the active site), the experimental observations do not allow the quantitative differences between the Friesner and Karplus calculations to be evaluated.

As noted previously, experimental data are unavailable regarding the stabilities of the enediol/enediolate intermediates. However, quantitation of the relative stabilities of these relative to the bound DHAP and G3P is important in evaluating the Knowles and Albery proposal that the evolution of enzyme function requires that all of the bound species, including the reactive intermediates, be isoenergetic. In the classical mechanism in which His 95 catalyzes tautomerization, no enediol intermediate is present on the reaction coordinate; in the criss-cross mechanism, an enediol intermediate occurs. Karplus and Cui calculated that the enediol is essentially isoenergetic with the bound DHAP and G3P, as originally hypothesized by Knowles and Albery. In contrast, Friesner and coworkers predict that all three intermediates in the preferred criss-cross mechanism are higher in energy than bound DHAP and G3P: the enediolates derived from DHAP and G3P are predicted to be 11.4 kcal mol^{-1} and 4.2 kcal mol^{-1} less stable than DHAP and G3P, respectively; the enediol obtained via the preferred criss-cross mechanism is 9.2 kcal mol^{-1} and 6.5 kcal mol^{-1} less stable than DHAP and G3P, respectively. Thus, the latter calculations do not support the Knowles and Albery hypothesis.

The different conclusions regarding the energies of the intermediates have an additional implication for the Knowles and Albery hypotheses, i.e., assessing the importance of "catalysis of elementary steps" to achieve a reduction in $\Delta G^{\ddagger}_{int}$. If the enediol/enediolate intermediates are approximately isoenergetic with DHAP and G3P, as predicted by Karplus, the value of $\Delta G^{\ddagger}_{int}$ is not significantly reduced from the value ($\sim$12 kcal mol^{-1}) found in nonenzymatic reactions. However, if the intermediates are significantly more unstable than DHAP and G3P, the value of $\Delta G^{\ddagger}_{int}$ must be decreased to account for the observed rates of proton abstraction from DHAP and G3P.

Thus, despite the apparent abundance of functional and structural information, important mechanistic issues remain to be resolved for a complete understanding of the strategy by which TIM catalyzes proton abstraction reactions.

6.4.2
Ketosteroid Isomerase

KSI

3-Oxo-Δ^5-ketosteroid isomerase (KSI), that catalyzes the cofactor-independent tautomeric interconversion of the α,β- and β,γ-unsaturated 3-oxo-steroids via a dienolate anion intermediate, has received considerable recent mechanistic and structural scrutiny by Pollack and Mildvan and their coworkers. The reaction catalyzed by KSI is arguably "simpler" than that catalyzed by TIM: the value of the pK_a of the unconjugated 5-androstene-3,17-dione is 12.7 whereas that of the conjugated 4-androstene-3,17-dione is 16.1 [53]; the value of the pK_a of Asp 38 that mediates the intramolecular 1,2-proton transfer reaction is $\sim$4.7 [54]. As a result of the "low" values for the pK_a of the substrate/product, the rate acceleration is "only" a factor of 10^7 (Table 6.1).

As a result of the relatively low values for the pK_as of the substrate/product and its stability due to the absence of competing side reactions, the dienolate anion intermediate can be chemically prepared by treatment of substrate/product with strong base and, after rapid neutralization, can be supplied to KSI as a "substrate" to allow an evaluation of the kinetics of processing of the intermediate to substrate/product [55]. Preparation of the enediolate intermediate in the TIM-catalyzed reaction is impossible because of its facile propensity to eliminate inorganic phosphate with the concomitant formation of methylglyoxal.

The likely importance of strong hydrogen bonding in stabilizing the dienolate anion intermediate was prominent in the formulation of both the Gerlt–Gassman and Cleland–Kreevoy proposals, even prior to the availability of high resolution

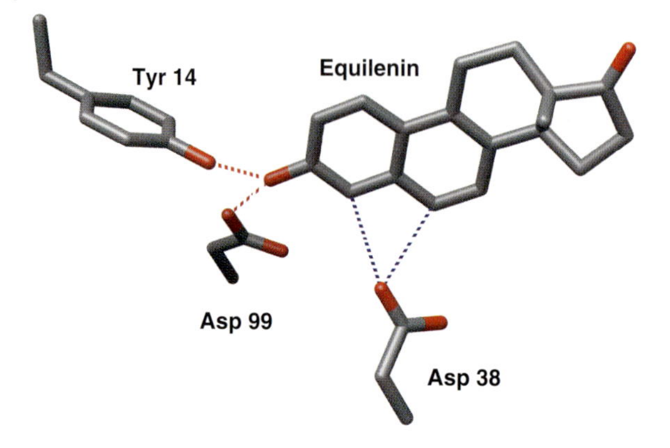

Figure 6.4. The active site of KSI.

structural information. When structures later became available, a high resolution NMR structure by Summers and Mildvan and coworkers [56] and high resolution crystal structures by Oh and coworkers [57–59], Asp 38 was located in a hydrophobic substrate binding cleft along with both Tyr 14, long known to be catalytically important, and Asp 99, whose presence but not identity had been suspected as a result of studies of the dependence of the kinetics constants on pH (Fig. 6.4). Tyr 14 and Asp 99, both potential hydrogen bond donors, were proximal to the suspected binding site for the 3-oxo group of the substrate/product; the remaining question was the geometry of the hydrogen bonding network involving these and the dienolate anion intermediate generated by proton abstraction from the substrate by Asp 38.

Two hydrogen bonding geometries were proposed [56, 60]: (i) both Asp 99 and Tyr 14 would hydrogen bond to the 3-oxo group, thereby providing two "independent" hydrogen bonds to stabilize the anionic intermediate; and (ii) Asp 99 and Tyr 14 would form an interacting dyad, with only Tyr 14 directly involved in a hydrogen bond to the substrate/intermediate. Prior to the report of the high resolution crystal structure, Mildvan reported evidence in favor of the second geometry based on the downfield NMR spectra of wild type KSI and mutants of Tyr 14 and Asp 99

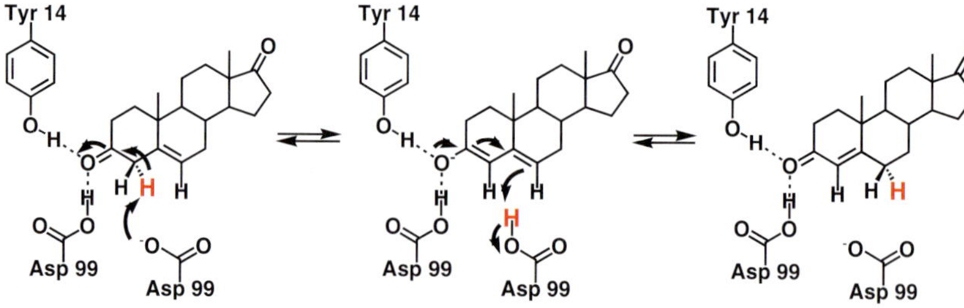

Figure 6.5. Mechanism for the reaction catalyzed by KSI.

[60]. However, the X-ray structure favored the first geometry, as did the observation that separate substitutions for both Tyr 14 and Asp 99 were accompanied by additive free energy effects on the values of the kinetic constants; as a result, each hydrogen bond was concluded to contribute one half, ~5 kcal mol^{-1}, of the total energy associated with the total rate acceleration [61]; the accepted mechanism is shown in Fig. 6.5. Although this dissection of the energetic importance of these hydrogen bonds allowed the conclusion that the intermediate is stabilized by strong hydrogen bonds rather than a single low barrier hydrogen bond, the important point is that the differential strengths of hydrogen bonds that can be achieved by large changes in the proton affinity of the carbon acid oxygen as it is transformed to an enolate anion intermediate are completely able to explain the rate acceleration for the KSI-catalyzed reaction. Prior to the formulation of the Gerlt–Gassman/Cleland–Kreevoy proposals and subsequent experimental confirmation by Herschlag, such changes in hydrogen bond strength were not expected in enzyme-catalyzed reactions.

Experiments reported by Pollack and his coworkers allow the conclusion that the dienolate anion intermediate is approximately isoenergetic with the more unstable unconjugated enone substrate/product, as proposed by Knowles and Albery in their theory for understanding optimization of catalytic efficiency [9]. Thus, based on the value of the rate constant for proton abstraction from the unconjugated enone, 1.7×10^5 s^{-1}, Pollack and coworkers calculated that the value of the $G^{\ddagger}{}_{int}$ for proton abstraction from carbon is 10 kcal mol^{-1}, a modest reduction from that expected (~ 13 kcal mol^{-1}) for the nonenzymatic reaction.

Various aspects of the reaction coordinate for the KSI-catalyzed reaction have been subjected to computational examination [62–64]. These are in accord with the experimental results, i.e., Tyr 14 and Asp 99 independently stabilize the dienolate anion intermediate via a hydrogen bond. Although the strengths of these hydrogen bonds each increase by ~5 kcal mol^{-1} as the anionic intermediate is formed, the hydrogen bonds are asymmetric with the protons associated with the donors.

6.4.3
Enoyl-CoA Hydratase (Crotonase)

CoAS–C(=O)–CH$_2$–CH(OH)–R $\underset{\pm H_2O}{\overset{ECH}{\rightleftharpoons}}$ CoAS–C(=O)–CH=CH–R

Enoyl-CoA hydratase (ECH; commonly known as crotonase), that catalyzes the cofactor-independent hydration of conjugated enoyl-CoA esters in β-oxidation, has been the subject of considerable debate regarding the timing of bond-making reactions and, therefore, the importance of a thioester enolate anion on the reaction coordinate. The active site contains Glu 144 and Glu 164 as the only possible acid–base catalysts. In the nonphysiological dehydration direction, the value of the pK_a

of the α-proton of the thioester is ~21 [65]. The pH-dependence of the values of the kinetic constants allows the suggestion that the value of the pK_a of one of the active site Glu residues is neutral and the other is unprotonated at physiological pH, with a value < 5 for the pK_a of Glu 164 that is thought to abstract the α-proton in the direction of dehydration [66]. The rate acceleration is estimated as $\sim 10^9$ (Table 6.1).

Several high-resolution structures are available for ECH, with the structure of the complex with 4-(*N*,*N*-dimethylamino)cinnamoyl-CoA arguably providing the most valuable insights into the mechanism [67]: 4-(*N*,*N*-dimethylamino)cinnamoyl-CoA is a substrate for hydration, but conjugation of the enoyl ester with the substituted aromatic substituent shifts the direction of the hydration reaction from favoring the hydrated product by a factor of 7 to favoring the enoyl-CoA substrate by a factor of >1000. Importantly, this structure also contains a water molecule, the second substrate, apparently "poised" for nucleophilic attack on the conjugated thioester by both Glu 144 and Glu 164 (Fig. 6.6).

Anderson and coworkers measured various substrate kinetic effects to deduce the timing of the C–H and C–O bond cleavage events in the ECH-catalyzed dehydration of 3-hydroxybutyrylpantetheine; the primary deuterium and oxygen-18 isotope effects were both significant, 1.60 and 1.053, respectively [68]. A further double isotope study examining the effect of solvent isotope substitution on the α-secondary deuterium isotope effect at carbon-3 for hydration of crotonylpantetheine revealed isotopic invariance, leading to the suggestion that incorporation of hydrogen at carbon-2 and the rehybridization at carbon-3 due to attack of the nucleophilic water occurred in the same transition state, i.e., the ECH-catalyzed reaction is concerted [69]. This interpretation is most simply explained by an E2 mechanism, in which Glu 144 functions as a general base to facilitate the attack of water (hydroxide) on carbon-3, and Glu 164 simultaneously delivers a solvent-derived hydrogen to carbon-2. However, as pointed out by Gerlt–Gassman in the formulation of their proposals [12], a concerted reaction would obviate the involvement of the thioester carbonyl group in facilitating the proton transfer events at carbon-2, suggesting an alternate explanation would be required to explain the measured isotope effects.

The precise catalytic roles of Glu 144 and Glu 164 remained uncertain, despite the availability of crystal structures of necessarily nonproductive complexes [31]. These structures pointed to the expected proximity of Glu 164 to carbon-2 and led to the suggestion that it mediates proton transfers to/from carbon. Accordingly, Glu 144 was expected to facilitate attack of water on carbon-3. In addition, the interpretation of the dependence of the kinetic constants for reactions catalyzed by wild type ECH as well as the E144Q and E164Q mutants is uncertain: although "common sense" would require that one be anionic, i.e., a general base, and the other be neutral, i.e., a general acid, no self-consistent, unequivocal support for this expectation could be obtained.

The complex with 4-(*N*,*N*-dimethylamino)cinnamoyl-CoA was solved at pH 7.3 by Bahnson, Anderson, and Petsko, conditions where ECH is active, at sufficiently high resolution (2.3 Å) that the hydrogen bonding interactions of the substrate, nu-

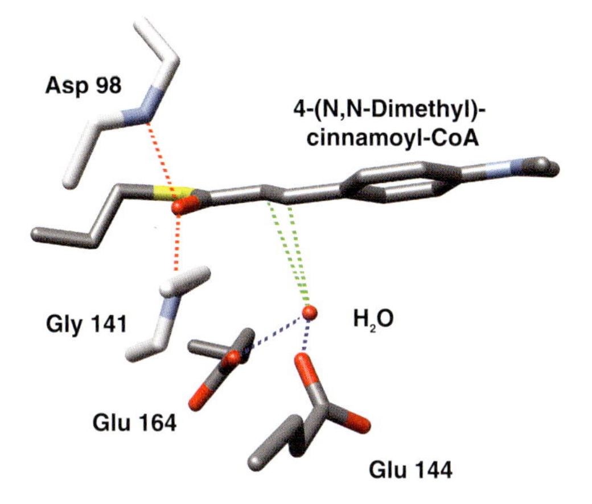

Figure 6.6. The active site of ECH.

cleophilic water, and both active site Glu residues could be unambiguously interpreted [67]. The thioester carbonyl oxygen is located in an oxyanion hole formed by the peptide NH groups of Ala 98 and Gly 141. As expected, the oxygen of the water was proximal to carbon-3. Surprisingly, the hydrogen bonding interactions of both Glu residues required that both be anionic and participating as hydrogen bond acceptors from the nucleophilic water, with the more basic *syn* orbital of Glu 144 and the less basic anti orbital of Glu 164 participating in the hydrogen bonds.

These unexpected details suggest a previously unrecognized E1cb mechanism for the hydration reaction (Fig. 6.7): Glu 144 is not an acid–base catalyst but orients the nucleophilic water molecule, and Glu 164 functions first as the general base that abstracts a proton from the water to catalyze rate-limiting addition to the bound enoyl-CoA with formation of a thioester enolate anion stabilized by hydrogen bonding interaction with the oxyanion hole. In the second, more rapid step, the conjugate acid of Glu 164 delivers the proton derived from the nucleophilic water to carbon-2. This mechanism is in accord with the results of the single and double kinetic isotope effect studies, summarized earlier, in which transfer of a

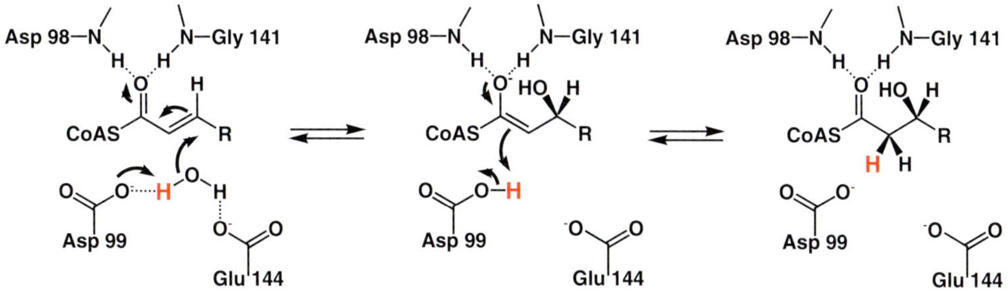

Figure 6.7. Mechanism for the reaction catalyzed by ECH.

solvent-derived proton to Glu 164, rather than Glu 144, occurs in the same transition state as hydroxide attacks carbon-3. It also provides a rationale for the observation that neither the E144Q nor E164Q mutant can catalyze exchange of the α-proton from the thioester substrate as observed for wild type ECH – their positions in the active site are not independent by virtue of their interactions with the same water molecule.

Other mechanisms are also consistent with this structure, including a cyclic four-membered transition state for a "concerted" reaction in which the bond forming reactions at carbons-2 and -3 are asynchronous, thereby allowing partial anionic charge to be localized on the carbonyl oxygen which it can be stabilized by the oxyanion hole.

The catalytic importance of the oxyanion hole has been confirmed. Tonge and coworkers constructed and kinetically characterized the G141P mutant of ECH [32]: Gly 141 provides one of the two peptidic NH groups in the oxyanion hole and is located at the N-terminus of an α-helix. The value of k_{cat} is reduced by a factor of 10^6 by the G141P substitution, providing persuasive evidence for the formation of a transiently stabilized thioester enolate anion intermediate; the reduction in k_{cat} corresponds to the loss of ~8 kcal mol^{-1} of stabilization, which, based on Herschlag' studies, is readily accommodated by the expected change in the pK_a of the thioester carbonyl oxygen as the enolate anion is formed.

No information is available concerning the concentration of the thioester enolate anion intermediate, by either experiment or computation, so the partitioning of the rate acceleration between reductions in ΔG^o and $\Delta G^{\ddagger}{}_{int}$ is not possible.

Further persuasive evidence in support of the expectation that the mechanism of the ECH-catalyzed reaction involves an E1cb mechanism with a stabilized thioester enolate anion intermediate is obtained from the membership of ECH in the mechanistically diverse enoyl-CoA hydratase superfamily [70]. Such superfamilies are derived from a common ancestor by divergent evolution; the members of these share a partial reaction, usually formation of a common intermediate, e.g., an enolate anion. The reactions catalyzed by members of the enoyl-CoA hydratase superfamily (almost) always utilize acyl esters of CoA as substrates; the reactions invariably can be rationalized with mechanisms that involve the formation of a thioester enolate anion intermediate, e.g., 1,3-proton transfer, 1,5-proton transfer, Dieckman and reverse Dieckman condensations, and β-decarboxylation. Although mechanisms with thioester enolate anion intermediates are plausible for each of these reactions, as in the ECH-catalyzed reaction, evidence for their existence on the reaction coordinate is circumstantial because the intermediates do not accumulate, thereby avoiding spectroscopic detection.

However, 4-chlorobenzoyl-CoA dehalogenase is also a member of the enoyl-CoA hydratase superfamily. The mechanism of its reaction involves nucleophilic aromatic substitution in which an active site Asp adds to the 4-position of the benzoyl ring to necessarily form a Meisenheimer complex; this Meisenheimer complex is an analog of a thioester enolate anion. Although the Meisenheimer complex cannot be observed for displacement of chloride from 4-chlorobenzoyl-CoA due to the rate constants for formation and decomposition of the intermediate, the Meisenheimer complex has been observed by Raman spectroscopy for displacement of fluoride from 4-fluorobenzoyl-CoA and nitrite from 4-nitrobenzoyl-CoA [71, 72]. The expectation is that the complex exists on the reaction coordinate for the reaction involving 4-chlorobenzoyl-CoA. By analogy, the reasonable assumption is that the structurally analogous thioester enolate anion is stabilized by the oxyanion hole in the reactions catalyzed by other members of the superfamily, including ECH.

6.4.4
Mandelate Racemase and Enolase

(R)-Mandelate ⇌ (MR) ⇌ (S)-Mandelate

PGA ⇌ (Enolase, $\pm H_2O$) ⇌ PEP

Both the 1,1-proton transfer reaction catalyzed by mandelate racemase (MR) and the dehydration catalyzed by enolase require Mg^{2+} for activity. The values of the pK_as for mandelate and 2-phosphoglycerate, the substrates for the MR- and enolase-catalyzed reactions, are estimated as 29 and 32, respectively [1]. The values of the pK_as of the general basic Lys residues are ~6 and ~9 in MR [6] and enolase [73], respectively. Thus, formation of a dienolate anion intermediate is extremely endergonic, unless the active site can stabilize the intermediate which is the obvious function of the essential Mg^{2+}. The rate accelerations for the MR- and enolase-catalyzed reactions are $\sim 10^{15}$ as a direct result of the values of the pK_as of the α-protons (Table 6.1).

The mechanisms of both reactions are reasonably well understood, so these serve as paradigms for understanding the mechanisms of divalent metal ion-assisted proton abstraction from carbon acids. In the case of MR [4], a single Mg^{2+} ion is located in the active site; one carboxylate oxygen of the substrate is coordinated to the Mg^{2+}, and the second is hydrogen-bonded to the carboxylate of Glu 317. One face of the active site contains Lys 166 that is positioned to mediate proton transfers to/from (*S*)-mandelate; the opposite face contains His 297 hydrogen-bonded to Asp 270 (His-Asp dyad) that is positioned to mediate proton transfers to/from (*R*)-mandelate (Fig. 6.8). The simplest mechanism based on this structure is that the enediolate anion obtained by proton abstraction from a substrate enantiomer is transiently stabilized by enhanced electrostatic (Mg^{2+}) and hydrogen bonding (Glu 317) (Fig. 6.9). Then, the conjugate acid of the catalyst on the opposite face of the active site protonates the intermediate to form the product

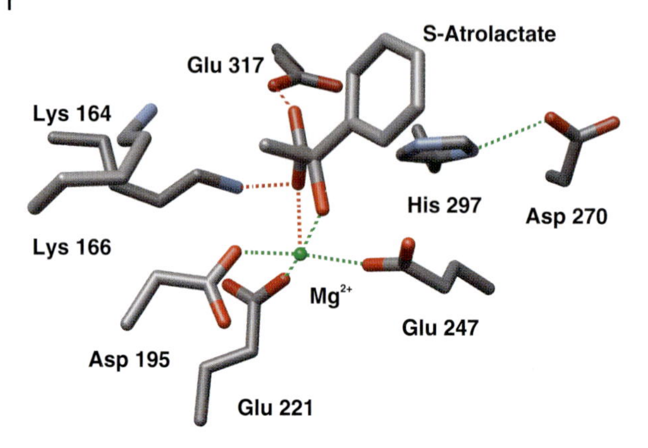

Figure 6.8. The active site of MR.

enantiomer. The observation that the H297N mutant is inactive as a racemase but active as an (*S*)-mandelate "exchangease", i.e., in D_2O solvent hydrogen is incorporated without racemization, provides the most persuasive evidence for the transient formation of a stabilized enediolate anion intermediate on the reaction coordinate [6]. Bearne and coworkers have quantitated the reaction coordinate for the MR-catalyzed reaction, although, as might be expected, the concentration of the enediolate anion intermediate remains elusive [10]. The transition states for substrate binding/product dissociation are approximately isoenergetic.

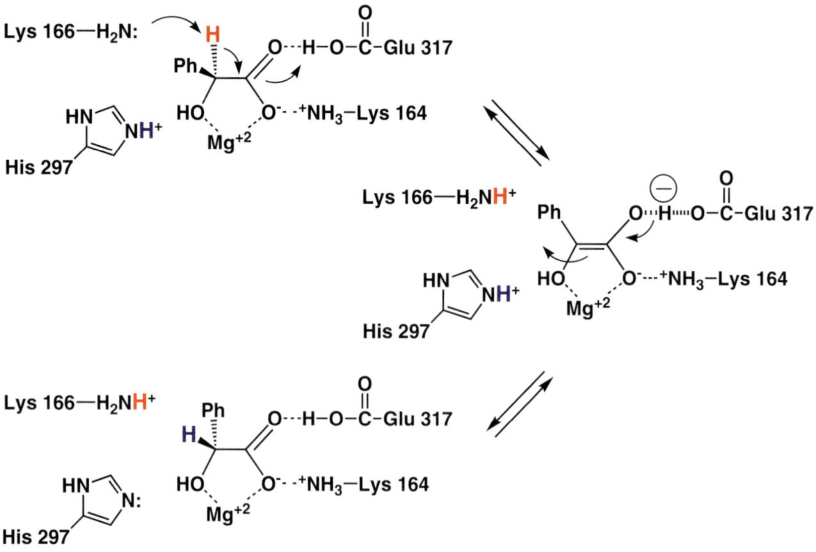

Figure 6.9. Mechanism for the reaction catalyzed MR.

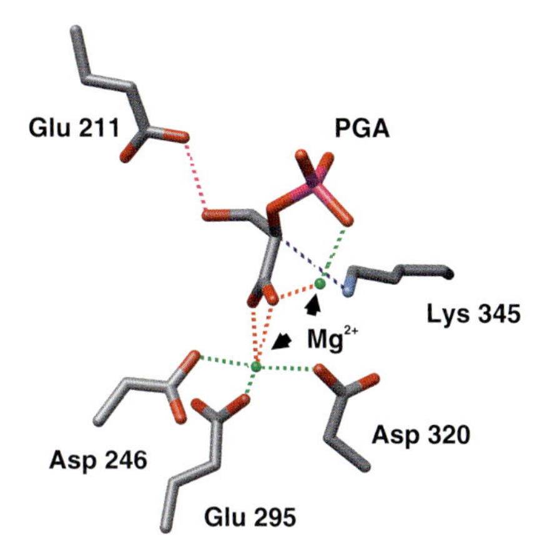

Figure 6.10. The active site of enolase.

In the case of enolase, two Mg^{2+} ions are located in the active site (Fig. 6.10) [74]. The carboxylate group of the substrate is a bidentate ligand of one Mg^{2+}; one carboxylate oxygen is also liganded to the second Mg^{2+} ion (a μ-oxo bridge ligand). One face of the active site contains Lys 345 that is positioned to abstract the α-proton from the 2-phosphoglycerate substrate to form a stabilized enediolate anion intermediate; the other face contains Glu 211 that is positioned to facilitate vinylogous departure of the 3-hydroxide leaving group from the intermediate by acid catalysis (Fig. 6.11). The coordination of the substrate carboxylate group to

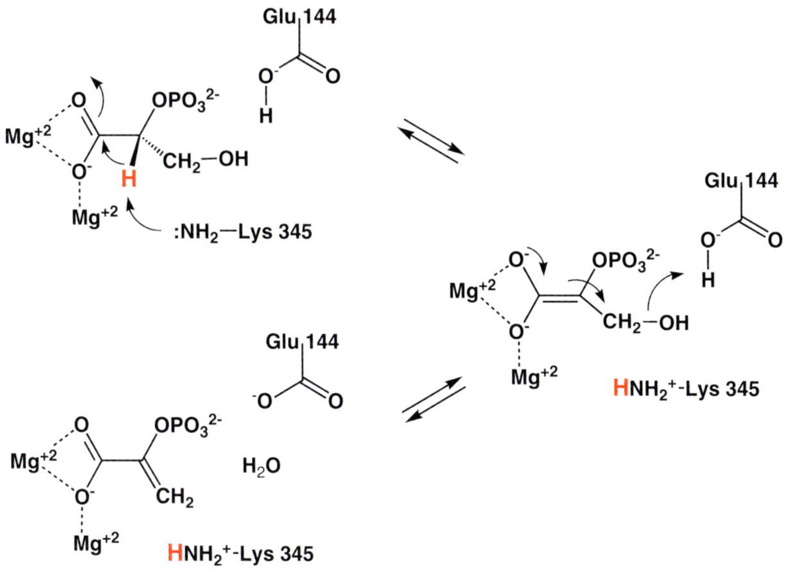

Figure 6.11. Mechanism for the reaction catalyzed by enolase.

two Mg^{2+} ions likely reflects the lower acidity of 2-PGA relative to the enantiomers of mandelate and the accompanying requirement for enhanced electrostatic stabilization of the intermediate.

Yang and coworkers have reported computational studies of the enolase-catalyzed reaction that are in accord with this mechanism and address the interesting, and likely general, problem of how an active site with a pair of divalent metal ions can both stabilize the accumulated anionic charge of the enediolate intermediate formed by α-proton abstraction by electrostatic interactions and promote the vinylogous elimination of the electron-rich β-hydroxide group [75]. Their conclusion is that longer range interactions involving a large number of active site functional groups and water molecules differentially effect the energies of the transition states for formation and breakdown of the enediolate anion intermediate. Although many of these proposed interactions have not been investigated experimentally, it is intuitive that a single principle, e.g., stabilization of an enolate anion intermediate, cannot be the only feature associated with catalysis.

Although not recognized until high-resolution structures were available for both, MR and enolase are homologous, i.e., members of the mechanistically diverse enolase superfamily [70, 76, 77]. The structures of both are composed of two domains: a $(\beta/\alpha)_7\beta$-barrel (a modified $(\beta/\alpha)_8$- or TIM-barrel) domain that provides the structural foundation for the ligands for the essential Mg^{2+} ion(s) as well as the acid–base catalysts; and a capping domain formed from polypeptides segments at both the N- and C-terminal ends of the polypeptide. With these structures and the abundance of sequence data now in the databases, MR and enolase contain three carboxylates at conserved positions at the ends of the third, fourth, and fifth β-strands of the barrel-domain as well as positionally, but not chemically, conserved acid–base catalysts on opposite faces of the active site. The evidence is persuasive that the MR and enolase, as well as other members of the superfamily that can be identified in the sequence databases, are derived from a common ancestor by divergent evolution. Whatever the functional identity of the ancestor, the reaction undoubtedly involved Mg^{2+} ion-assisted enolization of a carboxylate anion substrate.

6.5 Summary

A quantitative understanding of how enzymes catalyze rapid proton abstraction from weakly acidic carbon acids is necessarily achieved by dissecting the effect of active site structure on the values of ΔG°, the thermodynamic barrier, and $\Delta G^{\ddagger}{}_{int}$, the intrinsic kinetic barrier for formation of the enolate anion intermediate. The structural strategies by which ΔG° for formation of the enolate anion is reduced sufficiently such that these can be kinetically competent are now understood. In divalent metal ion-independent reactions, e.g., TIM, KSI, and ECH, the intermediate is stabilized by enhanced hydrogen bonding interactions with weakly acidic hydrogen bond donors; in divalent metal-dependent reactions, e.g., MR and enolase, the intermediate is stabilized primarily by enhanced electrostatic interactions with the directly coordinated metal ion. However, the influence of active site structure on $\Delta G^{\ddagger}_{int}$ for formation of the intermediate remains uncertain. Quantitation of the value of $\Delta G^{\ddagger}_{int}$ requires that the value ΔG° be known, but the concentrations of the intermediates usually cannot be measured, even in well-characterized reactions. Until that problem is solved, a complete understanding of catalytic strategies for proton abstraction from carbon acids remains elusive.

References

1 J. A. Gerlt, J. W. Kozarich, G. L. Kenyon, P. G. Gassman, *J. Am. Chem. Soc.*, **1991**, *113*, 9667–9669.
2 J. P. Richard, T. L. Amyes, *Curr. Opin. Chem. Biol.*, **2001**, *5*, 626–633.
3 A. Thibblin, W. P. Jencks, *J. Am. Chem. Soc.*, **1979**, *101*, 4963.
4 G. L. Kenyon, J. A. Gerlt, G. A. Petsko, J. W. Kozarich, *Acc. Chem. Res.*, **1995**, *28*, 178–186.
5 J. A. Landro, J. A. Gerlt, J. W. Kozarich, C. W. Koo, V. J. Shah, G. L. Kenyon, D. J. Neidhart, S. Fujita, G. A. Petsko, *Biochemistry*, **1994**, *33*, 635–643.
6 J. A. Landro, A. T. Kallarakal, S. C. Ransom, J. A. Gerlt, J. W. Kozarich, D. J. Neidhart, G. L. Kenyon, *Biochemistry*, **1991**, *30*, 9274–9281.
7 W. J. Albery, J. R. Knowles, *Biochemistry*, **1976**, *15*, 5631–5640.
8 W. J. Albery, J. R. Knowles, *Biochemistry*, **1976**, *15*, 5627–5631.
9 D. C. Hawkinson, T. C. Eames, R. M. Pollack, *Biochemistry*, **1991**, *30*, 10849–10858.
10 M. St Maurice, S. L. Bearne, *Biochemistry*, **2002**, *41*, 4048–58.
11 J. G. Belasco, W. J. Albery, J. R. Knowles, *Biochemistry*, **1986**, *25*, 2552–2558.
12 J. A. Gerlt, P. G. Gassman, *J. Am. Chem. Soc.*, **1993**, *115*, 11552–11568.
13 J. A. Gerlt, P. G. Gassman, *Biochemistry*, **1993**, *32*, 11943–11952.
14 A. Warshel, *J. Biol. Chem.*, **1998**, *273*, 27035–27038.
15 M. A. Spies, J. J. Woodward, M. R. Watnik, M. D. Toney, *J. Am. Chem. Soc.*, **2004**, *126*, 7464–7475.
16 S. Sun, M. D. Toney, *Biochemistry*, **1999**, *38*, 4058–4065.
17 D. C. Hawkinson, T. C. Eames, R. M. Pollack, *Biochemistry*, **1991**, *30*, 6956–6964.
18 A. R. Dewanti, B. Mitra, *Biochemistry*, **2003**, *42*, 12893–128901.
19 S. M. Lau, R. K. Brantley, C. Thorpe, *Biochemistry*, **1988**, *27*, 5089–5095.
20 J. A. Gerlt, P. G. Gassman, *J. Am. Chem. Soc.*, **1992**, *114*, 5928.
21 W. W. Cleland, M. M. Kreevoy, *Science*, **1994**, *264*, 1887–1890.
22 A. R. Fersht, J. P. Shi, J. Knill-Jones, D. M. Lowe, A. J. Wilkinson, D. M. Blow, P. Brick, P. Carter, M. M. Waye, G. Winter, *Nature*, **1985**, *314*, 235–238.
23 J. P. Guthrie, R. Kluger, *J. Am. Chem. Soc.*, **1993**, *115*, 11569–11572.
24 E. A. Komives, L. C. Chang, E. Lolis, R. F. Tilton, G. A. Petsko, J. R. Knowles, *Biochemistry*, **1991**, *30*, 3011–3019.
25 P. J. Lodi, J. R. Knowles, *Biochemistry*, **1991**, *30*, 6948–6956.
26 J. A. Gerlt, M. M. Kreevoy, W. Cleland, P. A. Frey, *Chem. Biol.*, **1997**, *4*, 259–267.
27 P. A. Frey, S. A. Whitt, J. B. Tobin, *Science*, **1994**, *264*, 1927–1930.
28 J. Lin, W. M. Westler, W. W. Cleland, J. L. Markley, P. A. Frey, *Proc. Natl. Acad. Sci. USA*, **1998**, *95*, 14664–14668.
29 S. O. Shan, S. Loh, D. Herschlag, *Science*, **1996**, *272*, 97–101.
30 S. O. Shan, D. Herschlag, *Proc. Natl. Acad. Sci. USA*, **1996**, *93*, 14474–14479.
31 C. K. Engel, T. R. Kiema, J. K. Hiltunen, R. K. Wierenga, *J. Mol. Biol.*, **1998**, *275*, 847–859.

32 A. F. Bell, J. Wu, Y. Feng, P. J. Tonge, *Biochemistry*, **2001**, *40*, 1725–1733.
33 D. Straus, R. Raines, E. Kawashima, J. R. Knowles, W. Gilbert, *Proc. Natl. Acad. Sci. USA*, **1985**, *82*, 2272–2276.
34 P. J. Lodi, L. C. Chang, J. R. Knowles, and E. A. Komives, *Biochemistry*, **1994**, *33*, 2809–14.
35 B. Plaut and J. R. Knowles, *Biochem J*, **1972**, *129*, 311–20.
36 S. De la Mare, A. F. Coulson, J. R. Knowles, J. D. Priddle, and R. E. Offord, *Biochem J*, **1972**, *129*, 321–31.
37 L. M. Fisher, W. J. Albery, J. R. Knowles, *Biochemistry*, **1976**, *15*, 5621–5626.
38 P. F. Leadlay, W. J. Albery, J. R. Knowles, *Biochemistry*, **1976**, *15*, 5617–5620.
39 S. J. Fletcher, J. M. Herlihy, W. J. Albery, J. R. Knowles, *Biochemistry*, **1976**, *15*, 5612–5617.
40 S. G. Maister, C. P. Pett, W. J. Albery, J. R. Knowles, *Biochemistry*, **1976**, *15*, 5607–5612.
41 J. M. Herlihy, S. G. Maister, W. J. Albery, J. R. Knowles, *Biochemistry*, **1976**, *15*, 5601–5607.
42 W. J. Albery, J. R. Knowles, *Biochemistry*, **1976**, *15*, 5588–5600.
43 A. C. O'Donoghue, T. L. Amyes, J. P. Richard, *Biochemistry*, **2005**, *44*, 2622–2631.
44 A. C. O'Donoghue, T. L. Amyes, J. P. Richard, *Biochemistry*, **2005**, *44*, 2610–2621.
45 Z. Zhang, S. Sugio, E. A. Komives, K. D. Liu, J. R. Knowles, G. A. Petsko, D. Ringe, *Biochemistry*, **1994**, *33*, 2830–2837.
46 G. Jogl, S. Rozovsky, A. E. McDermott, L. Tong, *Proc. Natl. Acad. Sci. USA*, **2003**, *100*, 50–55.
47 E. B. Nickbarg, R. C. Davenport, G. A. Petsko, J. R. Knowles, *Biochemistry*, **1988**, *27*, 5948–5960.
48 T. K. Harris, C. Abeygunawardana, A. S. Mildvan, *Biochemistry*, **1997**, *36*, 14661–14675.
49 T. K. Harris, R. N. Cole, F. I. Comer, A. S. Mildvan, *Biochemistry*, **1998**, *37*, 16828–16838.
50 V. Guallar, M. Jacobson, A. McDermott, R. A. Friesner, *J. Mol. Biol.*, **2004**, *337*, 227–239.
51 Q. Cui, M. Karplus, *J. Am. Chem. Soc.*, **2001**, *123*, 2284–2290.
52 Q. Cui, M. Karplus, *J. Am. Chem. Soc.*, **2002**, *124*, 3093–3124.
53 B. Zeng, R. M. Pollack, *J. Am. Chem. Soc.*, **1991**, *113*, 3838–3842.
54 D. C. Hawkinson, R. M. Pollack, N. P. Ambulos, Jr., *Biochemistry*, **1994**, *33*, 12172–12183.
55 B. F. Zeng, P. L. Bounds, R. F. Steiner, R. M. Pollack, *Biochemistry*, **1992**, *31*, 1521–1528.
56 Z. R. Wu, S. Ebrahimian, M. E. Zawrotny, L. D. Thornburg, G. C. Perez-Alvarado, P. Brothers, R. M. Pollack, M. F. Summers, *Science*, **1997**, *276*, 415–418.
57 S. W. Kim, S. S. Cha, H. S. Cho, J. S. Kim, N. C. Ha, M. J. Cho, S. Joo, K. K. Kim, K. Y. Choi, B. H. Oh, *Biochemistry*, **1997**, *36*, 14030–14036.
58 H. S. Cho, G. Choi, K. Y. Choi, B. H. Oh, *Biochemistry*, **1998**, *37*, 8325–8330.
59 H. S. Cho, N. C. Ha, G. Choi, H. J. Kim, D. Lee, K. S. Oh, K. S. Kim, W. Lee, K. Y. Choi, B. H. Oh, *J. Biol. Chem.*, **1999**, *274*, 32863–32868.
60 Q. Zhao, C. Abeygunawardana, A. G. Gittis, A. S. Mildvan, *Biochemistry*, **1997**, *36*, 14616–14626.
61 L. D. Thornburg, F. Henot, D. P. Bash, D. C. Hawkinson, S. D. Bartel, R. M. Pollack, *Biochemistry*, **1998**, *37*, 10499–10506.
62 I. Feierberg, J. Aqvist, *Biochemistry*, **2002**, *41*, 15728–15735.
63 D. Mazumder, K. Kahn, T. C. Bruice, *J. Am. Chem. Soc.*, **2003**, *125*, 7553–7561.
64 K. S. Kim, K. S. Oh, J. Y. Lee, *Proc. Nat. Acad. Sci. USA*, **2000**, *97*, 6373–6378.
65 T. L. Amyes, J. P. Richard, *J. Am. Chem. Soc.*, **1992**, *114*, 10297–10302.
66 H. A. Hofstein, Y. Feng, V. E. Anderson, P. J. Tonge, *Biochemistry*, **1999**, *38*, 9508–9516.
67 B. J. Bahnson, V. E. Anderson, G. A. Petsko, *Biochemistry*, **2002**, *41*, 2621–2629.
68 B. J. Bahnson, V. E. Anderson, *Biochemistry*, **1989**, *28*, 4173–4181.

69 B. J. Bahnson, V. E. Anderson, *Biochemistry*, **1991**, *30*, 5894–5906.

70 J. A. Gerlt, P. C. Babbitt, *Annu. Rev. Biochem.*, **2001**, *70*, 209–246.

71 J. Dong, P. R. Carey, Y. Wei, L. Luo, X. Lu, R. Q. Liu, D. Dunaway-Mariano, *Biochemistry*, **2002**, *41*, 7453–7463.

72 J. Dong, X. Lu, Y. Wei, L. Luo, D. Dunaway-Mariano, P. R. Carey, *Biochemistry*, **2003**, *42*, 9482–9490.

73 P. A. Sims, T. M. Larsen, R. R. Poyner, W. W. Cleland, G. H. Reed, *Biochemistry*, **2003**, *42*, 8298–8306.

74 G. H. Reed, R. R. Poyner, T. M. Larsen, J. E. Wedekind, I. Rayment, *Curr. Opin. Chem. Biol.*, **1996**, *6*, 736–743.

75 H. Liu, Y. Zhang, W. Yang, *J. Am. Chem. Soc.*, **2000**, *122*, 6560–6570.

76 P. C. Babbitt, M. S. Hasson, J. E. Wedekind, D. R. Palmer, W. C. Barrett, G. H. Reed, I. Rayment, D. Ringe, G. L. Kenyon, J. A. Gerlt, *Biochemistry*, **1996**, *35*, 16489–164501.

77 J. A. Gerlt, F. M. Raushel, *Curr. Opin. Chem. Biol.*, **2003**, *7*, 252–264.

78 A. Hall, J. R. Knowles, *Biochemistry*, **1975**, *14*, 4348–4353.

79 R. M. Pollack, B. F. Zeng, J. P. G. Mack, S. Eldin, *J. Am. Chem. Soc.*, **1989**, *111*, 6419–6423.

80 L. F. Mao, C. Chu, H. Schulz, *Biochemistry*, **1994**, *33*, 3320–3326.

81 S. L. Bearne, R. Wolfenden, *Biochemistry*, **1997**, *36*, 1646–1656.

82 S. L. Bearne, R. Wolfenden, *J. Am. Chem. Soc.*, **1995**, *117*, 9588–9589.

83 R. R. Poyner, L. T. Laughlin, G. A. Sowa, G. H. Reed, *Biochemistry*, **1996**, *35*, 1692–1699.

7
Multiple Hydrogen Transfers in Enzyme Action

M. Ashley Spies and Michael D. Toney

7.1
Introduction

The focus of this chapter is on enzyme mechanisms that employ multiple hydrogen transfers, where both transfers are mechanistically central steps. The exchange of hydrons with solvent often presents both challenges and opportunities to the kinetic analysis of enzyme systems that undergo multiple hydrogen transfers. The 1,1-proton transfer mechanisms of epimerases and racemases are prototypes for exploring multiple hydrogen transfers, and will thus be the focus of this chapter. Although simple deprotonation/reprotonation of a carbon center is used in the majority of epimerases and racemases, there is variation in the specifics of how this is accomplished (e.g., cofactor-dependent or cofactor-independent, from an activated or unactivated substrate, one- or two-base mechanism, etc.).

7.2
Cofactor-Dependent with Activated Substrates

Many substrates for epimerases and racemases are considered to be "activated", in the sense that the reactive carbon is adjacent to a carbonyl or carboxylate group. In addition to the intrinsic substrate activation, enzymes frequently achieve epimerization or racemization with the aid of a cofactor (organic or inorganic).

7.2.1
Alanine Racemase

Perhaps the best characterized organic cofactor-dependent racemase is alanine racemase, which employs pyridoxal 5′-phosphate (PLP) (Table 7.1). D-alanine is necessary for the synthesis of the peptidoglycan layer of bacterial cell walls in Gram negative and positive bacteria [1]. Alanine racemase is thus a ubiquitous enzyme in bacteria and an excellent drug target [2]. Both its crystal structure and mechanism have been well investigated. PLP reacts with amino acids to produce

Hydrogen-Transfer Reactions. Edited by J. T. Hynes, J. P. Klinman, H.-H. Limbach, and R. L. Schowen

ISBN: 978-3-527-30777-7

Table 7.1. Enzyme catalyzed racemization/epimerization.

Enzyme	Cofactor	Intermediates	Activated/ Unactivated substrate
alanine racemase; serine racemase; amino acid racemase of broad substrate specificity	PLP	cofactor-stabilized carbanion	activated
proline racemase; glutamate racemase; aspartate racemase; diaminopimelate (DAP) epimerase	none	carbanion	
phenylalanine racemase; actinomycin synthetase II (ACMSII); δ-L-(α-aminoadipoyl)-L-cysteinyl-D-valine (ACV)	PAN	enzyme-bound thioester	
mandelate racemase; *N*-acylaminoacid racemase	divalent metal (Mg^{2+}, Mn^{2+}, Co^{2+}, Ni^{2+}, Fe^{2+})	metal stabilized-enolate	
D-ribulose-5-phosphate 3-epimerase	none	ene-diol	
dTDP-L-rhamnose synthase (epimerase component)	none	two sequential enol intermediates	
methylmalonyl-coenzyme-A epimerase	divalent metal ($Co^{2+} > Mn^{2+} > Ni^{2+}$)	metal-bound enolate	
UDP-galactose-4-epimerase	NADH	keto-intermediate	unactivated
UDP-*N*-acetylglucosamine-2-epimerase	none	possible oxonium intermediate	
L-ribulose 5-phosphate 4-epimerase	divalent metal ($Mn^{2+} > Ni^{2+} > Ca^{2+} > Zn^{2+}$)	glycoaldehyde phosphate + metal-bound enolate; retro-aldol C–C bond cleavage	

an "external aldimine" intermediate (Fig. 7.1), which acidifies the Cα proton via resonance delocalization of negative charge in the resulting carbanionic intermediate [3]. Reprotonation of the carbanionic intermediate yields the antipodal aldimine. The deprotonation/reprotonation mechanism of alanine racemase is sup-

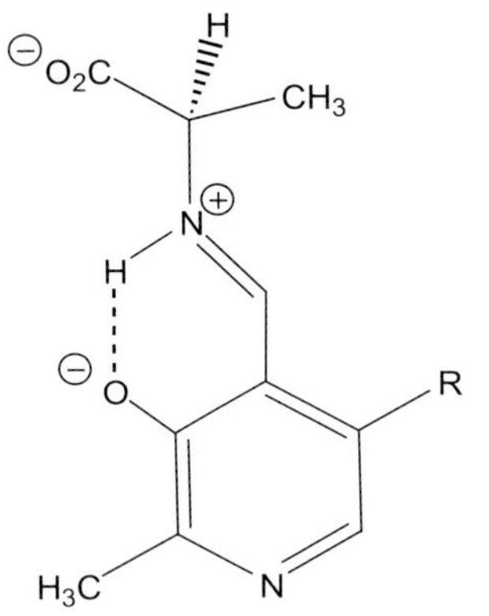

Figure 7.1. PLP external aldimine.

ported by the presence of solvent hydron at the Cα position of the racemized product [4].

A racemase (or epimerase) employing the deprotonation/reprotonation mechanism must be able to abstract a proton from one stereoisomer and reprotonate the other face of the ensuing carbanionic intermediate. This can either be accomplished by a "two-base" or "one-base" mechanism (Fig. 7.2). In the former, two bases flank Cα in the enzyme active site. In this case, the first base abstracts the Cα proton, while the conjugate acid of the second base donates a solvent-derived proton to Cα, generating the isomeric product. The reverse reaction is initiated by the second base abstracting the Cα proton, followed by the conjugate acid of the

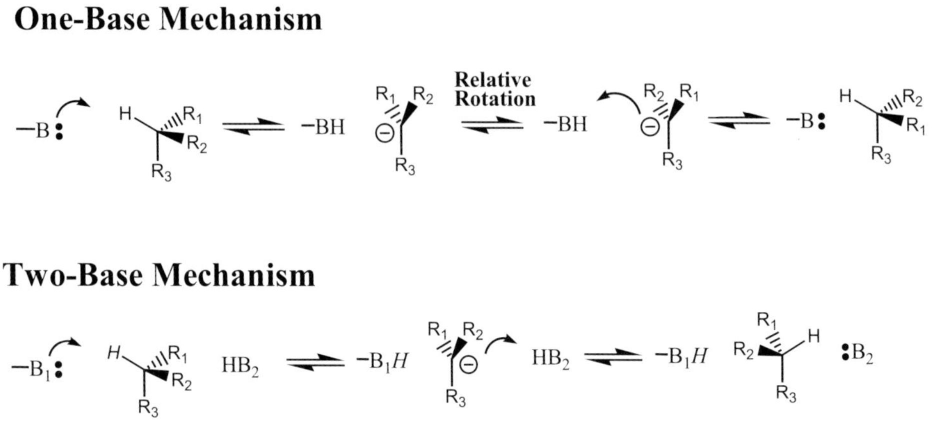

Figure 7.2. A one-base versus a two-base mechanism for deprotonation/reprotonation with a planar carbanionic intermediate. In a one-base mechanism, either the catalytic base or the substrate must reposition after the initial deprotonation, such that the opposite face of the substrate is reprotonated.

first base donating a solvent-derived proton. Alternatively, a one-base mechanism (Fig. 7.2) can, in principle, be operative, with a single active site base abstracting the Cα proton, followed by rotation of either the intermediate or the base catalyst, and subsequent reprotonation (also referred to as the "swinging door" mechanism) [5].

A frequently investigated characteristic of epimerases and racemases is the degree to which they exchange solvent hydron at the Cα position (e.g., racemization of [^{1}H]-substrates in D_2O solutions). The pattern of isotopic incorporation in the product and substrate is often used to assign a two-base mechanism (or eliminate a one-base mechanism) [6–10]. A classic two-base mechanism is consistent with solvent-derived (i.e., isotopic) hydron incorporated into the product, and an absence of solvent-derived hydron in the remaining substrate pool at low conversions. However, in the intuitively unlikely, but theoretically possible, event that hydron exchange occurs between the two bases, "internal return" of the abstracted hydron into the Cα position of the product could occur. Although improbable, such a scenario would yield an ambiguous isotopic exchange pattern. In the classic interpretation of the one-base mechanism, deprotonation and reprotonation are thought to occur in identical environments, yielding identical isotopic incorporation patterns in the substrate and product [5]. A more realistic description of the one-base mechanism would involve base migration to the opposite face of the planar intermediate after deprotonation, such that reprotonation may occur in a distinctly different microenvironment, effectively giving a two-base mechanism as far as hydron exchange is concerned. In short, isotopic incorporation patterns are not sufficient for the absolute diagnosis of one- versus two-base mechanisms, but can be highly suggestive of one or the other.

Alanine racemase has been found to have an asymmetry with regard to the rates of Cα proton exchange relative to racemization, with a smaller conversion/exchange ratio in the D → L than in the L → D direction [11]. This is consistent with a two-base mechanism in which the two bases are in different environments. However, as noted above, this is also consistent with a one-base mechanism, in which the single base reacts from two distinct environments. Furthermore, polyprotic bases such as lysine will exchange with solvent hydron more rapidly than monoprotic bases, which also contributes to the conversion/exchange asymmetry. Asymmetry was also exhibited with regard to the substrate kinetic isotope effects (KIEs) [11, 12]. Note that this asymmetry does not apply to the magnitude of the k_{cat}/K_M values themselves, which must be equivalent according to the Haldane relationship (i.e., the equilibrium constant for any racemase is equal to one).

Another measure of the asymmetric kinetic properties of the two bases in the alanine racemase mechanism is the qualitative behavior of the equilibrium "overshoots" observed. Overshoots are often observed in reaction progress curves run in deuterium oxide that are initiated with a single stereoisomer that is protiated at the Cα position (Fig. 7.3). The optical activity is monitored by polarimetry or circular dichroism (CD). At equilibrium, the signal is zero, since the product is a racemic mixture of D- and L-isomers. However, when there is a significant substrate-derived KIE on the reverse direction (product being fully deuterated in a two-base mecha-

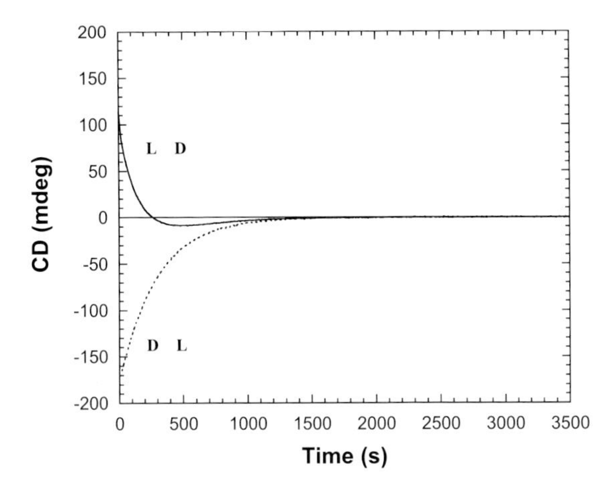

Figure 7.3. Racemization progress curves for protiated D- and L-alanine in a D_2O solution. The progress curve for the L isomer, but not the D isomer, shows an "overshoot" of the stereoisomeric equilibrium (i.e., CD = 0). (Reproduced with permission, © 2004 American Chemical Society.)

nism), the progress curve overshoots the initial point at which there is an equal concentration of stereocenters if the remaining substrate pool retains a substantial amount of protium at Cα. This is because the system is not at "isotopic" equilibrium (i.e., there is still protium-containing substrate, but all of the product is deuterated). This causes the reverse (deuterated) direction to be slower than the forward (protiated) direction, until all of the protium is washed out of the substrate, resulting in a transient excess accumulation of the deuterated product, i.e. an overshoot. The progress curve asymptotically returns to the zero point as the substrate protium is washed out, and both forward and reverse directions have an equivalent rate. The overshoot phenomenon was first characterized by Cardinale and Abels for proline racemase [13].

Progress curves for *B. stearothermophilus* alanine racemase catalyzed [^{1}H]-Ala racemization/washout in both directions are shown in Fig. 7.3 [14]. The L → D direction exhibits a clear overshoot, while there is no detectable overshoot in the D → L direction. This is in accordance with the smaller conversion/exchange ratio seen in the D → L direction [11].

Several *B. stearothermophilus* alanine racemase crystal structures support the asymmetric two-base model [15–17]. The crystal structure of the alanine phosphonate external aldimine suggests that the Tyr265 hydroxyl group, from the adjacent monomer of a homodimer, is one of the catalytic bases (Fig. 7.4) [17]. The other base is thought to be Lys39, which forms the internal aldimine with the PLP cofactor [18]. These two residues are completely conserved in all known alanine racemases. Lys39 is proposed to abstract the Cα proton in the D → L direction, while the Tyr265 would act as the base in the L → D direction (Fig. 7.5) [12, 18–20]. This agrees with both the asymmetry in the overshoots and in the isotopic

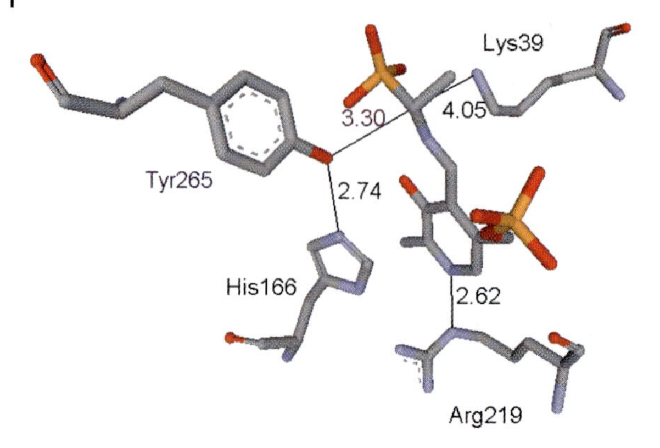

Figure 7.4. Active site of the *B. stearothermophilus* crystal structure with D-alanine phosphonate PLP-aldimine [17]. All distances are in Å.

conversion/exchange ratios. Table 7.2 compares a number of different epimerases and racemases with regard to the identity of their two bases and the symmetry of their overshoots.

Site-directed mutagenesis studies by Esaki and coworkers established that Lys39 acts as a base in the D → L direction and an acid in the L → D direction, while Tyr265 acts as a base in the L → D direction and an acid in the D → L direction

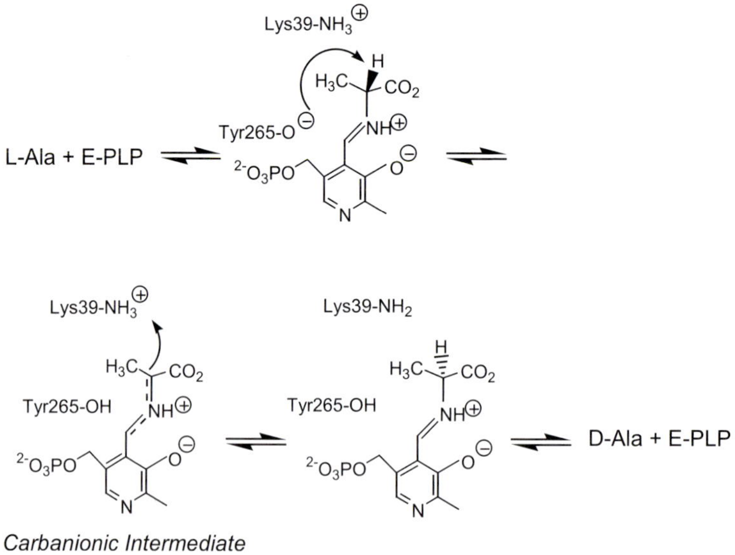

L-Ala + E-PLP ⇌ [Lys39-NH_3^+, Tyr265-O^-, PLP-aldimine] ⇌ [Lys39-NH_3^+, Tyr265-OH] *Carbanionic Intermediate* ⇌ [Lys39-NH_2, Tyr265-OH] ⇌ D-Ala + E-PLP

Figure 7.5. Racemization of L- to D-alanine by alanine racemase. (Reproduced with permission, © 2004 American Chemical Society.)

Table 7.2. Permutations of the two-base mechanism for racemization/epimerization.

Enzyme	Base 1	Base 2	Overshoots
mandelate racemase	His	Lys	asymmetric
alanine racemase	Tyr	Lys	asymmetric
proline racemase; aspartate racemase	Cys[a]	Cys[a]	symmetric
glutamate racemase	Cys	Cys	symmetric
DAP-epimerase[b]	Cys	Cys	†asymmetric
N-acylamino acid racemase	Lys	Lys	?
methylmalonyl CoA racemase[c]	Glu	Glu	?
ribulose 5-phosphate 3-epimerase[c]	Asp	Asp	?

[a] Identical cysteine residues from homodimer. [b] Double overshoot in the D,L → L,L direction; single overshoot in L,L → D,L direction. [c] Based on crystal structure.

[18–20]. The *B. stearothermophilus* K39A mutant had no detectible activity, but addition of methylamine restored approximately 0.1% of the wild-type activity [18]. Furthermore, the mutant exhibited an increase in the $^{D}(V)$ substrate-derived KIE value (in the reaction rescued by methylamine) when [^{2}H]-D-Ala was used as substrate, but not when [^{2}H]-L-Ala was used, while only the L → D direction yielded an increase in the $^{D2O}(V)$. The Y265A mutant had about 0.01% of the racemization activity of the wild-type [19]. Esaki and coworkers hypothesized that the identity of the base involved in racemization and transamination (a side reaction), in the L → D direction, is the same. Accordingly, the Y265A mutant completely lacked transamination activity in the L → D direction only. The *apo*-Y265A mutant also exhibited the ability to abstract stereospecifically tritium from the *R*-isomer of 4′-[^{3}H]-PMP in the presence of pyruvate, in contrast to the *apo*-wild-type enzyme, which abstracts hydron nonspecifically from both (*R*)- and (*S*)-[^{3}H]-PMP.

The active site from the alanine phosphonate crystal structure (Fig. 7.4) shows the pyridine ring nitrogen of PLP hydrogen bonded to the highly conserved Arg219 [17]. This interaction with Arg219 prohibits protonation of the pyridine nitrogen, thereby preventing full utilization of the "electron sink" potential of the pyridine ring. This suggests that a fully stabilized quinonoid intermediate is not employed in the mechanism of alanine racemases. In fact, no quinonoid intermediate can be detected spectroscopically in the wild-type enzyme [14, 21]. However, replacement of Arg219 with a glutamate via site-directed mutagenesis resulted in a spectroscopically detectable quinonoid intermediate in the mutant enzyme, which suffered a drop in activity of three orders of magnitude [12].

The absence of a detectable quinonoid intermediate and the site-directed mutagenesis studies on Arg219 suggested that the alanine racemase mechanism might

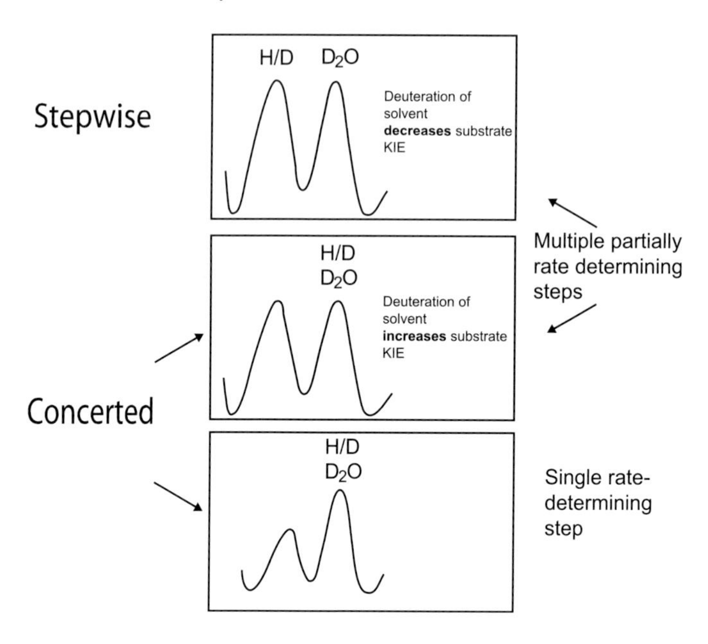

Figure 7.6. Multiple hydrogen kinetic isotope effects used to diagnose a concerted versus stepwise mechanisms.

proceed via a concerted double proton transfer. Multiple kinetic isotope effects (MKIEs) are the method of choice for discerning stepwise versus concerted mechanisms [22]. For example, consider the free energy profile for a stepwise double proton transfer. There are two internal free energy barriers, one for abstraction of a substrate-derived proton and the other for donation of the solvent-derived proton. One may monitor, for example, a substrate KIE in one direction in the presence and absence of deuterated solvent. For systems exhibiting normal KIEs, the presence of isotope in the solvent will increase the energy barrier for $C\alpha$ reprotonation, causing it to become more rate-determining. This results in a reduction in the observed substrate KIE (Fig. 7.6). Alternatively, a concerted mechanism has only a single internal barrier for both isotopically sensitive atom transfers (Fig. 7.6). Thus, introduction of deuterium at the solvent exchangeable position would either increase the expression of the substrate isotope effect if other steps are partially rate-determining, or result in no change if the isotopically sensitive step is fully rate-determining. The equations that describe the expression of MKIE values have been reviewed by O'Leary [22].

A method for obtaining very precise multiple hydrogen kinetic isotope effects was developed in order to determine whether alanine racemase catalyzes a concerted or a stepwise process [21]. The method employs an equilibrium perturbation-type analysis to deduce the substrate KIE in H_2O and in D_2O, thus enabling the MKIE to be determined. Cleland and coworkers first reported the equilibrium perturbation technique, which involves adding enzyme to an equilibrium mixture of substrate and product with only one of these being isotopically labeled [23]. Equilibrium perturbations are typically monitored with an optical spectroscopy. For normal isotope effects, the direction containing the heavy isotope will react more slowly than the opposite direction, producing a transient perturbation in the optical signal due to the transient accumulation of the slower reactant (i.e., the same phenomenon as the equilibrium overshoots described earlier). Cleland and coworkers derived the equations for extracting the $^{D}(V/K)$ KIE values from the magnitude of the mole fraction of the perturbation [23]. However, the case for a two-base mechanism, as described for alanine racemase, is more complicated due to the irreversible loss of substrate hydron into the solvent pool. Bahnson and Andersen derived an expression for obtaining substrate KIE values from equilibrium perturbation-type deuteron washout traces, which was applied to the case of the crotonase-catalyzed dehydration of 3-hydroxybutyrylpantetheine [24].

The entire scheme for a deuterium washout equilibrium perturbation is described in Fig. 7.7A. The two reactants initially present are boxed. The starting substrates for the perturbation are [^{2}H]-D-Ala (lower manifold) and [^{1}H]-L-Ala (upper manifold). All hydrons on the upper manifold are considered to have the same identity as solvent. An equilibrium perturbation-type washout of the [^{2}H]-D-Ala in H_2O proceeds by abstraction of Cα deuteron by a protiated enzyme (lower manifold), followed by donation of a proton, to yield the protiated L-isomer. The enzyme rapidly and irreversibly exchanges the deuteron for proton, moving from the lower to the upper manifold. The contemporaneous racemization of the L-isomer on the upper manifold occurs more rapidly than the racemization from the lower manifold. This transient accumulation of the slower species (in this case D-isomer) produces the perturbation in the optical signal, from which $^{D}(V/K)$ for the D $\rightarrow$ L direction may be determined.

The combination of stereoismers needed to obtain the $^{D}(V/K)$ value for the D $\rightarrow$ L direction *with the solvent exchangeable site being deuterated* is not immediately obvious. After considerable thought, it was determined that it is necessary to perform a perturbation starting with [^{1}H]-D-Ala and [^{2}H]-L-Ala in D_2O [21]. A complete protium washout in D_2O is described in Fig. 7.7B. The lower (washout) manifold is now faster than the upper (solvent) manifold, resulting in a transient accumulation of the [^{2}H]-L-Ala, instead of the isomer on the washout manifold. This yields a perturbation of the opposite direction (relative to the deuterium washout perturbation), with a magnitude that can be used to calculate $^{D}(V/K)_{D2O}$ (Fig. 7.8). The $^{D}(V/K)_{H2O}$ and $^{D}(V/K)_{D2O}$ values allow one to determine if the double proton transfer takes place in a concerted or stepwise mechanism, as described above. There was a significant reduction in the $^{D}(V/K)_{D2O}$ value, relative to the $^{D}(V/K)_{H2O}$ value, which is only consistent with a stepwise mechanism.

A recent global kinetic analysis of racemization progress curves for alanine racemase allowed the definition of the enzymatic free energy profile (Fig. 7.9) [14]. Nu-

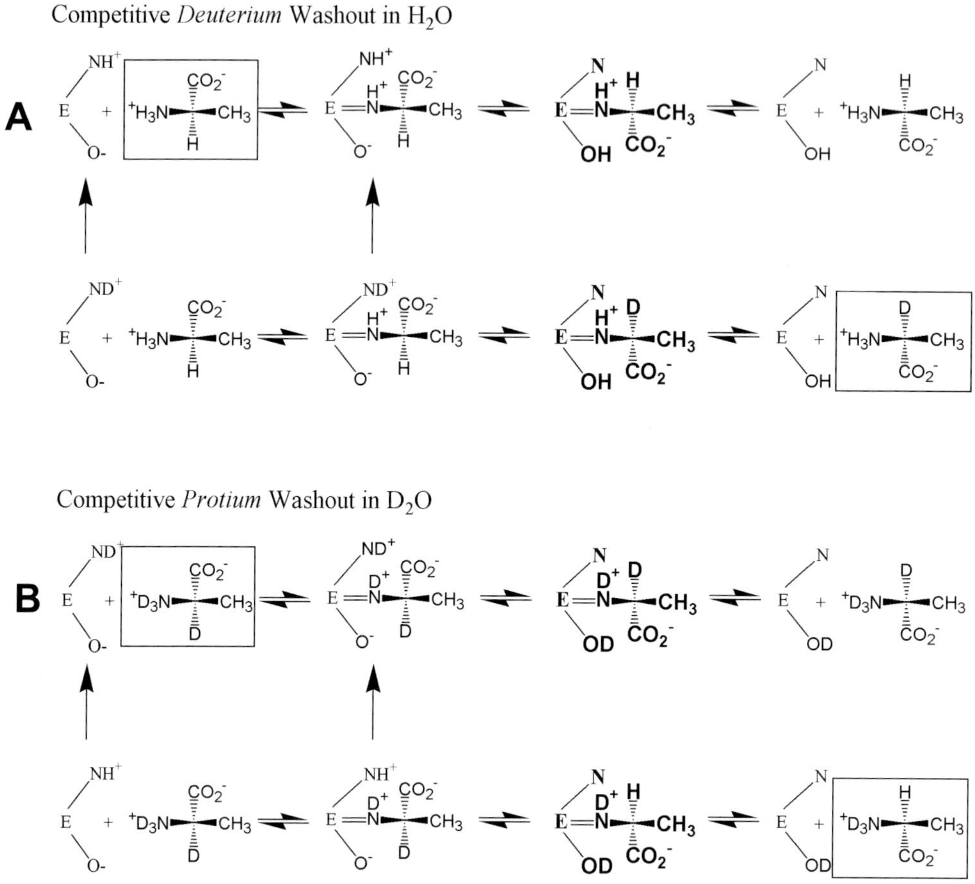

Competitive *Deuterium* Washout in H_2O

A

Competitive *Protium* Washout in D_2O

B

Figure 7.7. Schematic representation of a D-hydron washout perturbation. The upper panel describes the washout of deuterated D-alanine in H_2O (equal starting concentrations of [^{2}H]-D-alanine and [^{1}H]-L-alanine). Upon initiation of the perturbation, [^{2}H]-D-alanine–enzyme complex (lower manifold) and the [^{1}H]-L-alanine–enzyme complex (upper manifold) dominate, with transient accumulation of the former, due to its slower racemization. Upon racemization of the [^{2}H]-D-alanine–enzyme complex, the deuteron is washed out into the solvent pool. At equilibrium only the upper manifold exists, in which forward and reverse racemization rates are equivalent. The substrate-derived KIE is obtained from the difference in rates between the racemization of the [^{2}H]- and [^{1}H]-D-alanine–enzyme complexes. The same logic may be extrapolated to the washout of a proton in an all deuterated system. The lower panel describes the washout of a protiated D-alanine in D_2O (equal starting concentrations of [^{1}H]-D-alaine and [^{2}H]-L-alanine). One may obtain a multiple kinetic isotope effect by comparing the magnitudes of the KIEs from the H_2O and D_2O perturbations (i.e., the effect of solvent on the substrate derived KIE). (Reproduced with permission, © 2003 American Chemical Society.)

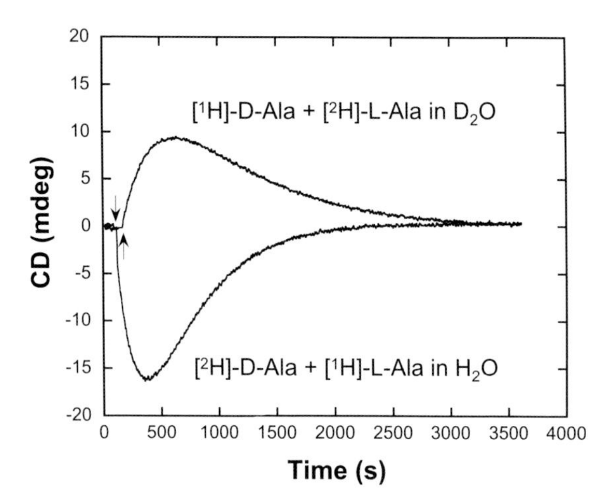

Figure 7.8. Equilibrium perturbation-type washout experiments. The isotopic compositions of the reactants are indicated. At 25 °C, H_2O solutions were pH 8.48 and D_2O solutions were pD 8.90, which gives enzyme in the same ionic state. (Reproduced with permission, © 2003 American Chemical Society.)

merical integration was used to simulate progress curves that correspond to the stepwise double proton transfer catalyzed by alanine racemase. Nonlinear regression was then used in global fits that reduced the mean square of the difference between the simulated and observed progress curves (Fig. 7.10). A series of global

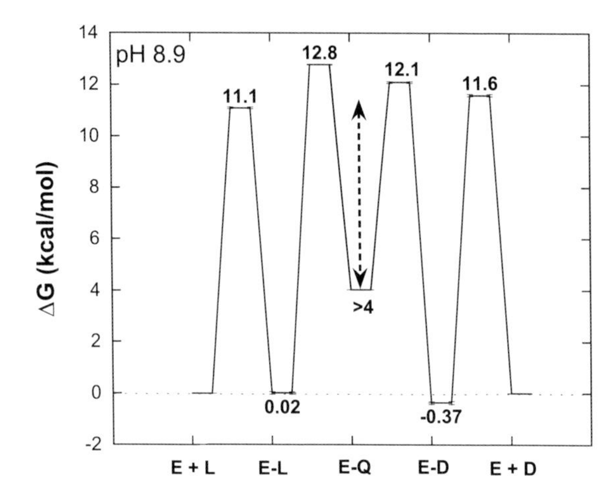

Figure 7.9. Free energy profile obtained from global analysis of racemization progress curves at pH 8.9. Standard state is 5 mM alanine. The double arrow represents the region of uncertainty for the quinonoid intermediate, which extends to a lower limit of approximately 4 kcal mol^{-1}. All other ground and transition state energies have uncertainties of less than 0.06 kcal mol^{-1}. (Reproduced with permission, © 2004 American Chemical Society.)

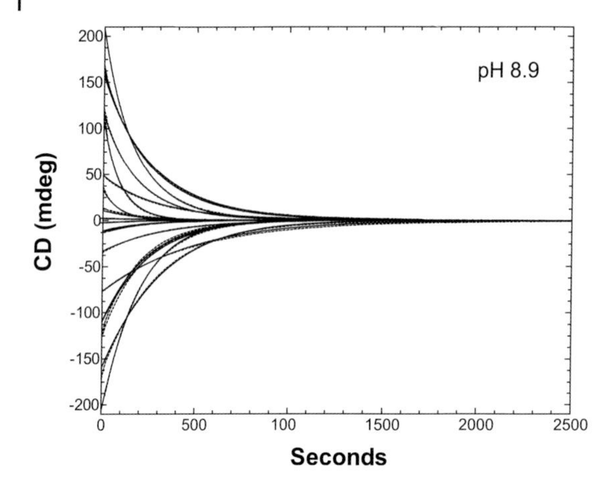

Figure 7.10. Global fit of a stepwise double proton transfer model to racemization progress curves at pH 8.9. Dashed lines, experimental data; solid lines, fitted curves. Positive and negative CD signals correspond to L- and D-alanine, respectively. Alanine concentrations range from 0.2 to 22 mM, yielding 17 progress curves, providing information about the enzyme in the unsaturated and saturated states. (Reproduced with permission, © 2004 American Chemical Society.)

fits commenced from randomized sets of rate constants. The microscopic rate constants from the global fits were used to calculate the steady-state parameters, which were in good agreement with the experimental values. The free-energy profile was also consistent with viscosity variation studies, KIE values and overshoots. The two internal proton transfer steps are mostly rate-determining (88%) at the pH optimum (8.9). The asymmetry between the two internal barriers is expected, based on the larger substrate KIE and overshoot in the L → D direction. Further underscoring this asymmetry, the microscopic rate constants from the global analysis were used to simulate overshoot progress curves (Fig. 7.11), which were consistent with the experimental overshoots [14].

Another interesting feature of the alanine racemase free energy profiles is the relatively high energy of the resonance-stabilized carbanionic intermediate. The precise energy of the intermediate could not be determined, due to the small contribution to the overall rate of the two rate constants leading away from it. However, the global analysis was able to show that it lies >4 kcal mol^{-1} higher in energy than the flanking aldimine intermediates. The high energy of the carbanionic intermediate contributes to the poor catalytic efficiency of alanine racemase (having an efficiency factor, E_f, of about 1×10^{-3}, where unity represents a "perfect" enzyme) [25]. Although catalytically inefficient, this high energy intermediate may prevent unwanted side reactions, and thus contribute to enhanced racemization fidelity.

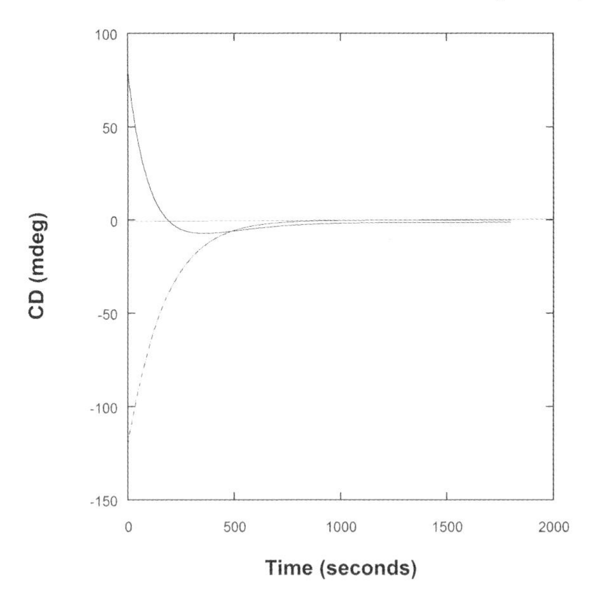

Figure 7.11. Simulated progress curves of protiated L- and D-alanine in D_2O, using rate constants obtained from the global fits in Fig. 7.10, and the published substrate derived $^D(V/K)$ values. An overshoot occurs in the L → D direction only, as in the experimental overshoots shown in Fig. 7.3.

7.2.2
Broad Specificity Amino Acid Racemase

The "amino acid racemase of broad substrate specificity" is an alanine racemase-like enzyme that has, as the name implies, the ability to racemize a number of amino acids (Lys > Arg > Met > Leu > Ala > Ser), preferring positively charged side chains, and having no detectable activity with aromatic or negatively charged amino acids [26, 27]. It is a PLP-dependent homodimer, derived from a number of organisms: *Pseudomonas putida* (previously known as *Pseudomonas striata*, but re-classified to *P. putida* [28]; *Pseudomonas taetrolens* and *Aeromonas caviae* [29]. The *P. putida* genome sequence was recently published [30]. The *P. putida* racemase has 23% sequence identity with that from *B. stearothermophilus* (unpublished observation), retaining the two catalytic bases and the arginine that hydrogen bonds to the pyridine ring nitrogen. A salient difference between the broad substrate specificity racemase and most alanine racemases is the absence of a tyrosine (Tyr354, in the *B. stearothermophilus*), which partially controls access to the active site. In the broad substrate specificity racemase from *P. putida*, the enzyme has an alanine at this position. The importance of this residue to substrate specificity was further illustrated by site-directed mutagenesis studies on *B. stearothermophilus*, in which a

Y354A mutant was shown to have significantly enhanced serine racemase activity [31].

Although the sequence alignments suggest that these amino acid racemases employ a two-base mechanism, there are conflicting isotope incorporation data. Internal transfer of labeled Cα proton was detected in the D $\rightarrow$ L direction of the *P. putida* enzyme, employing alanine as the substrate, which is difficult to reconcile with a two-base mechanism [9]. However, later studies measuring the rates of solvent incorporation into the substrate and product pools of each enantiomer of methionine showed an asymmetry, which is more difficult to explain with a one-base mechanism [32]. However, a one-base mechanism cannot entirely be eliminated, since the two faces of the carbanion may be exposed to significantly different protein environments. It may be that the mechanism changes between one and two bases as the structure of the substrate changes. One could imagine that a substrate that is not very tightly bound and thus able to rotate easily might default to a one-base mechanism while one that is tightly bound might be unable to rotate and require a two-base mechanism.

7.2.3 Serine Racemase

The presence of D-serine in mammalian brain tissue was first reported in 1989 [33, 34]. It has recently been established that D-serine is employed in the mammalian forebrain as a co-agonist for the *N*-methyl-D-asparate (NMDA) excitatory amino acid receptor [35, 36]. A PLP-dependent serine racemase has been cloned and purified from mammalian brain, and found to be a homodimer, which has a number of nonessential cofactors that enhance its activity, including Ca^{2+}, Mg^{2+} and ATP [37–40]. The mouse brain enzyme has also been shown to catalyze elimination from L-serine, to form pyruvate, with an activity comparable to that for racemization [41]. Interestingly, the first instance of this class of racemase was discovered by Esaki and coworkers in the silkworm, *Bombyx mori* [42]. D-serine concentration in the blood of *B. mori* larvae is thought to play a role in metamorphosis.

7.2.4 Mandelate Racemase

The mechanism of madelate racemase is very thoroughly characterized. The reaction catalyzed, stereoinversion of (*S*)- and (*R*)-mandelate, is shown in Fig. 7.12. The enzyme employs a metal cofactor (preferably Mg^{2+}, but also accepting Mn^{2+}, Co^{2+}, Ni^{2+}, and Fe^{2+}) [43, 44], as indicated in Table 7.1, and exhibits a high structural homology with the muconate lactonizing enzyme, which also employs a metal cofactor [45–47]. Both of these enzymes are members of an emerging class of enzymes, the "vicinal oxygen chelate" (VOC) superfamily, which includes methylmalonyl CoA epimerase and *N*-acylamino acid racemase (Table 7.1) [48–50]. The metal binding site of enzymes in the VOC superfamily is located at a conserved site within a TIM barrel, in which Mg^{2+} and Mn^{2+} are typically the preferred

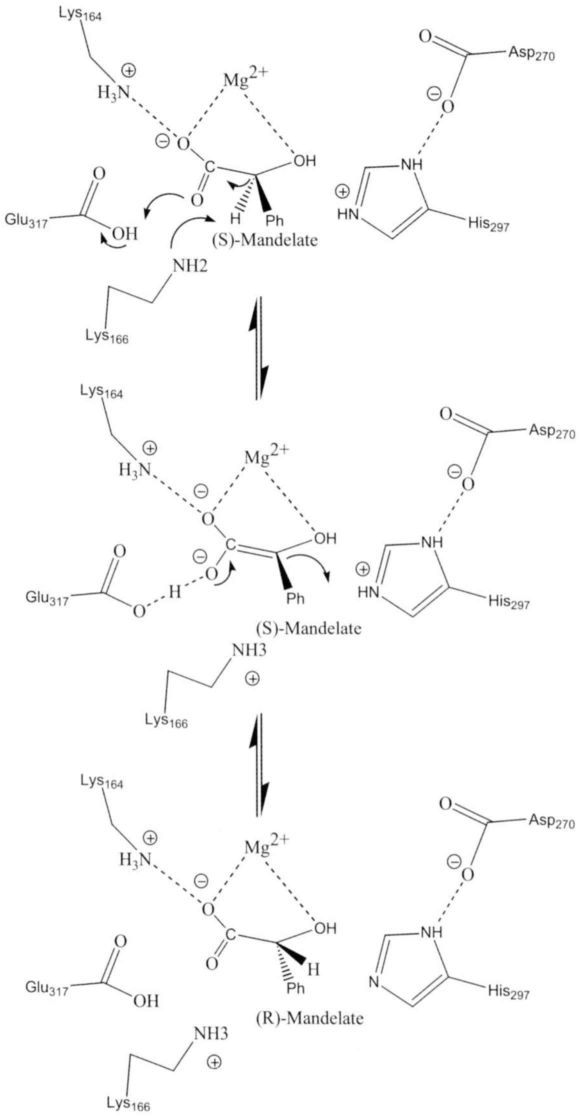

Figure 7.12. Mechanism for the stereoinversion of (*S*)- to (*R*)-mandelate catalyzed by mandelate racemase [10].

metals. The crystal structure of the complex of the K166R mutant (Lys166 is one of the catalytic bases) madelate racemase with (*R*)-mandelate bound in the active site shows that the metal cofactor is bound to both an oxygen from the carboxylate and to the Cα hydroxyl (Fig. 7.13) [51].

Early studies on madelate racemase demonstrated that the substrate Cα proton fully exchanges with D_2O solvent during racemization, and that the proton abstrac-

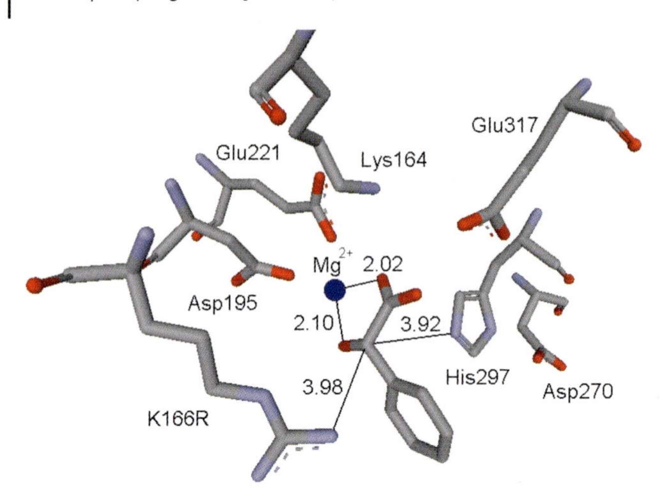

Figure 7.13. Active site of the K166R mutant of mandelate racemase from the crystal structure of the Michaelis complex with (*R*)-mandelate. All distances are in Å [51].

tion is partially rate-determining, in both directions [8]. Also, linear free energy relationships were demonstrated by substitution of electron withdrawing substituents on the phenyl ring of the substrate [52]. These gave the expected result that stabilization of the Cα carbanion leads to a more efficient substrate. Further studies yielded an isotope exchange pattern for both directions that is consistent with a two-base mechanism [10].

The crystal structure indicated that the two likely bases flanking the Cα carbon are Lys166 and His297 (Fig. 7.13) [46, 51]. From the active site architecture, it is thought that Lys164 plays an important role in lowering the pK_a ($\sim$4 units) of Lys166, such that it is present in the catalytically active basic form at physiological pH [53]. Similarly, the second catalytic base, His297 is hydrogen bonded to Asp270, which may act to increase its pK_a.

This asymmetry with respect to the catalytic bases in mandelate racemase is formally analogous to alanine racemase, and indeed the overshoots with mandelate racemase are highly asymmetric as with alanine racemase (Table 7.2) [10]. The $(S) \rightarrow (R)$ direction exhibits a much smaller overshoot than the $(R) \rightarrow (S)$ direction, and shows significant exchange of solvent deuteron into the substrate pool even when the extent of racemization is low. The $(R) \rightarrow (S)$ direction shows much less exchange of deuterium into the substrate pool.

7.2.5
ATP-Dependent Racemases

Many peptide-based antibiotics contain D-amino acids. D-phenylalanine is a component of gramacidin S, and is produced by phenylalanine racemase, which is a member of a family of ATP-dependent racemases and epimerases that also require the 4′-phosphopantethein (PAN) cofactor for activity (Table 7.1) [54, 55]. Adenylation is used to activate the amino acid for transfer to the thiol group of PAN, yielding a thioester with an acidified Cα proton [56, 57]. The stereoinversion is catalyzed directly on the thioesterified substrate, producing an equilibrated mixture of the enantiomers. The overall reaction (Fig. 7.14) is referred to as a "thiol-template" mechanism [54]. The coupling to ATP produces an overall reaction that is irreversible. Interestingly, reactions initiated with ATP and a single stereoisomer result in

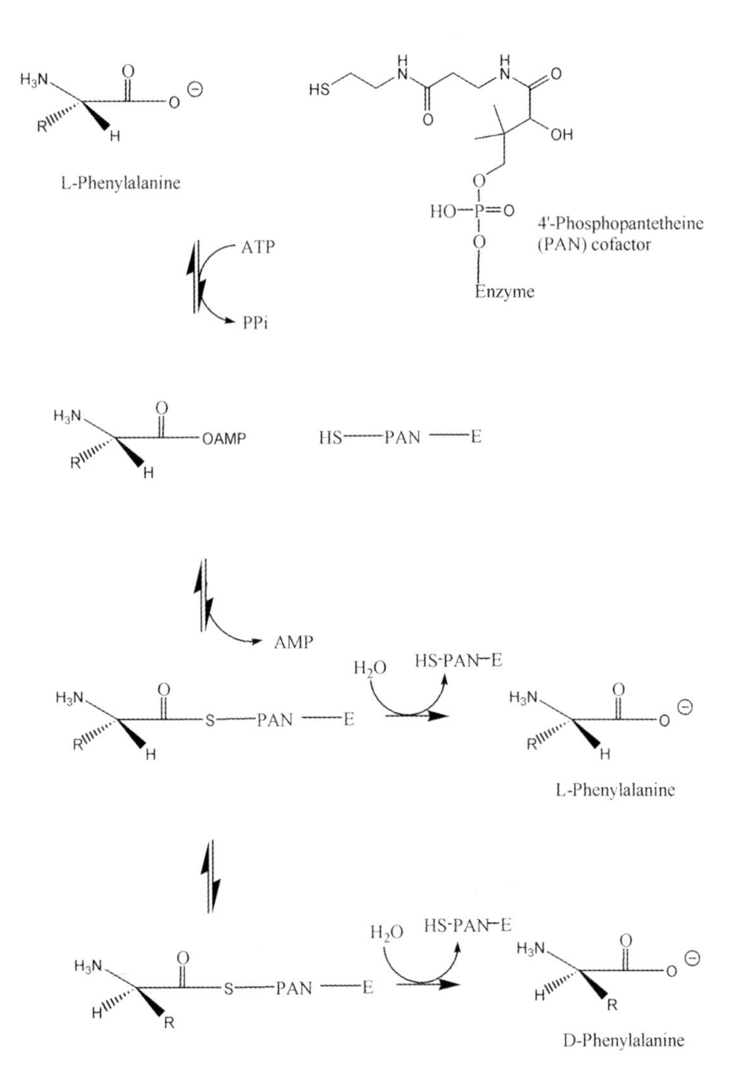

Figure 7.14. The thiol template mechanism for the stereoinversion of L- to D-phenylalanine catalyzed by phenylalanine racemase [54].

4:1 (D:L) mixtures of stereoisomers that are determined by the different rates of hydrolysis of the thioester intermediates [56, 57].

The thiol-template mechanism is utilized in other enzymes involved in production of peptide-based antibiotics. Actinomycin synthetase II (ACMSII) and δ-L-(α-aminoadipolyl)-L-cysteinyl-D-valine (ACV) synthetase catalyze the stereoinversion of valine residues within peptide-based antibiotics, and employ ATP and the PAN cofactor in a mechanism similar to that depicted in Fig. 7.14 [58, 59]. ACMSII catalyzes the stereoinversion of a valine within the tripeptide 4-MHA-L-Thr-D-Val (MHA, 4-methyl-3-hydroxyanthranilic acid), which is a precursor for the antibiotic actinomycin D. ACV synthetase catalyzes the stereoinversion of the valine within ACV, which is a precursor for penicillin and cephalosporin [60–63]. ACV synthetase has been shown to have much broader substrate specificity, also accepting non-natural substrates [64, 65].

A number of epimerases act at carbon centers that are α to thioester linkages with coenzyme-A (CoA). These enzymes are similar to mandelate racemase, in that they employ a metal cofactor (Table 7.1), which is thought to stabilize an enolate intermediate. They also belong to the VOC superfamily of enzymes, whose members all involve proton abstraction, proton transfer and metal cofactors that stabilize anionic intermediates [48–50].

7.2.6 Methylmalonyl-CoA Epimerase

In animals, the breakdown of lipids involves conversion of propionyl-CoA to succinyl-CoA. Methylmalonyl-CoA is a metabolic intermediate in this process. *In vivo*, it is necessary to convert the 2-(*S*)-form of methylmalonyl-CoA to the 2-(*R*)-form, for reaction with methylmalonyl-CoA mutase. This reaction is catalyzed by methylmalonyl-CoA epimerase (MMCE) [4, 66–68]. Methylmalonate is also employed in polyketide antibiotic biosynthesis, in the form of methylmalonate units, although less is known about the stereochemical requirements of these processes [69, 70].

MMCE is found in both animals and bacteria [67, 71]. The best characterized MMCE is from *Propionibacterium shermanii*, whose crystal structure with the Co^{2+} cofactor has been published [72, 73]. Modeling of the 2-methylmalonate substrate into the active site of the crystal structure shows that the Cα of the 2-(*R*)-epimer is ~3 Å from Glu48, while the 2-(*S*)-epimer is ~3 Å from Glu141. It is possible that these two residues are the catalytic bases employed in the stereoinversion. Early studies using 2-(*R*)-[^{3}H]-methylmalonlyl-CoA as the substrate showed total washout of the label in the product [74, 75]. A similar experiment using unlabeled 2-(*R*)-epimer in tritiated water showed no return of label into the substrate pool. This provides strong evidence for, but not definitive proof of, a two-base mechanism. In the event that a similar result is obtained for the $2(S) \rightarrow 2(R)$ direction, a one-base mechanism would be highly unlikely. The lack of internal return of substrate-derived hydron, in both directions, can only be explained by rapid exchange of the abstracted hydron with environmental solvent. However, rapid exchange of hydron is not consistent with the observation that there is no incorporation of solvent hydron into the substrate pool.

7.3 Cofactor-Dependent with Unactivated Substrates

Some epimerases act on substrates that are not activated (i.e., carbon centers not adjacent to carbonyls or carboxylates). This may be achieved by creating a transiently activated species, which is the actual target for stereoinversion. This is the strategy used by a number of NADH/NAD^+-dependent sugar epimerases. These may be further subdivided into enzymes that transiently oxidize the hydroxyl on the carbon adjacent to the stereogenic center, and those enzymes that transiently oxidize the stereogenic center itself. In the former case, the stereogenic center is activated for a 1,1-proton transfer, which precedes reduction of the keto-intemediate, to yield the epimeric product. In the latter case, a 1,1-hydride transfer directly results in epimerization about the stereogenic center. Figure 7.15 illustrates these two pathways.

UDP-galactose 4-epimerase utilizes the 1,1-hydride transfer route for sugar epimerization, yielding a 4-hexose intermediate [76]. The ketohexose intermediate is produced by hydride transfer from C-4 to the B-face of the nicotinamide ring. The ketohexose then moves such that the opposite face of the keto group is reduced by NADH, yielding the antipodal enantiomer (Fig. 7.15). This is formally analogous to Henderson and Johnston's "swinging door" one-base mechanism for 1,1 proton transfer [5], which has yet to be definitively demonstrated in a racemase.

Another group of sugar epimerases, which uses a metal cofactor instead of NADH/NAD^+, takes an entirely different approach to epimerization. L-ribulose 5-phosphate 4-epimerase, which is involved in the bacterial metabolism of arabinose, performs a retro-aldol cleavage of a C–C bond to yield a metal-stabilized enolate of dihydroxyacetone and glycoaldehyde phosphate, similar to the reaction catalyzed by class II aldolases [77–79]. The glycoaldehyde phosphate is thought to rotate, such that addition of the enolate generates the isomeric product.

7.4 Cofactor-Independent with Activated Substrates

7.4.1 Proline Racemase

Proline racemase is a member of a broad family of cofactor-independent epimerases and racemases, and has been very well characterized mechanistically. The proline racemase from *Clostridium sticklandii* was the first of the cofactor-independent racemases to be characterized [13, 80]. The enzyme participates in the catabolism of L-proline, producing D-proline as a substrate for D-proline oxidase [4]. Early

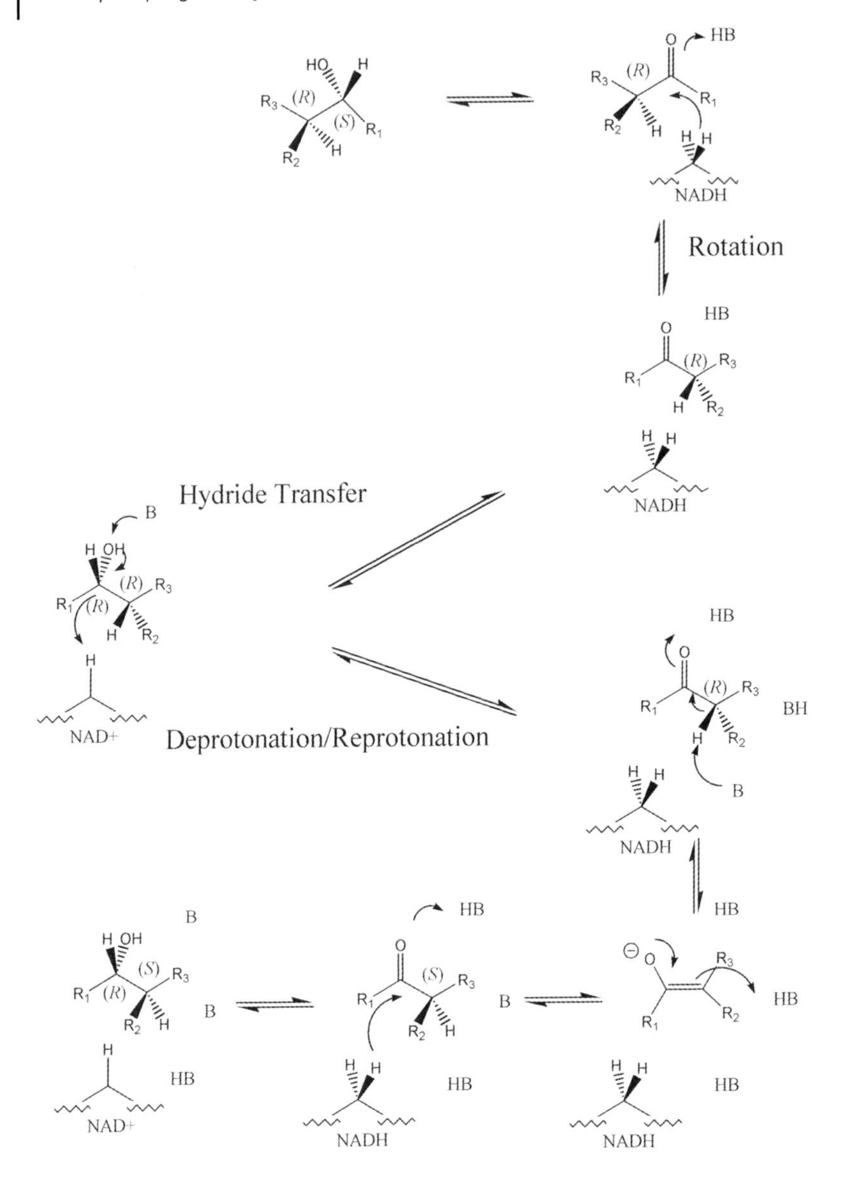

Figure 7.15. Two possible pathways for the stereoinversion of sugars by sugar epimerases utilizing the NADH/NAD$^+$ cofactor.

studies showed that both directions fully incorporate isotopic label in the product, with no label returning to the substrate pool [13]. This is strongly suggestive of a two-base mechanism with a planar carbanionic intermediate. Investigations on deuterium incorporation and primary KIE values led Cardinale and Ables to observe the first overshoot phenomenon (discussed above) [13]. The enzyme is a homodimer, in which the same cysteine from each monomer contributes to a single active site [80]. This is qualitatively different than the other cofactor-independent racemases (glutamate racemase and aspartate racemase) and epimerases (DAP-epimerase), which are monomers with pseudo-symmetry (i.e., the Cys active site bases are not at identical positions in two different subunits). This symmetry of proline racemase catalytic bases leads to a symmetry in the overshoots (Table 7.2) and KIE values. Global amino acid sequence alignments indicate that proline racemase is in a distinctly different protein family than aspartate/glutamate racemase (unpublished observation using the Pfam database [81]).

The enzyme exists in two different protonation states of the active site cysteines, each binding a different enantiomer. Conversion between enantiomers can be through the racemization path (upper manifold of Fig. 7.16) or through direct proton exchange with water (lower manifold of Fig. 7.16). Knowles and coworkers found that interconversion of enzyme protonation states was kinetically significant [82]. This was determined by measuring rates of tritiated proline washout as a function of the proline concentration. It was found that higher concentrations of proline promote slower washout of the Cα proton. Additional support for the rela-

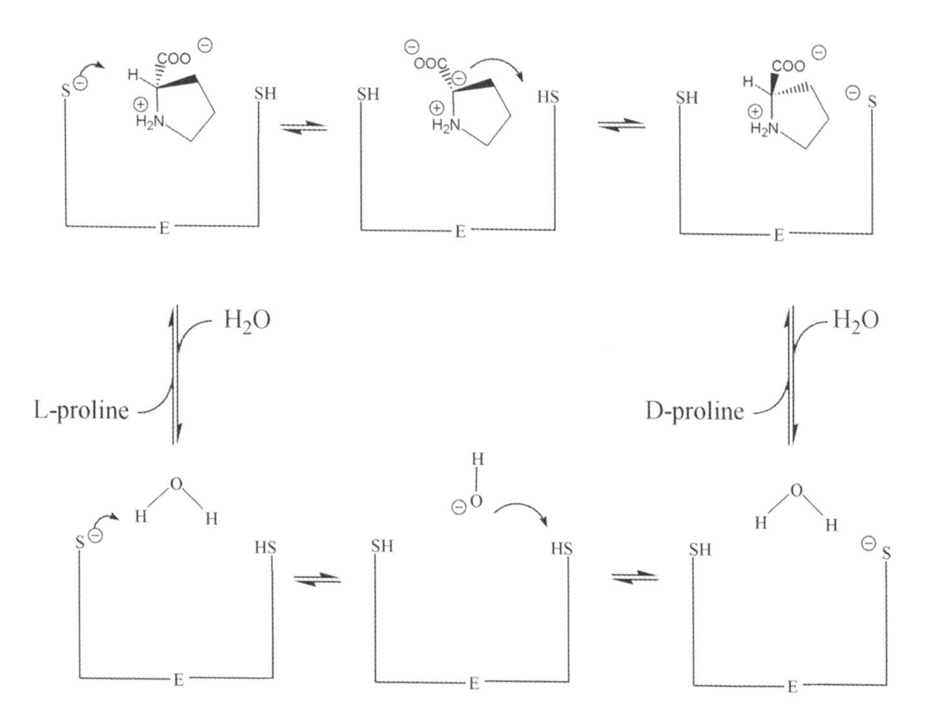

Figure 7.16. Mechanism for the stereoinversion of L- to D-proline catalyzed by proline racemase (upper manifold) and water catalyzed proton exchange of the free enzyme (lower manifold).

tively slow interconversion of the two protonation states was provided by conceptually similar experiments in which "oversaturation" by substrate was observed. Thus, product release is followed by relatively slow interconversion of the "substrate" and "product" protonation states, which Northrop has termed an "iso-mechanism" [83].

The free energy profile for proline racemase, including the energetic barriers for interconversion of the enzyme protonation states, was described in a series of papers by Knowles and coworkers in 1986 [82, 84–88]. A "tracer-perturbation" experiment was employed to estimate the energy barriers for the proton exchange [84]. This is performed by adding an excess of unlabeled L-proline to an equilibrated mixture of [^{14}C]-D,L-proline, promoting the formation of the enzymic form that binds D-proline. The enzyme-D-proline complex releases D-proline and either (i) undergoes proton exchange to convert to the form that binds L-proline, or (ii) binds labeled D-proline and converts it to L-proline. This experiment results in a net flux of label into the L-proline pool since the high concentration of unlabeled L-proline effectively competes against the conversion of labeled L-proline to labeled D-proline. It is the net flux of label into the L-proline pool that allowed Knowles and coworkers to estimate the rate constants for proton exchange between the two enzymic forms ($\sim 1 \times 10^5$ s^{-1}).

Oversaturation, mentioned above, is another phenomenon resulting from the competition between solvent catalyzed conversion of enzymic forms and the conversion via substrate racemization [82]. A consequence of this competition is that the net rate of racemization decreases, under reversible conditions, as the concentration of a proline increases. The loss of productivity in the racemization manifold is due to the product form of the enzyme binding substrate, and the substrate form of the enzyme binding product (i.e., substrate inhibition).

Knowles and coworkers also performed "competitive deuterium washouts" (i.e., an equilibrium perturbation-type washout experiments), using deuterated substrates in H_2O solutions, which yielded the $^{D}(V/K)$ values for both directions [85]. Further confirmation of these KIE values was validated by a "double competitive deuterium washout" experiment, in which both substrates are Cα deuterated, which yielded a ratio of the two $^{D}(V/K)$ values. The authors were also able to perform competitive deuterium washout experiments where direct proton exchange between free enzyme forms is rate-limiting (i.e., at high substrate concentration the lower manifold of Fig. 7.16 is dominant). This experiment indicated that interconversion of free enzyme forms is very similar to the racemization manifold, in that loss of proton from one form yields the other free enzyme form, with water acting as the catalyst, Fig. 7.16.

Isotope discrimination studies were employed to deduce if the double proton transfer of proline racemase is concerted or stepwise [88]. Isotope discrimination is an alternative manifestation of the multiple kinetic isotope effect techniques previously discussed, wherein racemization is conducted in mixed isotopic solvents of H_2O and D_2O and the discrimination in the incorporation of solvent deuterium is measured. If the double proton transfer is stepwise, deuteration of the substrate causes the solvent incorporation step to be less rate-determining, resulting in a decrease in the isotope effect (i.e., a decrease in discrimination). For a concerted mechanism, deuteration of the substrate position would not affect the solvent incorporation, resulting in no change in the isotope effect (i.e., no change in discrimination). In conjunction with other studies, the isotope discrimination studies led Knowles and coworkers to favor a stepwise mechanism in proline racemase.

7.4.2 Glutamate Racemase

D-Glutamate, like D-alanine, is a constituent of the peptidoglycan layer of bacterial cell walls [1, 2]. Glutamate racemase is a member of the cofactor-independent family of epimerases and racemases, with high sequence homology to aspartate racemase [89]. The enzyme belongs to the Aspartate/Glutamate Racemase superfamily, with an ATC-like fold, consisting of two similar domains, related by pseudo-dyad symmetry. The enzyme showed no dependence on PLP or metal cofactors, and was shown to be inactivated by thiol-specific and oxidizing reagents [89–92]. In *E. coli* and *Aquifex pyrophilus* the enzyme is thought to be a dimer with two active sites, while in *Lactobacillus fermenti* it is monomeric with a single active site [89, 91, 93].

The crystal structure for glutamate racemase from *Aquifex pyrophilus* has been determined with D-glutamine bound in the active site [93]. The authors hypothesized that the two catalytic bases are Cys70 and Cys178. However, the Cα of the D-glutamine ligand is not positioned for proton abstraction or donation from either of these groups, being ~7 Å away. The authors hypothesize, based on modeling studies, that the D-glutamine ligand is flipped 180° from the catalytic position assumed by the glutamate substrate. Each cysteine base is within about 4 Å of a carboxylate-containing residue (Cys70 is proximal to Asp7, Cys178 is proximal to Glu147) and the D7S and E147N mutants were found to have activity losses between 1 and 2 orders of magnitude relative to wild-type. The role of these acidic residues may be to increase the pK_a of the two catalytic cysteines.

Isotope incorporation studies show that racemization in D_2O results in Cα deuterium incorporation in the product, from both directions, which is suggestive of a two-base mechanism [94, 95]. The $^{D}(V/K)$ values and overshoots have been determined for both directions [96]. There is significant symmetry in the overshoots (Table 7.2) and KIE values, as one would expect, based on the identity of the catalytic bases. Unlike proline racemase, there is a single active site per monomer. Mutant enzymes lacking either of the two conserved cysteines (Cys to Ala in *E. coli*, and Cys to Thr in *L. fermenti* enzyme) residues exhibit a complete loss of activity [96]. Furthermore, mutants lacking one of the catalytic cysteines can eliminate HCl from *threo-β*-chloroglutamate, with each mutant being specific for one of the stereoisomers [96]. This suggests that the residues are on opposing sides of Cα. Unlike proline racemase, no oversaturation phenomenon was observed, indicating that interconversion of protonation states is kinetically insignificant [95].

7.4.3
DAP Epimerase

In addition to D-alanine and D-glutamate, many bacterial cell walls also contain *meso*-diaminopimelate (DAP) [2]. DAP is produced by epimerization from L,L-DAP to D,L-DAP by the cofactor independent diaminopimelate epimerase [97, 98]. The structure of this enzyme has been solved and two cysteines in the active site were proposed to be the acid–base catalysts [99]. The pattern of label incorporation from tritiated water is consistent with a two-base mechanism [97]. The enzyme has been shown to be stoichiometrically inhibited by the thiol alkylating agent aziDAP [97]. Interestingly, DAP epimerase has an equilibrium constant of 2 (K_{eq} = [D,L]/[L,L]) due to the statistically expected higher concentration of the [D,L] form at equilibrium between these species [100].

Although both catalytic bases are cysteines, the $^{D}(V/K)$ values for both directions are apparently not identical: 4.3 ± 0.7 for the L,L → D,L direction, and 5.4 ± 1.1 for the D,L → L,L direction [100]. These $^{D}(V/K)$ values have been ascribed to differences in the intrinsic KIE values for abstraction of the Cα protons for the respective directions. The ^{D}V values are significantly smaller than the $^{D}(V/K)$ values. Koo and Blanchard suggest this to be the result of a kinetically significant interconversion of the two protonation states, as observed with proline racemase [100]. The $^{D2O}(V/K)$ values for DAP-epimerase are inverse (L,L → D,L = 0.83 ± 0.08, D,L → L,L = 0.73 ± 0.09), which can be ascribed to the low fractionation factor of the thiol groups of the two catalytic bases. However, the ^{D2O}V values (L,L → D,L = 1.8 ± 0.1, D,L → L,L = 1.5 ± 0.1) are not inverse. Koo and Blanchard hypothesize that this may be due to a large SIE in the interconversion of the different protonation states of the enzyme.

DAP-epimerase yields an unusual overshoot pattern: a normal overshoot is seen in the L,L → D,L direction, but an unprecedented double-overshoot is seen in the D,L → L,L direction [100]. A simulation (using the program DynaFit [101]) of the DAP-epimerase double overshoot, based on rate constant values used in simulations by Koo and Blanchard, is shown in Fig. 7.17A. Koo and Blanchard proposed that the double overshoot is due to the fact that two stereocenters undergo exchange, but only one is racemized. The full reaction scheme, as presented by Koo and Blanchard, is illustrated in Fig. 7.18. The D,L-substrate initially reacts faster than the L,L-substrate, and enters an isotopically sensitive branch point. One observes a classic overshoot in both directions due to the fact that the substrate-derived KIE for the reverse direction results in a transient accumulation of the product (the orthodox source of an overshoot). However, the additional overshoot in the D,L → L,L direction was attributed to accumulation of [^{2}H]-D,[^{1}H]-L-DAP in the isotopically sensitive branch pathway, which results in a transient accumulation of the D,L-isomer, even though the reaction commenced with [^{1}H]-D,[^{1}H]-L-isomer (i.e., in the opposite direction from an orthodox overshoot). Surprisingly, removal of the isotopically sensitive branch point, such that only the bold species in Fig. 7.18 are present, yields an effectively identical simulated double overshoot

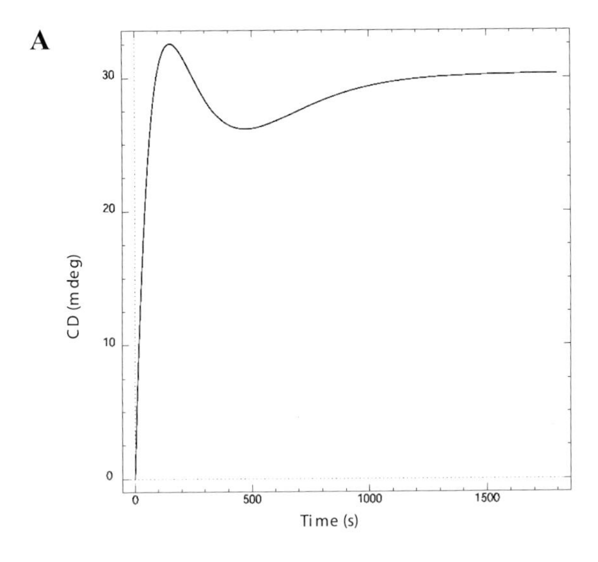

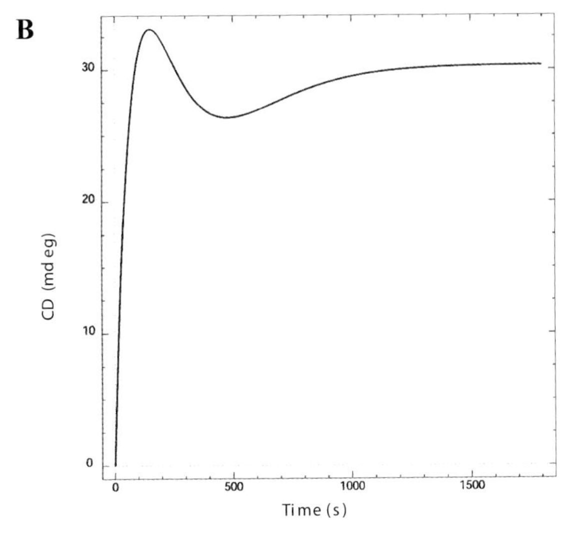

Figure 7.17. A, Simulated double overshoot for DAP-epimerase using the program DynaFit [101] and the rate constant values from Koo and Blanchard [100]. B, Simplified simulated double overshoot, in which the isotopically sensitive branch pathway from Fig. 7.18 is removed.

(Fig. 7.17B). This indicates that the source of the double overshoot phenomenon is simpler than previously thought. Figure 7.19 shows the concentrations of the various isotopic species of DAP during the double overshoot simulation shown in Fig. 7.17B. The peaks of the two overshoots are indicated in Fig. 7.19, demonstrating

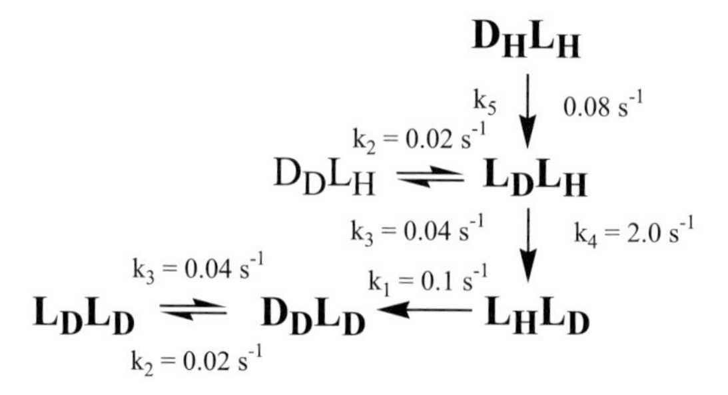

Figure 7.18. Kinetic scheme for DAP-epimerase, as described in Koo and Blanchard [100], used to generate the simulated overshoot in Fig. 7.17A. The species in bold type represent the simplified overshoot used to generate simulated overshoot in Fig. 7.17B.

that the source of the double overshoot is simply due to sequential transient accumulations of first L_HL_D and then D_DL_D, and a lag phase in the formation of L_DL_D. At the peak of the first overshoot there is an equal concentration of L_HL_D (positive optical signal) and D_HL_H/D_DL_D (no optical signal). The peak of the second overshoot occurs when there is an equal concentration of D_DL_D and L_DL_D/L_HL_D.

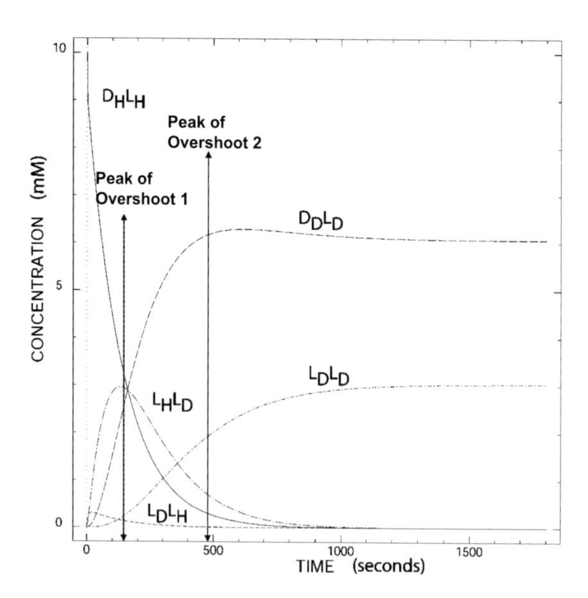

Figure 7.19. Concentrations of products, substrates and intermediates from the double overshoot of DAP-epimerase. The time points that correspond to the peaks of the two overshoots of the double overshoot are indicated.

7.4.4 Sugar Epimerases

There are a number of cofactor independent carbohydrate epimerases that act on activated substrates, such as keto-sugars and keto-sugar nucleotides, although there is a paucity of details about their mechanisms. D-ribulose-5-phosphate 3-epimerase catalyzes the stereoinversion of substrate about the C-3 carbon to form D-xylulose 5-phosphate (as in Fig. 7.15) [102, 103]. Solvent hydron is completely incorporated into the product at the C-3 carbon, during epimerization in the D-xylulose 5-phosphate to D-ribulose 5-phosphate direction [102]. This was taken as evidence for a two-base mechanism.

The keto-sugar nucleotide dTDP-L-rhamnose is synthesized from dTDP-4-keto-6-deoxy-D-glucose by dTDP-L-rhamnose synthase [104, 105]. The enzyme consists of two components, a cofactor independent epimerase and an NADH-dependent reductase. The epimerase component is inactive without the reductase component. The mechanism involves epimerization of two stereocenters flanking a carbonyl group, via sequential deprotonation/reprotonation, with two enol intermediates. Complete solvent isotope incorporation into both epimerized stereocenters was observed, and primary substrate-derived KIEs have been determined [104].

7.5 Cofactor-Independent with Unactivated Substrates

UDP-*N*-acetylglucosamine (UDP-GlcNAc) epimerase catalyzes a mechanistically exotic stereoinversion of UDP-GlcNAc to UDP-ManNAc, which is used in the synthesis of some bacterial cell walls [106, 107]. The epimerase has the unusual requirement for a small amount of substrate UDP-GlcNAc for activity (i.e., UDP-ManNAc and enzyme alone will not react). It is thought that UDP-GlcNAc may bind in a modulating site, separate from the active site [106].

The absence of an activated stereogenic center or a cofactor makes simple deprotonation/reprotonation prohibitively difficult. Solvent hydron incorporation in the target carbon of the product has been observed in both directions [108]. However, ^{18}O positional isotope exchange studies show evidence of C–O anomeric carbon cleavage during the reaction [109]. The ^{18}O label became distributed into both the anomeric position and into the pyrophosphate during catalysis, indicating C–O cleavage. This is consistent with a glycal mechanism, as proposed by Sala et al. (Fig. 7.20), in which trans-elimination of UDP yields a 2-acetamidoglucal enzyme-bound intermediate, followed by *syn*-addition of UDP to generate the isomeric product [109]. Thus, the enzyme avoids the energetically unfavorable E1cb reaction, and proceeds by either an oxonium intermediate E1 or a concerted E2 reaction. Limited kinetic isotope effect studies have been carried out. The value of 1.8 for $^{D}(V)$ in the forward direction indicates that C–H bond cleavage is at least partially rate-determining [109].

The 2-acetamidoglucal intermediate generated by UDP-*N*-acetylglucosamine 2-

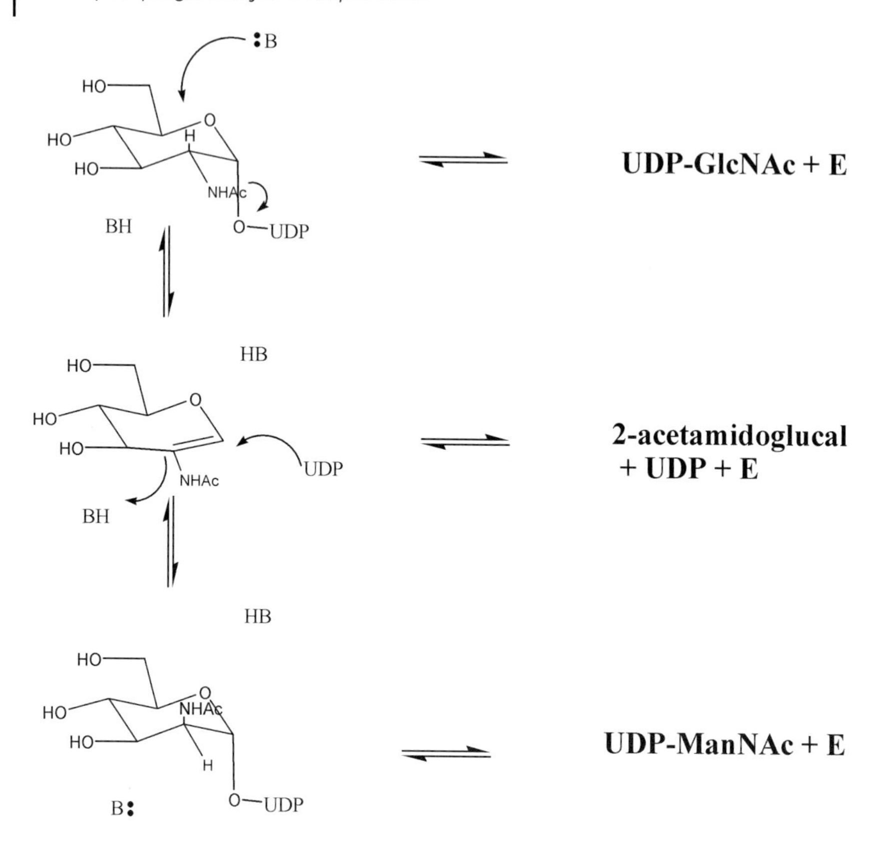

Figure 7.20. The glycal mechanism for the conversion of UDP-*N*-acetylglucosamine (UDP-GlcNAc) to UDP-*N*-acetylmannosamine (UDP-ManNAc) catalyzed by UDP-GlcNAc epimerase [109].

epimerase is thought to be more stable than the reactants or products. Thus, the enzyme has the unusual challenge of trying to prevent the release of a thermodynamically stable intermediate (relative to the free reactants and products), as opposed to protecting a higher energy intermediate from undesirable side reactions. This is precisely the opposite scenario faced by alanine racemase, which has a highly destabilized carbanionic intermediate, which may serve to enhance the fidelity of its racemization reaction.

7.6 Summary

The stereoinversion of carbon centers catalyzed by racemases and epimerases is an archetypal enzyme catalyzed reaction for studying multiple hydrogen transfers.

Epimerases and racemases may or may not employ enzyme cofactors (organic or inorganic) to activate the stereogenic center of the substrate. Common cofactor-stabilized intermediates include resonance-stabilized carbanions and metal-stabilized enolates. The substrate itself can be intrinsically activated if the stereogenic center is adjacent to a carbonyl or carboxylate group. A preponderance of racemases and epimerases act on activated substrates. A number of sugar and sugar nucleotide epimerases act on unactivated substrates. Double proton transfers may proceed, in principle, by either a one- or two-base mechanism. However, only two-base mechanisms have been observed for racemases.

References

1 Neidhart, F. C. (1999) *Escherichia coli and Salmonella*, 2nd edn., Blackwell Publishing, London.

2 Walsh, C. T. (1989) *J. Biol. Chem. 264*, 2393–2396.

3 Dixon, J. E., Bruice, T. C. (1973) *Biochemistry 12*, 4762–4766.

4 Adams, E. (1976) *Adv. Enzymol. Relat. Areas. Mol. Biol. 44*, 69–138.

5 Henderson, L. L., Johnston, R. B. (1976) *Biochem. Biophys. Res. Commun. 68*, 793–798.

6 Adams, E., Mukherjee, K. L., Dunathan, H. C. (1974) *Arch. Biochem. Biophys. 165*, 126–132.

7 Ahmed, S. A., Esaki, N., Tanaka, H., Soda, K. (1986) *Biochemistry 25*, 385–388.

8 Kenyon, G. L., Hegeman, G. D. (1979) *Adv. Enzymol. Relat. Areas. Mol. Biol. 50*, 325–360.

9 Shen, S. J., Floss, H. G., Kumagai, H., Yamaka, H., Esaki, N., Soda, K., Wasserman, S. A., Walsh, C. (1983) *J. Chem. Soc., Chem. Commun.* 82–83.

10 Powers, V. M., Koo, C. W., Kenyon, G. L., Gerlt, J. A., Kozarich, J. W. (1991) *Biochemistry 30*, 9255–9263.

11 Faraci, W. S., Walsh, C. T. (1988) *Biochemistry 27*, 3267–3276.

12 Sun, S. X., Toney, M. D. (1999) *Biochemistry 38*, 4058–4065.

13 Cardinale, G. J., Abeles, R. H. (1968) *Biochemistry 7*, 3970–3978.

14 Spies, M. A., Woodward, J. J., Watnik, M. R., Toney, M. D. (2004) *J. Am. Chem. Soc. 126*, 7464–7475.

15 Watanabe, A., Yoshimura, T., Mikami, B., Hayashi, H., Kagamiyama, H., Esaki, N. (2002) *J. Biol. Chem. 277*, 19166–19172.

16 Morollo, A. A., Petsko, G. A., Ringe, D. (1999) *Biochemistry 38*, 3293–3301.

17 Stamper, C. G. F., Morollo, A. A., Ringe, D. (1998) *Biochemistry 37*, 10438–10445.

18 Watababe, A., Kurokawa, Y., Yoshimura, T., Kurihara, T., Soda, K., Esaki, N. (1999) *J. Biol. Chem. 274*, 4189–4194.

19 Watanabe, A., Yoshimura, T., Mikami, B., Esaki, N. (1999) *J. Biochem. 126*, 781–786.

20 Watanabe, A., Kurokawa, Y., Yoshimura, T., Esaki, N. (1999) *J. Biochem. 125*, 987–990.

21 Spies, M. A., Toney, M. D. (2003) *Biochemistry 42*, 5099–5107.

22 O'Leary, M. H. (1989) *Annu. Rev. Biochem. 58*, 377–401.

23 Schimerlik, M. I., Rife, J. E., Cleland, W. W. (1975) *Biochemistry 14*, 5347–5354.

24 Bahnson, B. J., Anderson, V. E. (1991) *Biochemistry 30*, 5894–5906.

25 Albery, W. J., Knowles, J. R. (1976) *Biochemistry 15*, 5631–5640.

26 Inagaki, K., Tanizawa, K., Tanaka, H., Soda, K. (1984) *Progr. Clin. Biol. Res. 144A*, 355–63.

27 Soda, K., Osumi, T. (1969) *Biochem. Biophys. Res. Commun. 35*, 363–368.

28 Buchanan, R. E., Gibbons, N. E. (1974) *Bergey's Manual of Determinative Bacteriology*, 8th edn., The Williams and Wilkins Co., Baltimore.

29 Inagaki, K., Tanizawa, K., Tanaka, H., Soda, K. (1987) *Agric. Biol. Chem. 51*, 173–180.
30 Nelson, K. E., Weinel, C., Paulsen, I. T., Dodson, R. J., Hilbert, H., Martins dos Santos, V. A., Fouts, D. E., Gill, S. R., Pop, M., Holmes, M., Brinkac, L., Beanan, M., DeBoy, R. T., Daugherty, S., Kolonay, J., Madupu, R., Nelson, W., White, O., Peterson, J., Khouri, H., Hance, I., Chris Lee, P., Holtzapple, E., Scanlan, D., Tran, K., Moazzez, A., Utterback, T., Rizzo, M., Lee, K., Kosack, D., Moestl, D., Wedler, H., Lauber, J., Stjepandic, D., Hoheisel, J., Straetz, M., Heim, S., Kiewitz, C., Eisen, J. A., Timmis, K. N., Dusterhoft, A., Tummler, B., Fraser, C. M. (2002) *Environ. Microbiol. 4*, 799–808.
31 Patrick, W. M., Weisner, J., Blackburn, J. M. (2002) *Chembiochem 3*, 789–792.
32 Reynolds, K., Martin, J., Shen, S. J., Esaki, N., Soda, K., Floss, H. G. (1991) *J. Basic Microbiol. 31*, 177–188.
33 Nagata, Y., Konno, R., Yasumura, Y., Akino, T. (1989) *Biochem. J. 257*, 291–292.
34 Nagata, Y., Akino, T., Ohno, K. (1989) *Experientia 45*, 330–332.
35 Mothet, J. P., Parent, A. T., Wolosker, H., Brady, R. O., Jr., Linden, D. J., Ferris, C. D., Rogawski, M. A., Snyder, S. H. (2000) *Proc. Natl. Acad. Sci. USA 97*, 4926–4931.
36 Schell, M. J., Molliver, M. E., Snyder, S. H. (1995) *Proc. Natl. Acad. Sci. USA 92*, 3948–3952.
37 Cook, S. P., Galve-Roperh, I., Martinez del Pozo, A., Rodriguez-Crespo, I. (2002) *J. Biol. Chem. 277*, 27782–27792.
38 De Miranda, J., Santoro, A., Engelender, S., Wolosker, H. (2000) *Gene 256*, 183–188.
39 Wolosker, H., Sheth, K. N., Takahashi, M., Mothet, J. P., Brady, R. O., Jr., Ferris, C. D., Snyder, S. H. (1999) *Proc. Natl. Acad. Sci. USA 96*, 721–725.
40 Wolosker, H., Blackshaw, S., Snyder, S. H. (1999) *Proc. Natl. Acad. Sci. USA 96*, 13409–13414.
41 Strisovsky, K., Jiraskova, J., Barinka, C., Majer, P., Rojas, C., Slusher, B. S., Konvalinka, J. (2003) *FEBS Lett. 535*, 44–48.
42 Uo, T., Yoshimura, T., Shimizu, S., Esaki, N. (1998) *Biochem. Biophys. Res. Commun. 246*, 31–34.
43 Fee, J. A., Hegeman, G. D., Kenyon, G. L. (1974) *Biochemistry 13*, 2533–2538.
44 Fee, J. A., Hegeman, G. D., Kenyon, G. L. (1974) *Biochemistry 13*, 2528–2532.
45 Petsko, G. A., Kenyon, G. L., Gerlt, J. A., Ringe, D., Kozarich, J. W. (1993) *Trends Biochem. Sci. 18*, 372–376.
46 Neidhart, D. J., Howell, P. L., Petsko, G. A., Powers, V. M., Li, R. S., Kenyon, G. L., Gerlt, J. A. (1991) *Biochemistry 30*, 9264–9273.
47 Hoffman, M. (1991) *Science 251*, 31–32.
48 Armstrong, R. N. (2000) *Biochemistry 39*, 13625–13632.
49 Babbitt, P. C., Gerlt, J. A. (1997) *J. Biol. Chem. 272*, 30591–30594.
50 Bernat, B. A., Laughlin, L. T., Armstrong, R. N. (1999) *Biochemistry 38*, 7462–7469.
51 Kallarakal, A. T., Mitra, B., Kozarich, J. W., Gerlt, J. A., Clifton, J. G., Petsko, G. A., Kenyon, G. L. (1995) *Biochemistry 34*, 2788–2797.
52 Kenyon, G. L., Hegeman, G. D. (1970) *Biochemistry 9*, 4036–4043.
53 Kenyon, G. L., Gerlt, J. A., Petsko, G. A., Kozarich, J. W. (1995) *Acc. Chem. Res. 28*, 178–186.
54 Lipmann, F. (1973) *Acc. Chem. Res. 6*, 361–367.
55 Gevers, W., Kleinkauf, H., Lipmann, F. (1968) *Proc. Natl. Acad. Sci. USA 60*, 269–276.
56 Gevers, W., Kleinkauf, H., Lipmann, F. (1969) *Proc. Natl. Acad. Sci. USA 63*, 1335–1342.
57 Yamada, M., Kurahashi, K. (1969) *J. Biochem. (Tokyo) 66*, 529–540.
58 Stindl, A., Keller, U. (1994) *Biochemistry 33*, 9358–9364.

59 Stindl, A., Keller, U. (1993) *J. Biol. Chem. 268*, 10612–10620.
60 van Liempt, H., von Dohren, H., Kleinkauf, H. (1989) *J. Biol. Chem. 264*, 3680–3684.
61 Banko, G., Demain, A. L., Wolfe, S. (1987) *J. Am. Chem. Soc. 109*, 2858–2860.
62 Baldwin, J. E., Bird, J. W., Field, R. A., O'Callaghan, N. M., Schofield, C. J., Willis, A. C. (1991) *J. Antibiot. (Tokyo) 44*, 241–248.
63 Jensen, S. E., Westlake, D. W., Wolfe, S. (1983) *Can. J. Microbiol. 29*, 1526–1531.
64 Zhang, J., Wolfe, S., Demain, A. L. (1992) *Biochem. J. 283*(3), 691–698.
65 Baldwin, J. E., Shiau, C. Y., Byford, M. F., Schofield, C. J. (1994) *Biochem. J. 301*(2), 367–372.
66 Mazumder, R., Sasakawa, T., Kaziro, Y., Ochoa, S. (1962) *J. Biol. Chem. 237*, 3065–3068.
67 Leadlay, P. F. (1981) *Biochem. J. 197*, 413–419.
68 Allen, S. H., Kellermeyer, R., Stjernholm, R., Jacobson, B., Wood, H. G. (1963) *J. Biol. Chem. 238*, 1637–1642.
69 Pfeifer, B. A., Admiraal, S. J., Gramajo, H., Cane, D. E., Khosla, C. (2001) *Science 291*, 1790–1792.
70 Marsden, A. F., Caffrey, P., Aparicio, J. F., Loughran, M. S., Staunton, J., Leadlay, P. F. (1994) *Science 263*, 378–380.
71 Stabler, S. P., Marcell, P. D., Allen, R. H. (1985) *Arch. Biochem. Biophys. 241*, 252–264.
72 McCarthy, A. A., Baker, H. M., Shewry, S. C., Patchett, M. L., Baker, E. N. (2001) *Structure (Camb) 9*, 637–646.
73 McCarthy, A. A., Baker, H. M., Shewry, S. C., Kagawa, T. F., Saafi, E., Patchett, M. L., Baker, E. N. (2001) *Acta Crystallogr. Sect. D 57*, 706–708.
74 Fuller, J. Q., Leadlay, P. F. (1983) *Biochem. J. 213*, 643–650.
75 Leadlay, P. F., Fuller, J. Q. (1983) *Biochem. J. 213*, 635–642.
76 Frey, P. A. (1996) *Faseb J. 10*, 461–470.
77 Deupree, J., Wood, W. A. (1975) *Methods Enzymol. 41*, 412–419.
78 Deupree, J. D., Wood, W. A. (1972) *J. Biol. Chem. 247*, 3093–3097.
79 Salo, W. L., Fossitt, D. D., Bevill, R. D., 3rd, Kirkwood, S., Wood, W. (1972) *J. Biol. Chem. 247*, 3098–3100.
80 Rudnick, G., Abeles, R. H. (1975) *Biochemistry 14*, 4515–4522.
81 Bateman, A., Birney, E., Cerruti, L., Durbin, R., Etwiller, L., Eddy, S. R., Griffiths-Jones, S., Howe, K. L., Marshall, M., Sonnhammer, E. L. (2002) *Nucleic Acids Res. 30*, 276–280.
82 Fisher, L. M., Albery, W. J., Knowles, J. R. (1986) *Biochemistry 25*, 2529–2537.
83 Northrop, D. B., Rebholz, K. L. (1994) *Anal. Biochem. 216*, 285–290.
84 Fisher, L. M., Albery, W. J., Knowles, J. R. (1986) *Biochemistry 25*, 2538–2542.
85 Fisher, L. M., Belasco, J. G., Bruice, T. W., Albery, W. J., Knowles, J. R. (1986) *Biochemistry 25*, 2543–2551.
86 Belasco, J. G., Bruice, T. W., Fisher, L. M., Albery, W. J., Knowles, J. R. (1986) *Biochemistry 25*, 2564–2571.
87 Belasco, J. G., Bruice, T. W., Albery, W. J., Knowles, J. R. (1986) *Biochemistry 25*, 2558–2564.
88 Belasco, J. G., Albery, W. J., Knowles, J. R. (1986) *Biochemistry 25*, 2552–2558.
89 Gallo, K. A., Knowles, J. R. (1993) *Biochemistry 32*, 3981–3990.
90 Yagasaki, M., Iwata, K., Ishino, S., Azuma, M., Ozaki, A. (1995) *Biosci. Biotechnol. Biochem. 59*, 610–614.
91 Nakajima, N., Tanizawa, K., Tanaka, H., Soda, K. (1986) *Agric. Biol. Chem. 50*, 2823–2830.
92 Nakajima, N., Tanizawa, K., Tanaka, H., Soda, K. (1988) *Agric. Biol. Chem. 52*, 3099–3104.
93 Hwang, K. Y., Cho, C. S., Kim, S. S., Sung, H. C., Yu, Y. G., Cho, Y. (1999) *Nat. Struct. Biol. 6*, 422–426.
94 Choi, S. Y., Esaki, N., Ashiuchi, M., Yoshimura, T., Soda, K. (1994) *Proc. Natl. Acad. Sci. USA 91*, 10144–10147.
95 Gallo, K. A., Tanner, M. E., Knowles, J. R. (1993) *Biochemistry 32*, 3991–3997.

96 Tanner, M. E., Gallo, K. A., Knowles, J. R. (1993) *Biochemistry 32*, 3998–4006.
97 Wiseman, J. S., Nichols, J. S. (1984) *J. Biol. Chem. 259*, 8907–8914.
98 Antia, M., Hoare, D. S., Work, E. (1957) *Biochem. J. 65*, 448–459.
99 Cirilli, M., Zheng, R., Scapin, G., Blanchard, J. S. (1998) *Biochemistry 37*, 16452–16458.
100 Koo, C. W., Blanchard, J. S. (1999) *Biochemistry 38*, 4416–4422.
101 Kuzmic, P. (1996) *Anal. Biochem. 237*, 260–273.
102 Davis, L., Lee, N., Glaser, L. (1972) *J. Biol. Chem. 247*, 5862–5866.
103 McDonough, M. W., Wood, W. A. (1961) *J. Biol. Chem. 236*, 1220–1224.
104 Melo, A., Glaser, L. (1968) *J. Biol. Chem. 243*, 1475–1478.
105 Glaser, L., Zarkowsky, H., Ward, L. (1972) *Methods Enzymol. 28*, 446–454.
106 Kawamura, T., Ishimoto, N., Ito, E. (1979) *J. Biol. Chem. 254*, 8457–8465.
107 Kawamura, T., Kimura, M., Yamamori, S., Ito, E. (1978) *J. Biol. Chem. 253*, 3595–3601.
108 Salo, W. L. (1976) *Biochim. Biophys. Acta 452*, 625–628.
109 Sala, R. E., Morgan, P. M., Tanner, M. E. (1996) *J. Am. Chem. Soc. 118*, 3033–3034.

8
Computer Simulations of Proton Transfer in Proteins and Solutions

Sonja Braun-Sand, Mats H. M. Olsson, Janez Mavri, and Arieh Warshel *

8.1
Introduction

Proton transfer (PT) reactions play a major role in many enzymatic and other biological processes. Thus it is important to quantify the nature of such reactions by reliable computer modeling approaches. This chapter will review the advances in the field and present a unified way of modeling and analyzing PT reactions in proteins and solutions. We will start by considering the current options for reliable simulations. We will focus on the empirical valence bond (EVB) approach that has been used in studies that paved the way for the modern microscopically based treatments of PT in solutions and proteins (e.g. Refs. [1, 2]). It will be argued that the EVB presents currently the most effective strategy for exploring and modeling different aspects of such processes, ranging from hydrogen bonding to PTs in enzymatic reactions and to proton translocations along a chain of donors and acceptors. We will demonstrate the effectiveness of the EVB in quantifying the trend in PT reactions and in analyzing linear free energy relationships (LFER). We will also clarify misunderstandings about the nature of LFER that involve PT reactions. The issue of proton translocations (PTR) along conduction chains will be discussed, considering some misconceptions about the role of proton wires and the orientations of the neutral water molecules. Finally we will address the role of dynamics and nuclear quantum mechanical effects in PT in enzyme catalysis.

8.2
Simulating PT Reactions by the EVB and other QM/MM Methods

The rates of proton transfer reaction in solutions and proteins are determined by the corresponding rate constants (e.g. Ref. [3]).

* Corresponding author.

Hydrogen-Transfer Reactions. Edited by J. T. Hynes, J. P. Klinman, H.-H. Limbach, and R. L. Schowen

ISBN: 978-3-527-30777-7

$$k = \kappa\left(\frac{1}{2}\langle|\dot{x}|\rangle_{TS}/\Delta x^{\ddagger}\right)\exp[-\Delta G^{\ddagger}\beta] \quad (8.1)$$

Where κ is the transmission factor, $\langle|\dot{x}|\rangle_{TS}$ is the average of the absolute value of the velocity along the reaction coordinate at the transition state (TS), and $\beta = 1/k_B T$ (where k_B is the Boltzmann constant and T the absolute temperature). The term $\Delta G^{\ddagger}$ designates the multidimensional activation free energy that expresses the probability that the system will be in the TS region. The free energy reflects enthalpic and entropic contributions and also includes nonequilibrium solvation effects [4] and, as will be shown below, nuclear quantum mechanical effects. It is also useful to comment here on the common description of the rate constant as

$$k = A \cdot \exp[-\Delta E^{\ddagger}\beta] \quad (8.2)$$

This Arrhenius expression is of course useful for experimental analysis, but it may lead to unnecessary confusion about the factors that determine the rate constant. That is, as is now recognized by the chemical physics community [5–7], Eq. (8.1) provides an accurate description of the rate constant when all the dynamical effects are cast into the transmission factor and all the probabilistic effects are expressed by $\Delta G^{\ddagger}$. Of course, $\Delta G^{\ddagger}$ includes the activation entropy while the use of Eq. (8.2) places this crucial effect in A and makes it hard to separate the dynamical and probabilistic factors.

With the above background, we start the discussion of the evaluation of $\Delta G^{\ddagger}$, which in fact is the most important step. We would also like to emphasize that the ability to calculate $\Delta G^{\ddagger}$ (and the corresponding free energy profiles) for enzyme reactions and the corresponding reference solution reaction is crucial for any attempt to obtain a quantitative understanding of enzyme reactions.

The common prescription of obtaining potential surfaces for chemical reactions involves the use of quantum chemical computational approaches, and such approaches have become quite effective in treating small molecules in the gas phase (e.g. Ref. [8]). However, here we are interested in chemical reactions in very large systems, which cannot be explored at present by *ab initio* methods. Similarly, molecular mechanics simulations (e.g. Ref. [9]) that have been proven to be very effective in exploring protein configurational space cannot be used to describe bond breaking and bond making reactions in proteins or solutions. The generic solution to the above problem has been provided by the development of the hybrid quantum mechanics/molecular mechanics (QM/MM) approach [10]. This approach divides the simulation system (for example, the enzyme/substrate complex) into two regions. The inner region, region I, contains the reacting fragments which are represented quantum mechanically. The surrounding protein/solvent region, region II, is represented by a molecular mechanics force field.

Molecular orbital (MO) QM/MM methods are now widely used in studies of complex systems in general, and enzymatic reactions in particular, and we can only mention several works (for example, Refs. [11–22]). Despite these advances, we are not yet at the stage where one can use MO-QM/MM approaches in fully quantitative studies of enzyme catalysis. The major problem is associated with the fact that a quantitative evaluation of the potential surfaces for the reacting fragment should involve *ab initio* electronic structure calculations, and such calculations are too expensive to allow for the configurational averaging needed for proper free energy calculations. Specialized approaches can help one move toward *ab initio* QM/MM free energy calculations (see Ref. [23]), but even these approaches are still in a development stage. Fortunately, one can use approaches that are calibrated on the energetics of the reference solution reaction to obtain reliable results with semiempirical QM/MM studies, and the most effective and reliable way of doing this is the EVB method described below.

During our search for reliable methods for studies of enzymatic reactions it became apparent that, in studies of chemical reactions, it is more physical to calibrate surfaces that reflect bond properties (that is, valence bond-based, VB, surfaces) than to calibrate surfaces that reflect atomic properties (for example, MO-based surfaces). Furthermore, it appears to be very advantageous to force the potential surfaces to reproduce the experimental results of the broken fragments at infinite separation in solution. This can be effectively accomplished with the VB picture. The resulting empirical valence bond (EVB) method has been discussed extensively elsewhere [3, 24], but its main features will be outlined below, because it provides the most direct microscopic connection to PT processes.

The EVB is a QM/MM method that describes reactions by mixing resonance states (or more precisely diabatic states) that correspond to valence-bond (VB) structures, which describe the reactant, intermediate (or intermediates), and product states. The potential energies of these diabatic states are represented by classical MM force fields of the form:

$$\varepsilon_i = \alpha^i_{\text{gas}} + U^i_{\text{intra}}(\mathbf{R}, \mathbf{Q}) + U^i_{\text{Ss}}(\mathbf{R}, \mathbf{Q}, \mathbf{r}, \mathbf{q}) + U_{\text{ss}}(\mathbf{r}, \mathbf{q}) \tag{8.3}$$

Here $\mathbf{R}$ and $\mathbf{Q}$ represent the atomic coordinates and charges of the diabatic states, and $\mathbf{r}$ and $\mathbf{q}$ are those of the surrounding protein and solvent. α^i_{gas} is the gas-phase energy of the ith diabatic state (where all the fragments are taken to be at infinite separation), $U_{\text{intra}}(\mathbf{R}, \mathbf{Q})$ is the intramolecular potential of the solute system (relative to its minimum); $U_{\text{Ss}}(\mathbf{R}, \mathbf{Q}, \mathbf{r}, \mathbf{q})$ represents the interaction between the solute (S) atoms and the surrounding (s) solvent and protein atoms. $U_{\text{ss}}(\mathbf{r}, \mathbf{q})$ represents the potential energy of the protein/solvent system ("ss" designates surrounding-surrounding). The ε_i of Eq. (8.3) forms the diagonal elements of the EVB Hamiltonian (H_{ii}). The off-diagonal elements of the Hamiltonian, H_{ij}, are represented typically by simple exponential functions of the distances between the reacting atoms. The H_{ij} elements are assumed to be the same in the gas phase, in solutions and in proteins. Since one may wonder about this assumption we note the following; (i) the assumption of constant H_{ij} is in fact the main reason for the empirical success of LFER approaches that correlate the changes of the diabatic energies with the activation barrier, and (ii) the validity of the assumption of a relatively small environmental effect on H_{ij} has been established in recent constraint DFT studies [25], which follow the prescription proposed by Eqs. (17)–(18) of Ref. [26] and obtained a very small change in the *ab initio* effective H_{ij} for S_N2 reaction in solution and in the gas phase.

The ground state energy, E_g, is obtained by solving

$$\boldsymbol{H}_{\mathrm{EVB}}\boldsymbol{C}_{\mathrm{g}} = E_{\mathrm{g}}\boldsymbol{C}_{\mathrm{g}} \tag{8.4}$$

Here, C_g is the ground state eigenvector and E_g provides the EVB potential surface.

For example, we can describe the reaction

$$\mathrm{XH} + \mathrm{Y} \rightarrow \mathrm{X}^- + \mathrm{HY}^+ \tag{8.5}$$

by three resonance structures

$$\begin{aligned}
\Psi_a &= [X- \quad H \qquad Y]\ \phi_a \\
\Psi_b &= [X^- \quad H- \quad Y^+]\phi_b \\
\Psi_c &= [X^- \quad H^+ \quad Y]\ \phi_c
\end{aligned} \tag{8.6}$$

where the ϕ_s are the wave functions for the solvent and for the solute electrons, which are not included in the XH Y system. For simplicity, it is convenient to treat the high energy state, ψ_c, by a perturbation treatment restricting ourselves to the two states ψ_a and ψ_b. The potential surface for the two-state VB model is obtained by solving the secular equation

$$\begin{vmatrix} H_{aa} - E & H_{ba} - S_{ba}E \\ H_{ab} - S_{ab}E & H_{bb} - E \end{vmatrix} \tag{8.7}$$

where the matrix elements of $\mathbf{H}$ can be obtained by performing gas-phase *ab initio* calculations or represented by semiempirical analytical potential functions (fitted to the potential surface and charge distribution obtained from experimental information and/or *ab initio* calculations). The solvent can then be incorporated in the Hamiltonian of the system by using the expression

$$\begin{aligned}
\varepsilon_a &= H_{aa} = H_{aa}^0 + U_{\mathrm{Ss}}^a + U_{\mathrm{ss}} \\
\varepsilon_b &= H_{bb} = H_{bb}^0 + U_{\mathrm{Ss}}^b + U_{\mathrm{ss}}
\end{aligned} \tag{8.8}$$

where U_{Ss} is the interaction between the solute (S) charges and the surrounding solvent (s), and U_{ss} is the solvent–solvent interaction. The overlap integral, S_{ab}, is usually absorbed into the semiempirical H_{ab} and the solute-solvent interactions are described by analytical potential functions as discussed in Ref. [27]. The matrix elements for the isolated solute can be represented by

$$
\begin{aligned}
H^0_{aa} &= \varepsilon^0_a = \Delta M(b_1) + U^{(a)}_{\mathrm{nb}} + \sum_{m(a)} \frac{1}{2} K^{(m)}_b (b_m - b_0)^2 + \sum_{m'(a)} \frac{1}{2} K^{(m')}_\theta (\theta_{m'} - \theta_0)^2 \\
H^0_{bb} &= \varepsilon^0_b = \Delta M(b_2) + U^{(b)}_{\mathrm{nb}} - 332/r_2 + \alpha^{(b)} \\
&\quad + \sum_{m(b)} \frac{1}{2} K^{(m)}_b (b_m - b_0)^2 + \sum_{m'(b)} \frac{1}{2} K^{(m')}_\theta (\theta_{m'} - \theta_0)^2 \\
H_{ab} &= A_{ab} \exp\{-\mu(r_2 - r^0_2)\}
\end{aligned}
\tag{8.9}
$$

where b_1, b_3 and r_2 are, respectively, the X–H, H–Y, and X–Y distances, ΔM is the value of the Morse potential for the indicated bond relative to its minimum value, the quadratic bonding terms describe all bonds in the solute system, which are connected to X or Y and the quadratic angle bonding term describes all angles defined by the given covalent bonding arrangement that includes the X, Y, or H atoms. U_{nb} is the nonbonded interaction between nonbonded atoms in the αth resonance structure. These interaction terms are represented by either $Ae^{-\mu r}$ or 6-12 van der Waals potential functions. The parameter $\alpha^{(b)}$ is the energy difference between ψ_a and ψ_b with the fragments at infinite separation in the gas phase. The off-diagonal term H_{ab} can be evaluated by the three-state EVB approach of Ref. [28] and fitted to the two-state model. Note that the same two-state model can be fitted to gas-phase *ab initio* calculations.

The EVB methodology provides a computationally inexpensive Born–Oppenheimer surface suitable for describing chemical reaction in an enzyme or in solution. Running such MD trajectories on the EVB surface of the reactant state can (in principle) provide the free energy function, Δg, that is needed to calculate the activation energy, $\Delta g^{\ddagger}$. However, since trajectories on the reactant surface will reach the transition state only rarely, it is usually necessary to run a series of trajectories on potential surfaces that gradually drive the system from the reactant to the product state [3]. The EVB approach accomplishes this by changing the system adiabatically from one diabatic state to another. In the simple case of two diabatic states, this "mapping" potential, ε_m, can be written as a linear combination of the reactant and product potentials, ε_1 and ε_2:

$$
\varepsilon_m = (1 - \eta_m)\varepsilon_1 + \eta_m \varepsilon_2 \quad (0 \le \eta_m \le 1) \tag{8.10}
$$

When η_m is changed from 0 to 1 in $n+1$ fixed increments ($\eta_m = 0/n, 1/n, 2/n, \ldots, n/n$), potentials with one or more of the intermediate values of η_m will force the system to fluctuate near the TS.

The free energy, ΔG_m, associated with changing η_m from 0 to m/n is evaluated by the well known free energy perturbation (FEP) procedure described elsewhere (see, for example, Ref. [3]). However, after obtaining ΔG_m we still need to obtain the free energy that corresponds to the adiabatic ground state surface (the E_g of Eq. (8.4)) along the reaction coordinate, x. This free energy (referred to as a "free energy functional") is obtained by the FEP-umbrella sampling (FEP/US) method [3, 27], which can be written as

$$\Delta g(x') = \Delta G_m - \beta^{-1} \ln\langle\delta(x - x') \exp[-\beta(E_g(x) - \varepsilon_m(x))]\rangle_{\varepsilon_m} \tag{8.11}$$

where ε_m is the mapping potential that keeps x in the region of x'. If the changes in ε_m are sufficiently gradual, the free energy functionals, $\Delta g(x')$, obtained with several values of m overlap over a range of x', and patching together the full set of $\Delta g(x')$ gives the complete free energy curve for the reaction. In choosing the general reaction coordinate, x, we note that the regular geometrical coordinate, used in gas-phase studies, cannot provide a practical way to model the multidimensional reaction coordinate of reactions in solution and protein. In modeling such processes, it is crucial to capture the effect of the solvent polarization and probably the best way to describe this effect microscopically is to follow our early treatment [1, 3] and to use the electronic energy gap as the general reaction coordinate ($x = \varepsilon_2 - \varepsilon_1$).

The FEP/US approach may also be used to obtain the free energy functional of the isolated diabatic states. For example, the diabatic free energy, Δg_1, of the reactant state can be calculated as

$$\Delta g_1(x') = \Delta G_m - \beta^{-1} \ln\langle\delta(x - x') \exp[-\beta(\varepsilon_1(x) - \varepsilon_m(x))]\rangle_{\varepsilon_m} \tag{8.12}$$

The diabatic free energy profiles of the reactant and product states provide the microscopic equivalent of the Marcus' parabolas [29, 30].

The EVB method satisfies some of the main requirements for reliable studies of enzymatic reactions. Among the obvious advantages of the EVB approach is the facilitation of proper configurational sampling and converging free energy calculations. This includes the inherent ability to evaluate nonequilibrium solvation effects [4]. Another important feature of the EVB method is the ability to capture correctly the linear relationship between activation free energies and reaction energies (LFER) observed in many important reactions (for example, Ref. [3]). Furthermore, the EVB benefits from the aforementioned ability to treat consistently and conveniently the solute–solvent coupling. This feature is essential not only in allowing one to properly model charge-separation reactions, but also in allowing a reliable and convenient calibration. Calibrating EVB surfaces using *ab initio* calculations was found to provide quite reliable potential surfaces.

The seemingly simple appearance of the EVB method may have led to the initial impression that this is an oversimplified qualitative model, rather than a powerful quantitative approach. However, the model has been eventually widely adopted by other groups as a general model for studies of reactions in large molecules and in condensed phase (for example, Refs. [31–34]). Several very closely related versions have been put forward with basically the same ingredients as in the EVB method (see Refs. [35, 36]). It might also be useful to clarify that our EVB approach included calibration on *ab initio* surface from quite an early stage [27] so that this element is not a new development. Furthermore, although early works (e.g. Refs. [37, 38]) have some relationship to the EVB, they were merely combinations of VB and MM treatments and thus miss the crucial QM/MM coupling obtained by adding the MM description of each state in the diagonal EVB Hamiltonian (Eq.

(8.12)). It is this coupling idea that made the EVB such a powerful way of modeling reactions in condensed phases.

Since we will be dealing with proton transport processes, it might be useful to clarify that the EVB and the so-called MS-EVB [32, 39] (that was so effective in studies of proton transport in water) are more or less identical. More specifically, the so-called MS-EVB includes typically 6 EVB states in the solute quantum mechanical (QM) region and the location of this QM region changes if the proton moves. The QM region is surrounded by classical water molecules (the molecular mechanics (MM)), whose effect is sometimes included inconsistently by solvating the charges of the gas-phase QM region (this leads to inconsistent QM/MM coupling with the solute charges as explained in, for example, Refs. [4, 9]. More recently, the coupling was introduced consistently by adding the interaction with the MM water in the diagonal solute Hamiltonian. Now our EVB studies are performed repeatedly with multi-state treatment (for example, 5 states in Ref. [40]) and this has always been done with a consistent coupling to the MM region. Thus, the only difference that we can find between the two versions is that our EVB studies did not change the identity of the atoms in the QM region during simulations of individual chemical steps (this was done only while considering different steps). Such treatment provides the optimal strategy when one deals with processes in proteins that involve relatively high barriers, rather than with low barrier transport processes (so that the identity of the reacting region has not changed during the simulations). Also note that the MS-EVB simulations in proteins, where we have a limited number of quantum sites, do not have to change the QM region (for example, Ref. [41]) during the simulations. Thus we conclude that the EVB and MS-EVB are identical methods, although we appreciate the elegant treatment of changing the position of the QM region during simulations, which is a very useful advance in EVB treatments of processes with a very low activation barrier.

8.3 Simulating the Fluctuations of the Environment and Nuclear Quantum Mechanical Effects

The EVB provides an effective way to explore the effect of the fluctuations of the environment on PT reactions. That is, the electrostatic potential from the fluctuating polar environment interacts with the charge distribution of each resonance structure and thus the fluctuations of the environment are directly reflected in the time dependence of the EVB Hamiltonian. This point emerged from our early studies [1, 2, 4, 42] and is illustrated in Figs. 8.1 and 8.2. Figure 8.1 shows how the fluctuations of the field from the environment change the energy of the ionic state, and thus the potential for a PT at a given fixed environment. The same effect is illustrated in Fig. 8.2 where we consider the time-dependent energetics of the two EVB diabatic states approximately (the front and back panels) in the PT step in the reaction of lysozyme. The figure also describes the time dependence of the adiabatic ground state and the corresponding barrier ($\Delta E^{\neq}$) for a PT process. As

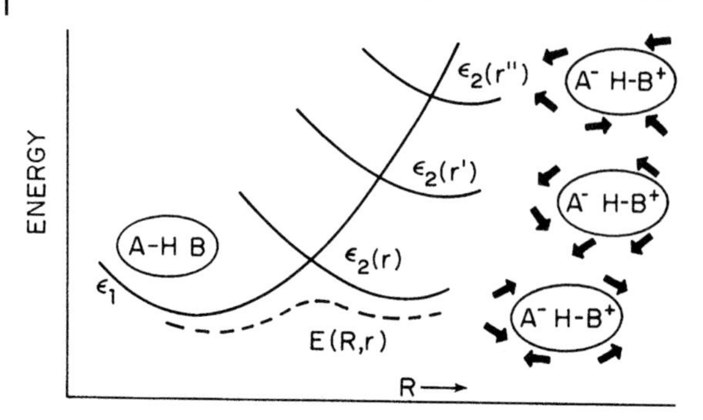

Figure 8.1. Energetics of the proton transfer between an acid (A) and a base (B) by valence bond resonance structures. The reaction is described in terms of a covalent resonance structure (A–H B) and an ionic resonance structure (A^- $H\text{–}B^+$). The energies of the valence bond structures are as given in the text and depend on the coordinates of the reacting atoms, **R**, and the coordinates of the rest of the system, **r**. Only the ionic resonance structure interacts strongly with the dipoles and charges of the surrounding solvent or protein cage. Thus, the energy of the ionic structure, ε_2, changes strongly with fluctuation of the surrounding dipoles. The reaction potential surface, $E(\mathbf{R}, \mathbf{r})$, obtained by mixing of the relevant resonance forms is also shown.

seen from the figure, the actual transfer would occur when the product state is stabilized and $\Delta E^{\neq}$ is reduced. As pointed out in our early studies, the fluctuations of the energy gap between the back and front panels of Fig. 8.2 tells us when $\Delta E^{\neq}$ will be reduced (the same point was adopted in Ref. [43], overlooking its origin). Furthermore, the fluctuations described in Fig. 8.1 can be used to evaluate dynamical effects by considering the autocorrelation of the time-dependent gap between the energies of the reactant and product states (see discussion in Ref. [44]).

The fluctuations of the electrostatic energy gap can also provide an interesting insight into nuclear quantum mechanical (NQM) effects. That is, we can use an approach that is formally similar to our previous treatment of electron transfer (ET) reactions in polar solvents [1] where we considered ET between the solute vibronic channels (for example, Ref. [1]). Our starting point is the overall rate constant

$$k_{ab} = \sum_{mm'} k_{am,bm'} \exp\{-E_{am}\beta\} \Big/ \sum_{m} \exp\{-E_{am}\beta\} \tag{8.13}$$

where $\beta = 1/(k_B T)$ (with k_B the Boltzmann constant) and E_{am} is the energy of the mth vibronic level of state a. By Eq. (8.13) we assume that the reactant well vibrational states are populated according to the Boltzmann distribution. The individual vibronic rate constant, $k_{am,bm'}$, is evaluated by monitoring the energy difference between E_{am} and E_{bm} as a function of the fluctuations of the rest of the molecule and therefore as a function of time. The time-dependent energy gap can be used to

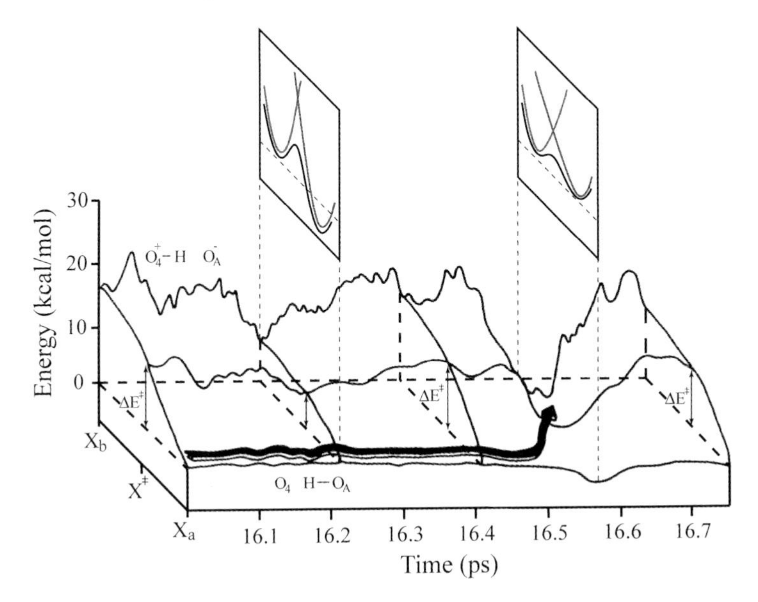

Figure 8.2. The time-dependent barrier for proton transfer from a carboxylic oxygen of Glu35 (O_A) to the glycosidic oxygen of the substrate (O_4) in lysozyme. The solid lines represent the potential energies of the reactant state ($O_4 \cdots H{-}O_A$), the product state ($O_4^- \cdots H{-}O_A^+$), and the TS, with the zero of the energy scale defined as the mean potential energy of the reactant state. The energy of the TS is also shown. The energies were calculated during an MD trajectory on the reactant surface. The effective activation barrier ($\Delta E^{\neq}$) is determined mainly by fluctuations of the electrostatic energy of the product state, in which the EVB structure has a large dipole moment. (From Ref. [2] in part).

evaluate the probability of surface crossing between the two states by adopting a semiclassical trajectory approach [45] to rate processes in condensed phases [1, 46]. This approach can be best understood and formulated by considering Fig. 8.3 and asking what is the probability that a molecule in state am will cross to state bm'. Treating the fluctuation of the vibronic state classically, one finds that the time-dependent coefficient for being in state $\psi_{bm'}$, while starting from state ψ_{am}, is given by

$$\dot{C}_{am,bm'}(t) = -(\mathrm{i}/\hbar) H_{ab} S_{m,m'} \exp\left\{(\mathrm{i}/\hbar)\int^{t} \Delta\varepsilon_{bm',am}\,\mathrm{d}t'\right\} \tag{8.14}$$

where the $S_{m,m'}$ is the Franck–Condon factor for transition from m to m' and H_{ab} is the off-diagonal electronic matrix element of the EVB Hamiltonian. As illustrated in Fig. 8.2, the energy gap, $\Delta\varepsilon$, is given by

$$\Delta\varepsilon_{bm',am} = \left((\varepsilon_b - \varepsilon_a) + \hbar\sum_r \omega_r(m'_r - m_r)\right) \tag{8.15}$$

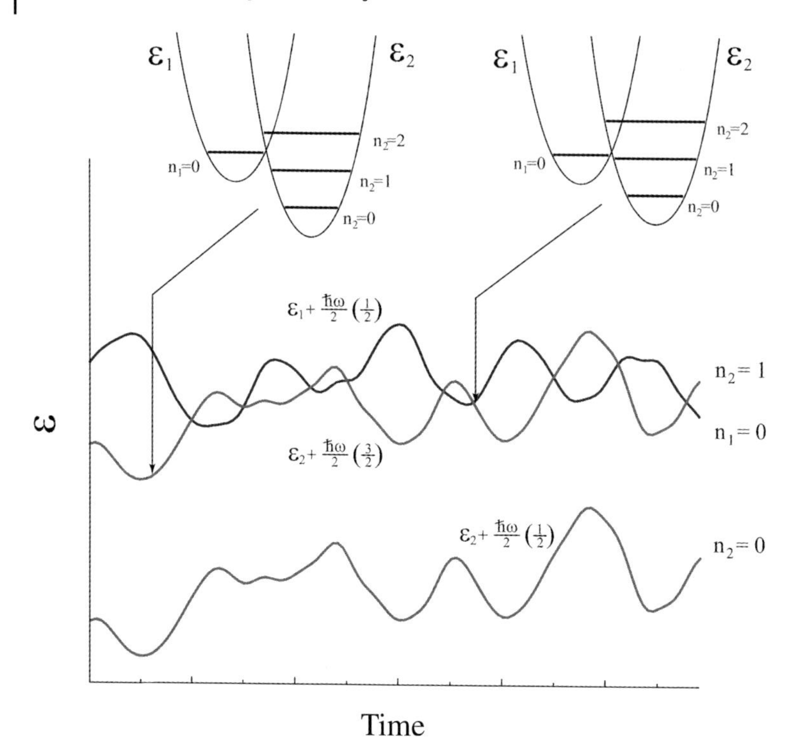

Figure 8.3. A semiclassical vibronic treatment of proton transfer. This model, which is valid only for small H_{12}, treats the carbon–proton stretching vibration quantum mechanically and the rest of the system classically. In this way, we monitor the energy gap between the vibronic states $\varepsilon_1 + h\omega_H/2(n_1 + 1/2)$ and $\varepsilon_2 + h\omega_H/(n_2 + 1/2)$ for trajectories of the system with a fixed X–H bond length (see Ref. [1] for a related treatment). The figure depicts the time dependence of ε_1, ε_2 and ε_2 plus single and double excitations of the X–H bond and also provides the energy levels at two points on the trajectory. A semiclassical surface hopping treatment of the crossing probability between the vibronic states, due to the fluctuating energy gap, leads to Eq. (8.21) (see Ref. [42]).

where ω_r and m_r are the frequency and quantum number of the rth mode. The corresponding rate constant is given by [46]

$$\begin{aligned} k_{am,bm'} &= \lim_{\tau\to\infty}(1/\tau|C_{bm',bm}(\tau)|) \\ &= |H_{ab}/\hbar|^2 \sum S^2_{mm'} \left| \int^t \mathrm{d}t \exp\left\{ \mathrm{i}\omega'_{ba}t + i\sum \omega_r(m'_r - m_r)t \right.\right. \\ &\qquad \left.\left. + (\mathrm{i}/\hbar)\int_0^t \mathrm{d}t'_a u(t'_a) \right\} \right|^2 \Big/ \tau \end{aligned} \tag{8.16}$$

where u is the electronic energy gap relative to its average values, given by

$$u = \varepsilon_b - \varepsilon_a - \langle \Delta\varepsilon_{ba} \rangle_a \tag{8.17}$$

and $\langle\ \rangle_a$ designates an average obtained over the fluctuations around the minimum of state a. The above rate constant can be treated using a cumulant expansion (see Ref. [46]) giving

$$k_{am,bm'} = |H_{ab}/\hbar|^2 \sum S^2_{mm'} \int_{-\infty}^{\infty} \exp[(i/\hbar)\langle\Delta\varepsilon_{bm',am}\rangle + \gamma(t)\,dt]$$

$$\gamma(t) = -(i/\hbar)^2 \int (t - t')\langle\Delta\varepsilon(0)\Delta\varepsilon(t')\rangle_a\,dt' \tag{8.18}$$

In the high temperature limit one obtains [42, 46]

$$k_{am,bm'} = |H_{ab}S_{mm'}/\hbar|^2(\pi\hbar^2/k_B T\lambda)^{1/2} \exp\{-\Delta g^{\neq}\beta\} \tag{8.19}$$

where λ is the "solvent reorganization energy" defined by

$$\lambda = \langle\Delta\varepsilon_{ba}\rangle_a - \Delta G_0 \tag{8.20}$$

Basically, this expression reflects the probability of vibronic transition from the reactant well to the product well (as determined by the vibrational overlap integrals (the $S_{mn'}$) modulated by the chance that $\Delta\varepsilon$ will be zero. This chance is determined by the activation free energy, $\Delta g^{\neq}$, whose value can be approximated by the activation free energy, $\Delta g^{\neq}$, which can be approximated by

$$\Delta g^{\neq}_{mm'} \approx \left[\Delta G^0 + \sum_r \hbar\omega r(m'_r - m_r) + \lambda\right]^2 \Big/ 4\lambda \tag{8.21}$$

This relationship is only applicable if the system can be described by the linear response approximation (see Ref. [46]), but this *does not* require that the system will be harmonic. The above vibronic treatment is similar to the expression developed by Kuznetsov and Ulstrup [47]. However, the treatment that leads to Eq. (8.21), which was developed by Warshel and coworkers [1, 42, 46], is based on a more microscopic approach and leads to much more consistent treatment of $\Delta g^{\neq}$ (see also below) where we can use rigorously ΔG^0 rather than ΔE. Furthermore, our dispersed polaron (spin boson) treatment [46] of Eq. (8.18) and if needed Eq. (8.21) gives a clear connection between the spectral distribution of the solvent fluctuations and the low temperature limit of Eq. (8.18). It is also useful to note that Borgis and Hynes [48] and Antoniou and Schwartz [49] have used a similar treatment but considered only the lowest vibrational levels of the proton.

Before considering the very serious limitation of the vibronic treatment, it is useful to comment about $\Delta g^{\neq}_{am,bm}$. That is, when the linear response approximation is not valid, we can obtain a more accurate estimate of $\Delta g^{\neq}$ using a free energy perturbation method. That is, $\Delta g^{\neq}_{am,bm}$ determines the probability that ε_{am} and $\varepsilon_{bm'}$ intersect by [46]

$$\Delta g^{\neq}_{mm'} \approx -k_B T \ln(\bar{n}^{\neq}_{mm'}/\bar{n}^{0}_{mm'}) \tag{8.22}$$

where $\bar{n}^{\neq}_{mm'}$ is the number of times the energy gap is between zero and ΔC and $\bar{n}^{0}_{mm'}$ is the number of times the energy gap is at the value $\Delta\varepsilon^0 \pm \Delta C$ that gives the largest value for $\bar{n}_{mm'}$ and C is an energy bin width. This probability can be determined in a direct way using molecular dynamics by running trajectories on the reactant surface with fixed A–H bond distance and monitoring the energy gap $\varepsilon_{bm'} - \varepsilon_{am}$ and counting the times this gap is zero. However, a direct calculation of such surface intersection events might require an extremely long computer time. Instead we can use a free energy perturbation (FEP) approach, propagating trajectories over a mapping potential of the form of Eq. (10)

$$\varepsilon(\eta_j) = \varepsilon_{am}(1 - \eta_j) + \varepsilon_{bm'}\eta_j \tag{8.23}$$

where the change of η from zero to one moves the system from ε_{am} to $\varepsilon_{bm'}$. The vibronic free energy function can be determined in analogy to Eq. (8.12) by

$$\begin{aligned} P(\Delta\varepsilon')_{\alpha m} &= \exp\{-\Delta g_{\alpha m}(\Delta\varepsilon_{am,bm'})\beta\} \\ &\approx \exp\{-\Delta G(\lambda_j)\beta\}\langle\delta(\Delta\varepsilon - \Delta\varepsilon')\exp\{-(\varepsilon_{am}(\Delta\varepsilon) - \varepsilon_j(\Delta\varepsilon))\}\beta\rangle_j \end{aligned} \tag{8.24}$$

Unfortunately, the use of the above vibronic treatment is valid only in the diabatic limit when $H_{ab}S^2_{mn}$ is sufficiently small. Now, H_{ab} in the case of PT processes is far too large to allow for a diabatic treatment. Thus, the question is, what is the magnitude of $H_{ab}S^2_{mn}$. Here we note that for $0 \to 0$ transitions, S_{mn} may be quite small since it is given by $S_{00} = \exp\{-\Delta^2/4\}$ where Δ is the dimensionless origin shift for the proton transfer ($\Delta_H \approx 6.5 \times (\Delta r/0.6)(\omega/3000)^{1/2}$ when Δr is given in Å, ω in cm^{-1}, Δ in dimensionless units, and this expression is based on the fact that at 0.6 Å, Δ is 6.5). However, S_{mn} approaches unity for hot transitions. At any rate, if the largest contribution to the rate constant comes from S_{00}, we may use Eq. (8.16) as a guide. Here, however, we face another problem, that is, the magnitude of the parameters in Eq. (8.16) is far from obvious. First, if we consider a colinear PT and treat all the coordinates except the X–H stretching frequency, then we have to deal with a large intramolecular reorganization energy. Second, the effective frequency for the X–H stretch can be very different than the typical frequency of about 3000 cm^{-1} once the X···Y distance starts to be shorter than 3.2 Å. In this range H_{12} starts to affect in a drastic way the ground state curvature (Fig. 8.4). Here one can use the idea introduced by Warshel and Chu [42] and modify the diabatic potential to make it close to the adiabatic potential. As long as the main contribution to the rate constant comes from the S_{00} term it is reasonable to represent this effect by using

$$\omega(\Delta) = \omega_0(1 - (4H_{12}/\lambda_H)^2\alpha) \tag{8.25}$$

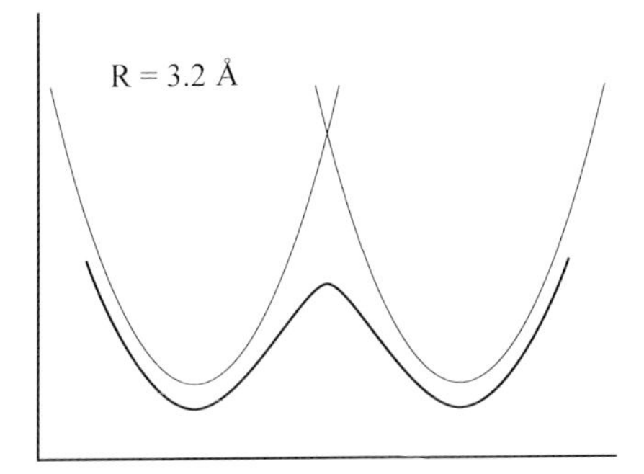

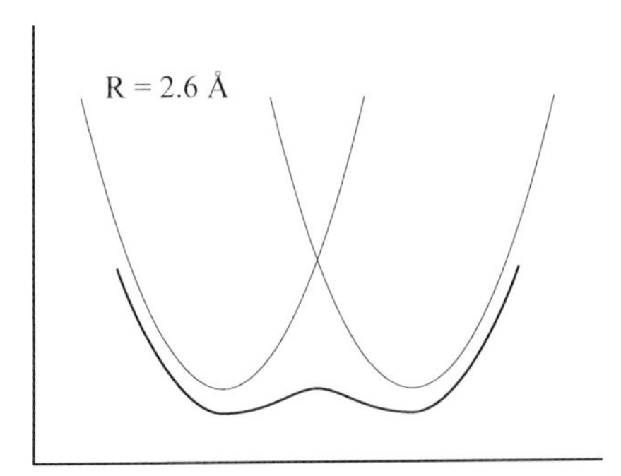

Figure 8.4. Dependence of adiabatic potential on the X···Y distance.

where α is approximately 0.5. It may also be useful to try to account for the complications due to the fact that the intramolecular solute coordinate contributes to the apparent reorganization energy and are also coupled to the PT coordinate. A part of this problem can be reduced by integrating the vibronic rate constant over the "soft" coordinates, and in particular the X···Y distance. This can be done by writing

$$\bar{k}_{ab} = \int k_{ab}(\Delta) \exp\{-U(R(\Delta)\beta\} \, d\Delta \tag{8.26}$$

The effectiveness of the above treatment will be examined in Section 8.7, but even if it can guide us with regards to the general trend it cannot be used in a phenomenological way to estimate actual molecular properties. Perhaps for this reason, microscopic estimates [50] of the parameters in Eq. (8.19) for the reaction catalyzed by lipoxygenase were found to be very different than those obtained by fitting Eq. (8.19) to the observed isotope effects [51–53].

The treatment of NQM effects can be accomplished on a much more quantitative level by including the adiabatic limit and modifying the centroid path integral approach [54–56]. The centroid path integral represents the unifying approach, which is valid both in the adiabatic and diabatic limits. This is done in a way that allows us to use classical trajectories as a convenient and effective reference for the corresponding centroid calculations. This QCP approach [57, 58] will be described briefly below.

In the QCP approach, the nuclear quantum mechanical rate constant is expressed as

$$k_{qm} = F_{qm} \frac{k_B T}{h} \exp(-\Delta g^{\neq}_{qm}) \tag{8.27}$$

where F_{qm}, k_B, T, and h are, respectively, the transmission factor, Boltzmann's constant, the temperature, Planck's constant, and $\beta = 1/k_B T$. The quantum mechanical activation barrier, $\Delta g^{\neq}_{qm}$, includes almost all the nuclear quantum mechanical effects, whereas only small effects come from the pre-exponential transmission factor in the case of systems with a significant activation barrier [31, 44].

The quantum mechanical free energy barrier, $\Delta g^{\neq}_{qm}$, can be evaluated by Feynman's path integral formulation [59], where each classical coordinate is replaced by a ring of quasiparticles that are subjected to the effective "quantum mechanical" potential

$$U_{qm} = \sum_{k=1}^{p} \frac{1}{2p} M\Omega^2 \Delta x_k^2 + \frac{1}{p} U(x_k) \tag{8.28}$$

Here, $\Delta x_k = x_{k+1} - x_k$ (where $x_{p+1} = x_1$), $\Omega = p/\hbar\beta$, M is the mass of the particle, and U is the actual potential used in the classical simulation. The total quantum mechanical partition function can then be obtained by running classical trajectories of the quasiparticles with the potential U_{qm}. The probability of being at the transition state is in this way approximated by a probability distribution of the center of mass of the quasiparticles (the centroid) rather than the classical single point.

Such calculations of centroid probabilities in the condensed phase reactions are very challenging since they may involve major convergence problems. The QCP approach offers an effective and rather simple way to evaluate this probability without significantly changing the simulation program. This is done by propagating classical trajectories on the classical potential surface of the reacting system and using the positions of the atom of the system to generate the centroid position for the quantum mechanical partition function. This treatment is based on the finding that the quantum mechanical partition function can be expressed as [57, 60]

$$Z_{qm}(\bar{x}) = Z_{cl}(\bar{x}) \left\langle \left\langle \exp\left\{ -(\beta/p) \sum_k U(x_k) - U(\bar{x}) \right\} \right\rangle_{fp} \right\rangle_U \tag{8.29}$$

where $\bar{x}$ is the centroid position, $\langle \cdots \rangle_{fp}$ designates an average over the free particle quantum mechanical distribution obtained with the implicit constraint that $\bar{x}$ coincides with the current position of the corresponding classical particle, and $\langle \cdots \rangle_U$

designates an average over the classical potential U. It is worth stressing that path integral calculations involving computationally expensive quantum chemical evaluation of forces and energies would benefit much from the QCP scheme. In quantum chemical calculations involving quasiparticles, one cannot realize exclusions between quasiparticles and therefore computational effort is proportional to the number of the quasiparticles in the necklace. In the present approach the quantum connection can be performed *a posteriori*, using the stored trajectory. It would be interesting to apply this approach to a Car-Parinello path integral scheme and demonstrate almost negligible increases of the CPU time relative to the classical treatment of the nuclei [61].

Using Eq. (8.27) we can obtain the quantum mechanical free energy surface by evaluating the corresponding probability by the same combined free energy perturbation umbrella sampling approach that has been repeatedly applied in our classical simulations as well as in our quantum mechanical simulations, but now we use the double average of Eq. (8.15) rather than an average over a regular classical potential. The actual equations used in our free energy perturbation (FEP) umbrella sampling calculations are given elsewhere, but the main point of the QCP is that the quantum mechanical free energy function can be evaluated by a centroid approach, which is constrained to move on the classical potential. This provides stable and relatively fast converging results that have been shown to be quite accurate in studies of well-defined test potentials where the exact quantum mechanical results are known.

8.4
The EVB as a Basis for LFER of PT Reactions

The approach used to obtain the EVB free energy functionals (the Δg_i of Eq. (8.12)) was originally developed in Ref. [1] in order to provide the microscopic equivalent of Marcus theory for electron transfer reactions [29]. This approach allows one to explore the validity of the Marcus formula on a microscopic molecular level [62]. While this point is now widely accepted by the ET community [44], the validity of the EVB as perhaps the most general tool in microscopic LFER studies of PT reactions is less appreciated. This issue will be addressed below.

In order to explore the molecular basis of LFER, we have to consider a one-step chemical reaction and to describe this reaction in terms of two diabatic states, ε_1 and ε_2, that correspond to the reactant and product states. In this case the ground state adiabatic surface is given by

$$E_g = \frac{1}{2}[(\varepsilon_1 + \varepsilon_2) - \sqrt{(\varepsilon_1 - \varepsilon_2)^2 + 4H_{12}^2}] \tag{8.30}$$

With this well-defined adiabatic surface we can explore the correlation between $\Delta g^{\ddagger}$ and ΔG^0. Now the EVB/umbrella sampling procedure (for example, Ref. [3]) allows one to obtain the rigorous profile of the free energy function, $\Delta\bar{g}$, that corresponds to E_g and the free energy functions, Δg_1 and Δg_2, that correspond to ε_1 and ε_2, respectively (see Fig. 8.5). It is important to point out here that such profiles have been evaluated quantitatively in many EVB simulations of chemical reactions in solutions and proteins (for reviews see Refs. [24, 63]). The corresponding profiles provide the activation free energy, $\Delta g^{\ddagger}$, for the given chemical step. The calculated activation barrier can then be converted (for example, see Ref. [3]) to the corresponding rate constant using transition state theory (TST):

$$k_{i\to j} \cong (RT/h)\exp\{-\Delta g^{\neq}_{i\to j}/RT\} \tag{8.31}$$

A more rigorous expression for $k_{i\to j}$ can be obtained by multiplying the TST expression by a transmission factor that can be calculated easily by running downhill trajectories [3]. However, the corresponding correction which takes into account barrier recrossing is an order of unity for reactions in aqueous solutions and enzymes [4].

In many case it is useful to estimate $\Delta\bar{g}$ and $\Delta g^{\neq}$ by an approximated expression. Here we note that with the simple two-state model of Eq. (8.11) we can obtain a very useful approximation to the $\Delta\bar{g}$ curve. That is, using the aforementioned free energy EVB/umbrella sampling formulation, we obtain the $\Delta\bar{g}$ that corresponds to the E_g and the free energy functions, Δg_i, that correspond to the ε_i surfaces. This leads to the approximated expression

$$\Delta\bar{g}(x) = \frac{1}{2}[(\Delta g_1(x) + \Delta g_2(x)) - \sqrt{(\Delta g_1(x) - \Delta g_2(x))^2 + 4H_{12}^2(x)}] \tag{8.32}$$

This relationship can be verified in the case of small H_{12} by considering our ET studies [62], while for larger H_{12} one should use a perturbation treatment. Now we can exploit the fact that the Δg_i curves can be approximated by parabolas of equal curvatures (this approximated relationship was found to be valid by many microscopic simulations (e.g. Ref. [24])). It is known empirically that for a series of chemical reactions those which are more exothermic typically proceed faster. This approximation can be expressed as

$$\Delta g_i(x) = \lambda\left(\frac{x - x_0^{(i)}}{x_0^{(j)} - x_0^{(i)}}\right)^2 \tag{8.33}$$

where λ is the so-called "solvent reorganization energy" (which is illustrated in Fig. 8.3).

Using Eqs. (8.32) and (8.33), one obtains the Hwang Åqvist Warshel (HAW) equation [27, 64], which is given in the general case by

$$\Delta g^{\neq}_{i\to j} = (\Delta G^0_{i\to j} + \lambda_{i\to j})^2/4\lambda - H_{ij}(x^{\neq}) + H_{ij}^2(x_0^{(i)})/(\Delta G^0_{i\to j} + \lambda_{i\to j}) + \Gamma_{ij} \tag{8.34}$$

where $\Delta G^0_{i\to j}$ is the free energy of the reaction, and H_{ij} is the off-diagonal term that mixes the two relevant states with the average value at the transition state, $x^{\neq}$, and at the reactant state, $x_0^{(i)}$. Γ_{ij} is the NQM correction that reflects the effect of tunneling and zero point energy corrections in cases of light atom transfer reactions. Γ therefore includes all effects associated with the quantum mechanical nature of the nuclei motion.

Repeated quantitative EVB studies of PT and other reactions in solutions and proteins (for example, Refs. [24, 65]) established the quantitative validity of Eq. (8.34). With this fact in mind we can take these equations as a quantitative correlation between $\Delta g^{\neq}_{i\to j}$ and $\Delta G^0_{i\to j}$. Basically, when the changes in ΔG^0 are small, we obtain a linear relationship between $\Delta g^{\neq}_{i\to j}$ and $\Delta G^0_{i\to j}$. This linear relationship, which can be obtained by simply differentiating the $\Delta g^{\neq}_{i\to j}$ of Eq. (8.34) with respect to $\Delta G^0_{i\to j}$, can be expressed in the form

$$\Delta\Delta g^{\neq}_{i\to j} = \theta\Delta\Delta G^0_{i\to j} \tag{8.35}$$

where $\theta = (\Delta G^0_{i\to j} + \lambda)/2\lambda$, and where the contribution from the last term of Eq. (8.34) is neglected. The linear correlation coefficient depends on the magnitude of ΔG^0 and λ. At any rate, more details about this linear free energy relationship (LFER) or free energy relationship (FER), and its performance in studies of chemical and biochemical problems are given elsewhere [3, 24, 27, 66–68].

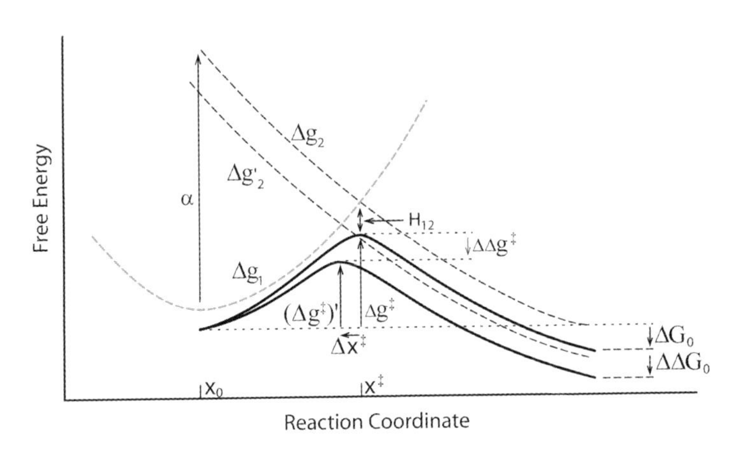

Figure 8.5. A schematic description of the relationship between the free energy difference ΔG_0 and the activation free energy $\Delta g^{\ddagger}$. The figure illustrates how a shift of Δg_2 by $\Delta\Delta G_0$ (that changes Δg_2 to $\Delta g_2{}'$ and ΔG_0 to $\Delta G_0 + \Delta\Delta G_0$) changes $\Delta g^{\ddagger}$ by a similar amount.

The main point of Eq. (8.36) and Fig. 8.3 is that the $\Delta G^0_{i\to j}$, which determines the corresponding $\Delta g^{\neq}_{i\to j}$, is correlated with the difference between the two minima of the $\Delta\bar{g}$ profile that correspond to states i and j, respectively.

While our ability to reproduce the observed LFER might not look like a conceptual advance, the fact that the EVB provides a rigorous basis for FER in condensed phases leads to a different picture than what has been assumed in traditional LFER studies. That is, as is clear from the HAW relationship, it is essential to take into account the effect of H_{ij} on LFER studies that involve actual chemical reactions (rather than ET reactions). In such cases, H_{ij} is frequently very significant, and its neglect leads to an incorrect estimate of the relevant reorganization energy. This point has not been widely appreciated because of the fact that the correlation between $\Delta g^{\neq}$ and ΔG does not depend so critically on H_{ij}. Thus, as long as one fits the experimentally observed relationship by phenomenological parameters, it is hard to realize that the relevant reorganization energies are underestimated in a drastic way. A case in point is provided by the systematic analyses of PT reactions in enzymes [24, 58], and of hydride transfer and S_N2 reactions [27, 66]. A specific example of a LFER analysis will be given in the next section.

The use of the EVB and Eq. (8.34) in studies of reactions in solutions has been extended to studies of LFERs in enzymes. The successes of this approach have been demonstrated in several important systems. At present, we view these studies as the most quantitative LFER studies of enzymes. It is also useful to point out the successes of our approach in LFER studies of electron transport in proteins (for example, Ref. [69]).

A recent study of Kiefer and Hynes [70] used an EVB formulation, with a dielectric continuum treatment of the solvent, in an attempt to derive a LFER for PT reactions. Although their derivation did not include the crucial effect of H_{ij}, and overlooked our earlier works it is encouraging to see a further realization of the effectiveness of the EVB in providing a molecular basis to LFER treatments.

8.5 Demonstrating the Applicability of the Modified Marcus' Equation

In order to illustrate our point about the difficulties associated with phenomenological LFER treatments of reactions in solutions and in enzymes, it is instructive to consider the studies of Human carbonic anhydrase III (which will be referred to here as CA III) [71]. Studies of this system [71, 72] demonstrated that the rate of PT in mutants of CA III is correlated with the pK_a difference between the donor and acceptor. It was found that the observed LFER follows a Marcus' type relationship. Although this study provided an excellent benchmark for studies of PT in proteins, it also raised the question about uniqueness of the parameters deduced from phenomenological LFER studies. This issue will be explored below.

The catalytic reaction of CA III can be described in terms of two steps. The first is attack of a zinc-bound hydroxide on CO_2 [73].

$$CO_2 + EZn^{+2}(OH^-) + H_2O \leftrightarrows HCO_3^- + EZn^{+2}(H_2O) \tag{8.36}$$

The reversal of this reaction is called the "dehydration step". The second step involves the regeneration of the OH^- by a series of PT steps [74, 75]

$$EZn^{+2}(H_2O) + B \underset{k_B}{\overset{k_{-B}}{\rightleftarrows}} EZn^{+2}(HO^-) + BH^+ \tag{8.37}$$

where $K_{-B} = k_{-B}/k_B$ (in the notation of Ref. [71]), BH^+ can be water, buffer in solution or the protonated form of Lys64 (other CAs have His in position 64). Previous experimental studies [71] have established a LFER that was fitted to the Marcus' equation using

$$\Delta g^{\ddagger} = w^r + \{1 + \Delta G^0/4\Delta G_0^{\ddagger}\}^2 \Delta G_0^{\ddagger} \tag{8.38}$$

where the observed reaction free energy is given by $\Delta G^0_{obs} = w^r + \Delta G^0 - w^p$, where w^r is the reversible work of bringing the reactants to their reacting configuration and w^p is the corresponding work for the reverse reaction. ΔG^0 is the free energy of the reaction when the donor and acceptor are at their optimal distance. $\Delta G_0^{\ddagger}$ is the so-called intrinsic activation barrier, which is actually $\frac{1}{4}$ of the corresponding reorganization energy, λ. Here we use $\Delta g^{\ddagger}$ rather than $\Delta G^{\ddagger}$ for the activation barrier, following the consideration of Ref. [3]. Equation (8.38) can also be written in the well-known form

$$\Delta g^{\ddagger} = w^r + (\Delta G^0 + \lambda)^2/4\lambda \tag{8.39}$$

The phenomenological fitting processes yielded $\lambda = 5.6$ kcal mol^{-1} and $w^r \cong 10.0$ kcal mol^{-1}. The estimated value of λ appears to be in conflict with the value deduced from microscopic computer simulation studies ($\lambda \cong 80$ kcal mol^{-1} in Ref. [64]). Furthermore, the large value of w^r is hard to rationalize, since the reaction involves a proton transfer between a relatively fixed donor and acceptor (residue 64 and the zinc bound hydroxide). The very small value of λ obtained by fitting Eq. (8.20) to experiment is not exclusive to CA III. Similarly, small values were obtained in analysis of other enzymes and are drastically different than the values obtained by actual microscopic computer simulations (note in this respect that λ cannot be measured directly).

As pointed out before [58, 76, 77], the above discrepancies reflect the following problems. First, the reaction under study may involve more than two intersecting parabolas and thus cannot be described by Eq. (8.20). Second, although Eq. (8.20) gives a proper description for electron transfer (ET) reactions where the mixing between the reactant and product state (H_{12}) is small, it cannot be used to describe proton transfer or other bond breaking reactions, where H_{12} is large. In such cases one should use the HAW equation, Eq. (8.34) [27, 64].

In order to obtain a proper molecular description of LFERs, it is essential to represent each reactant, product or intermediate by a parabolic free energy function [3]. In the case of CA III, we describe the proton transfer from residue 64 (Lys or His) to the zinc bound hydroxyl via a bridging water molecule (and alternatively two water molecules), by considering the three states

$$\begin{aligned}\psi_1 &= \mathrm{BH^+(H_2O)}_b\mathrm{(OH^-)}_a\mathrm{Zn^{+2}}\\ \psi_2 &= \mathrm{B(H_3O^+)}_b\mathrm{(OH^-)}_a\mathrm{Zn^{+2}}\\ \psi_3 &= \mathrm{B(H_2O)}_b\mathrm{(H_2O)}_a\mathrm{Zn^{+2}}\end{aligned} \tag{8.40}$$

where we denote by B the base at residue 64, and where ψ_1 and ψ_3 correspond, respectively, to the right and left sides of Eq. (8.37). The relative free energy of these states can be estimated from the corresponding pK_as, where the pK_as of $(H_2O)_a$ and B are known from different mutations [71], while the pK_a of $(H_2O)_b$ can be calculated by the PDLD/S-LRA approach [78]. Note that our three-state system can be easily extended to include one more water molecule and one more state.

The result of a HAW LFER analysis of the CA III system is illustrated in Fig. 8.6 and the overall dependence of $\Delta\Delta g^{\neq}$ on ΔG_{13} is presented in Fig. 8.7 (for more details see Ref. [77]). As seen from the figures our model reproduced the observed trend. However, the origin of the trend is very different than that deduced from the two-state Marcus' equation. That is, the flattening of the LFER at $\Delta pK_a > 0$, which would be considered in a phenomenological analysis of a two-state model as the beginning of the Marcus' inverted region (where $\Delta G_0 = -\lambda$), is due to the behavior of the three-state system (see Ref. [79]).

The extraction of λ from fitting Eq. (8.38) to the observed LFER requires that $\lambda = -\Delta G^0$ so that $\Delta G^0 < 0$ at the point where the LFER becomes flat. This means that we must have data from regions where $\Delta G^0 < 0$. However, at least for the cases when $\Delta g^{\ddagger}_{12}$ is rate limiting, ΔG^0_{12} cannot be negative, and the observation of the beginning of a flat LFER is actually due to other factors. It is also important to realize that λ_{out} cannot become too small and never approach zero, which is the continuum limit for a completely nonpolar environment (see discussion in Ref. [80]). The reason is quite simple; the protein cannot use a nonpolar environment, since this will decrease drastically the pK_a of $(H_3O^+)_b$. Instead, proteins use polar environments with partially fixed dipoles. However, no protein can keep its dipoles completely fixed (the protein is flexible) and thus gives a non-negligible λ_{out}. Of course, this reorganization energy is still smaller than the corresponding value for proton transfer in solutions, but it never approaches the low value obtained from fitting in a two-state Marcus' formula [81].

As long as we obtain the value of H_{ij} from fitting to observed LFERs, it is possible to argue that both Eqs. (8.39) and (8.34) reflect a phenomenological fitting with

Figure 8.6. Analysis of the energetics of PT in the i, ii, and iii mutants of CA III for the case where the transfer of a proton from residue 64 to the zinc-bound hydroxide involves two water molecules. The figure describes the three states of Eq. (8.40) and considers their change in each of the indicated mutants (relative to the native enzyme), and displays the changes in the diabatic potential surfaces and the corresponding changes in the adiabatic activation barriers. The figure also gives the changes in the diabatic activation energies. The final activation barrier is taken in each case as the highest adiabatic barriers (taken from Ref. [79]).

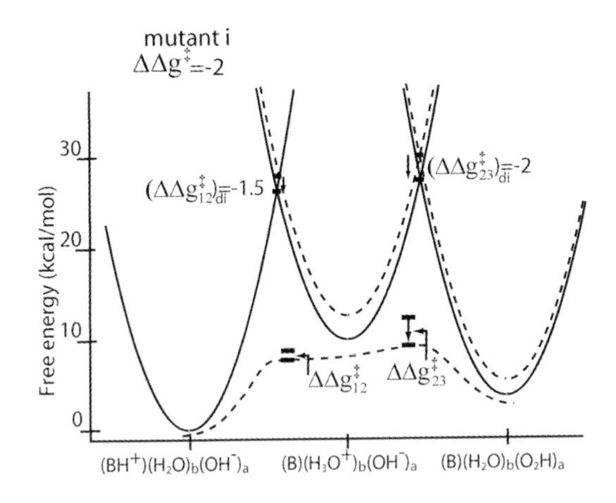
mutant i
ΔΔg‡=-2
Free energy (kcal/mol)
0
10
20
30
(ΔΔg‡12)di=-1.5
(ΔΔg‡23)di=-2
ΔΔg‡12
ΔΔg‡23
(BH+)(H2O)b(OH-)a
(B)(H3O+)b(OH-)a
(B)(H2O)b(O2H)a

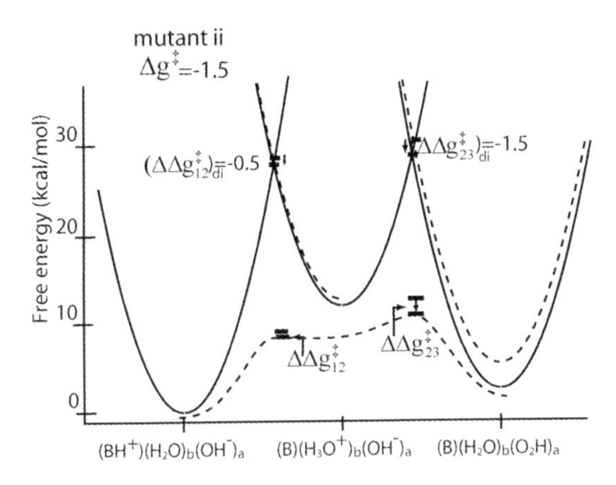
mutant ii
Δg‡=-1.5
Free energy (kcal/mol)
0
10
20
30
(ΔΔg‡12)di=-0.5
ΔΔg‡23)di=-1.5
ΔΔg‡12
ΔΔg‡23
(BH+)(H2O)b(OH-)a
(B)(H3O+)b(OH-)a
(B)(H2O)b(O2H)a

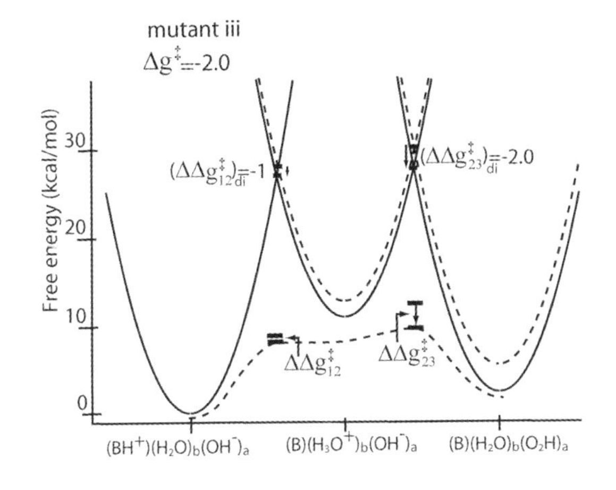
mutant iii
Δg‡=-2.0
Free energy (kcal/mol)
0
10
20
30
(ΔΔg‡12)di=-1
(ΔΔg‡23)di=-2.0
ΔΔg‡12
ΔΔg‡23
(BH+)(H2O)b(OH-)a
(B)(H3O+)b(OH-)a
(B)(H2O)b(O2H)a

a free parameter (ω and λ in the case of Eq. (8.39), and H_{ij} in the case of Eq. (8.34)). The difference, however, is that Eq. (8.34) and the use of three free energy functionals reflects much more realistic physics. This is evident, for example, from the fact that with Eq. (8.34) we do not obtain an unrealistically large w^r. Note in this respect that ΔG_{12}^0 in Eq. (8.23) might look like w in the phenomenological fitting to the Marcus' equation. It is also important to emphasize at this point that the present treatment is not a phenomenological treatment with many free parameters, as might be concluded by those who are unfamiliar with molecular simulations. That is, our approach is based on realistic molecular parameters obtained while starting from the X-ray structure of the protein and reproducing the relevant pK_as and reorganization energy. Reproducing the observed LFER by such an approach without adjusting the key parameters is fundamentally different than an approach that takes the observed LFER and adjusts free parameters in a given model to reproduce it. In such a case, one can reproduce experimental data by almost any model.

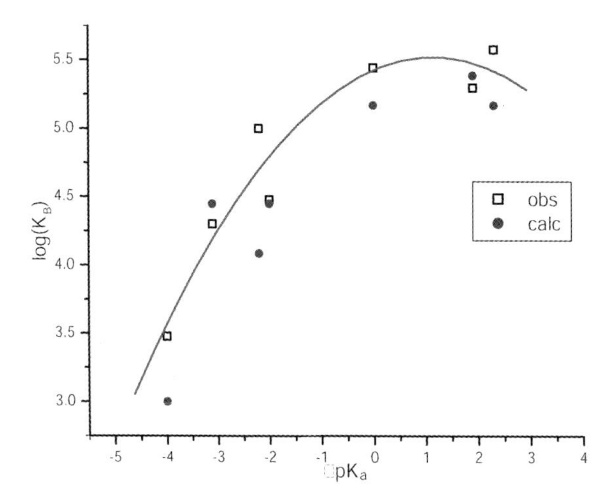

Figure 8.7. Calculated and observed FER for CA III. The different systems are marked according to the notation of Ref. [79]. The term ΔpK_a corresponds to the pK_a difference between the zinc-bound water and the pK_a of the given donor group ($\Delta pK_a = -\Delta G_{13}/2.3RT$).

Finally, we would like to address the validity of the general use of the HAW Equation (Eq. (8.34)) and the multistate procedure used for studies of the proton transport (PTR) in CA. The use of Eq. (8.34) for subsequent PT steps might look to some as an *ad hoc* approach, considering the assumption that PTR processes involve the Grotthuss mechanism, which is not sensitive to the ΔG_{ij}^0 values for the sequential transfer process. However, the assumption that the Grotthuss mechanism is a key factor underwent recently a major paradigm shift, where those who supported this idea started to attribute major importance to the electrostatic barrier

[82], in agreement with our view [83, 84]. Further support to this point is given below.

In order to further explore the validity of the stepwise modified Marcus' model, we developed recently [85] a simplified EVB model which represents the given conduction chain by an explicit EVB, while representing the rest of the environment (protein and solvent) implicitly. This is done by using the same type of solute surface as in Eq. (8.8), while omitting the explicit solute–solvent and solvent–solvent terms (the U_{Ss} and U_{ss} terms) and replacing them by implicit terms using:

$$\varepsilon_i = \varepsilon_i^0 + (\hbar\omega_Q/2)\lfloor(Q_{i,k(i)} + \delta)^2 + (Q_{i,k'(i)} - \delta)^2\rfloor + \Delta^{(i)} \tag{8.41a}$$

$$\bar{E}_{tot} = \bar{E}_g + \sum_i (\hbar\omega_Q/2)\lfloor(Q_{i,k(i)}^2 + B(Q_{i,k(i)} - Q_{i,k'(i)})^2)\rfloor \tag{8.41b}$$

Where the Qs are the solvent coordinates that are given by the electrostatic component of the energy gap $(-(\hbar\omega_Q\delta_Q)Q_{ij} = \varepsilon_j^{el} - \varepsilon_i^{el})$, δ is the dimensionless origin shift of the solvent coordinate, ω_Q is the effective vibrational frequency of the solvent, and $\bar{E}_g$ is the lowest eigenvalue of Eq. (8.7), and the B term represents the coupling between the solvent coordinates. Here $\bar{E}_g$ reflects the effect of the ε_i of Eq. (8.41a) and the other term in Eq. (8.41b) reflect the cost in solvent energy associated with moving the solvent coordinates from their equilibrium positions. Equation (8.41) is written for the case of a chain of water molecules, so that we assign one solvent coordinate to each pair of oxygens. In this case, the index i corresponds to a proton on the ith oxygen inside the chain, while $k'(i)$ and $k(i)$ correspond to the oxygens before and after i, respectively. When i is the first oxygen, there is no $Q_{i,k'}$. The free energy, $\bar{g}_g$, associated with the energy surface, $\bar{E}_g$, (here the free energy accounts for the average over the coordinates of the active space) is treated as the effective free energy surface that includes implicitly the rest of the system. That is, we use

$$g(\mathbf{r})_{eff} = \bar{g}_g(\mathbf{r}) \tag{8.42}$$

where $\mathbf{r}$ are the coordinates of the active space. In doing so, we note that in this simplified expression we treat the environment implicitly by adjusting the $\Delta^{(i)}$s to $\bar{\Delta}^{(i)}$ while imposing the requirement

$$\begin{aligned}(\Delta G_{i\to j})_{eff} &= (\Delta G_{i\to j})_{complete}\\ (\Delta g_{i\to j}^{\ddagger})_{eff} &= (\Delta g_{i\to j}^{\ddagger})_{complete}\end{aligned} \tag{8.43}$$

where $(\ )_{eff}$ represents the quantity obtained with the effective EVB potential and $(\ )_{complete}$ designates the results obtained when the EVB of the entire system is included explicitly. For convenience we usually determine $(\Delta G_{i\to j})_{complete}$ (and the corresponding $\Delta^{(i)}$ values of the effective model) by the semimacroscopic electrostatic calculations outlined below. The simplified system has therefore identical free energy of activation and identical reaction free energy as the full system, and it allows for much longer simulations.

With the effective potential defined above it is possible to examine the time dependence of PTR processes by Langevin dynamics (LD) simulations (considering the fluctuation of the missing parts of the system by using an effective friction), and such simulations were used to study the PTR in CA. The simulation established that the rate of the PTR process is determined by the energetics of the proton along the conduction chain, once the energy of the proton in two successive sites is significantly higher than the energy of the proton in the bulk water. The model was also applied to PTR in the K64H-F198D mutant of CA III and reproduced the observed rate constant. Typical simulations for the case where the energy of the proton on His64 is raised by 1.2 kcal mol^{-1}, in order to accelerate the calculations, are described in Fig. 8.8. The calculated average time for PTR from His64 to the Zn-bound hydroxide is about 5×10^{-6} s. Correcting this result for the energy shift and the effect of using overdamped rather than underdamped Brownian dynamics (BD) simulation gives a result that is close to the observed k_B ($k_B = 3 \times 10^{-6}$ s). The simulation provides an additional major support for the use of the HAW model.

8.6
General Aspects of Enzymes that Catalyze PT Reactions

Computer simulation studies have been used since 1976 to explore the origin of the catalytic power of different enzymes (for a recent review see Ref. [9]). Basically, the EVB studies, as well as the consistent QM/MM calculations identified electrostatic transition state stabilization (TSS) as the key catalytic factor (a discussion of

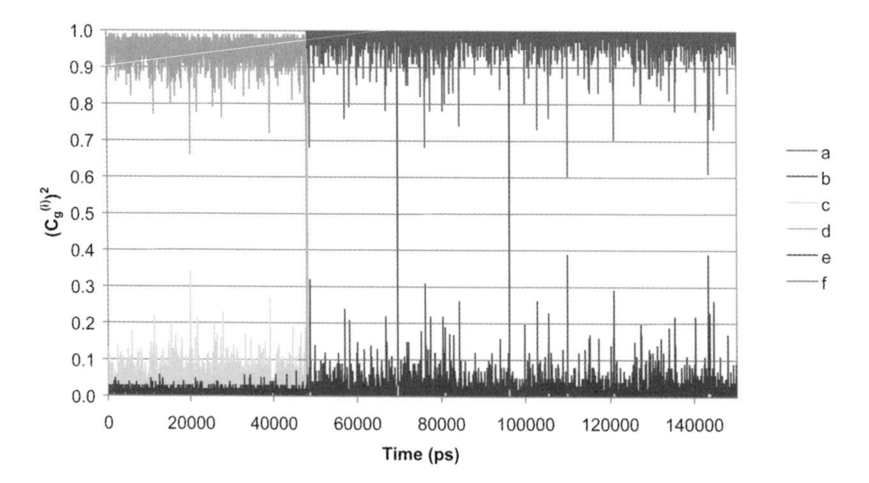

Figure 8.8. The time dependence of the probability amplitude of the transferred proton for a LD trajectory for a PTR that starts at His64 and ends at OH^- in the overdamped version of model S/A of the K64H-F198D mutant of CA III. The calculations were accelerated by considering a case where the minimum at site d is raised by 1.2 kcal mol^{-1} (taken from Ref. [85]).

inconsistent analyses has been given elsewhere [9, 63]. In many cases, it was found that enzymes catalyze PT steps by reducing the pK_a differences between the donor and acceptor [86]. However, in other cases (for example, the reaction of serine proteases [87]) the catalytic effect is exerted at the highest transition state, which is not a PT step, and the PT step is not catalyzed. As far as the present work is concerned, it is interesting to note that when a PT or hydride transfer step is catalyzed (for example, Refs. [3, 4, 88]) it is always accomplished by electrostatic stabilization of the corresponding change in charge distribution. This is done by the preorganized polar environment of the protein, which reduces the reorganization energy during the PT step (see discussion in Ref. [89]).

In considering the catalytic effect in PT processes, it is important to avoid common confusion associated with the so-called low barrier hydrogen bond (LBHB) proposal (for example, Ref. [90]). The LBHB proposal assumes that hydrogen bonding to the TS involves covalent interactions with special catalytic power. EVB considerations have shown that this assumption is incorrect since covalent interactions lead to delocalized charge distribution, which is less stable than localized charge in the protein polar environment (see for example Ref. [91]). In the polar environment the existence of a localized charge is more favorable than a delocalized charge. Since the latter case corresponds to the LBHB that corresponds to the transition state, the LBHBs in this sense are anticatalytic rather than catalytic. Unfortunately, in addition to the clear inconsistency of the LBHB proposal, it was also invoked in discussing PT steps. Now the transition states of PT steps do involve covalent delocalized TSs, but this has nothing to do with the LBHB proposal or with catalysis, since the same delocalization is involved in the reference reaction in aqueous solution.

Regardless of the exact nature of the PT step and the corresponding catalytic effect, we feel that our approach offers an extremely robust way of examining the role of the enzyme in these steps. All that is needed is to compare the energetics as to "reactant" and "product" states in the enzyme and in solution. This reduces to a simple electrostatic problem that can be addressed by the EVB/FEP calculations.

8.7 Dynamics, Tunneling and Related Nuclear Quantum Mechanical Effects

Recent studies (see Ref. [92] and references in Ref. [4]) have suggested that vibrationally enhanced tunneling (VET) of PT, hydride transfer, and hydrogen transfer reactions plays a major role in enzyme catalysis. According to this interesting proposal, nature created by evolution protein vibrational modes that are strongly coupled to the hydrogen atom motion. Some workers (for example, Ref. [93]) assumed that there exists here an entirely new phenomenon that makes TST inapplicable to enzymatic reactions. However, the VET effect is not new and is common to many chemical reactions in solution [1, 48, 94]. Moreover, the VET is strongly related to TST. That is, when the solvent fluctuates and changes the energy gap (see Refs. [1, 2]) the light atom sees a fluctuating barrier that allows, in some cases, for a larger rate of tunneling. As shown in Ref. [2], these fluctuations are taken into account in the statistical factor of the classical TST and the same is true when quantum effects are taken into account. Thus, the recent realization that the solvent coordinates should be considered in tunneling studies is not new, nor does it mean that this effect is important in catalysis.

Warshel and Chu [42] and Hwang et al. [60] were the first to calculate the contribution of tunneling and other nuclear quantum effects to PT in solution and enzyme catalysis, respectively. Since then, and in particular in the past few years, there has been a significant increase in simulations of quantum mechanical-nuclear effects in enzyme and in solution reactions [16]. The approaches used range from the quantized classical path (QCP) (for example, Refs. [4, 58, 95]), the centroid path integral approach [54, 55], and variational transition state theory [96], to the molecular dynamics with quantum transition (MDQT) surface hopping method [31] and density matrix evolution [97–99]. Most studies of enzymatic reactions did not yet examine the reference water reaction, and thus could only evaluate the quantum mechanical contribution to the enzyme rate constant, rather than the corresponding catalytic effect. However, studies that explored the actual catalytic contributions (for example, Refs. [4, 58, 95]) concluded that the quantum mechanical contributions are similar for the reaction in the enzyme and in solution, and thus, do not contribute to catalysis.

Interestingly, the MDQT approach of Hammes-Schiffer and coworkers [31] allowed them to explore the quantum mechanical transmission factor. It was found that even with quantum mechanical considerations, the transmission factor is not so different from unity, and thus, we do not have a large dynamical correction to the TST rate constant.

It is important to clarify here that the description of PT processes by curve crossing formulations is not a new approach nor does it provide new dynamical insight. That is, the view of PT in solutions and proteins as a curve crossing process has been formulated in early realistic simulation studies [1, 2, 42] with and without quantum corrections and the phenomenological formulation of such models has already been introduced even earlier by Kuznetsov and others [47]. Furthermore, the fact that the fluctuations of the environment in enzymes and solution modulate the activation barriers of PT reactions has been demonstrated in realistic microscopic simulations of Warshel and coworkers [1, 2]. However, as clarified in these works, the time dependence of these fluctuations does not provide a useful way to determine the rate constant. That is, the electrostatic fluctuations of the environment are determined by the corresponding Boltzmann probability and do not represent a dynamical effect. In other words, the rate constant is determined by the inverse of the time it takes the system to produce a reactive trajectory, multiplied by the time it takes such trajectories to move to the TS. The time needed for generation of a reactive trajectory is determined by the corresponding Boltzmann probability, and the actual time it takes the reactive trajectory to reach the transition state (of the order of picoseconds), is more or less constant in different systems.

It is also important to clarify that the solvent reorganization energy, which determines the amplitude of the solvent fluctuations, is not a "static dynamical effect" (as proposed by some) but a unique measure of the free energy associated with the reorganization of the solvent from its reactant to its product configuration (see Ref. [62] for a more rigorous definition). In fact, the reorganization energy, λ, and the reaction energy, ΔG, determine the activation free energy ($\Delta G^{\ddagger}$) and the corresponding Boltzmann probability of reaching the transition state (see discussion in Refs. [2, 4]). Now, since the $\Delta G^{\ddagger}$ is a probability factor it can be determined by Monte Carlo simulations without any dynamical considerations. Furthermore, since the transmission factor (κ in Eq. (8.1)) is the only rigorous dynamical part of the rate constant and since κ is close to unity in enzymes and solutions, see e.g. Ref. [4], the corresponding rate constants do not show significant dynamical effects. Furthermore, attempts to define dynamical catalytic effect in a different way and to include in such factor nonequilibrium solvation effects [100] have been shown to be very problematic (e.g. Ref. [4]). Similarly, we have shown that the reasonable definition of dynamical effects by the existence of special vibrations that lead coherently to the TS does correspond to the actual simulation in enzyme and solution.

Before concluding this section, we find it useful to discuss the specific case of lipoxygenase, which has been brought recently [101] as an example of the role of NQM in catalysis. The catalytic reaction of lipoxygenase involves a very large isotope effect (~80) and thus it is tempting to suggest that the enzyme catalyzes this reaction by enhancing the tunneling effect. This proposal [53] has been examined by several theoretical approaches ranging from phenomenological models [31, 53, 102] to continuum representations of the protein [103] and to complete microscopic treatment of the system using the QCP approach [50]. The phenomenological studies were able to fit the observed kinetics with an unrealistically large reorganization energy ($\lambda \approx 20$ kcal mol^{-1}) using the vibronic formulation of Eq. (8.16), while the microscopic reorganization energy was actually found to be around 2 kcal mol^{-1} [50]. The unrealistic phenomenological parameters may reflect several major problems, including the fact that the vibronic formula is invalid in the adiabatic limit and the other problems discussed in Section 8.3. Some of these fundamental problems can be reduced by modifying the diabatic potential to reflect the enormous effect of H_{12} (which is not considered in any of the vibronic treatments except in that of Warshel and Chu [42]). This type of treatment, which leads to Eq. (8.25) has been used recently [104] in a very qualitative examination of the KIE in lipoxygenase that compares the vibronic and the much more rigorous QCP simulations that will be considered below.

Hammes-Schiffer and coworkers [103] have progressed beyond the phenomenological vibronic treatment by using much more realistic potentials and a semi-microscopic treatment. However, the protein effect was modeled macroscopically with an arbitrarily low dielectric constant. Such a treatment makes it hard to explore the actual role of the protein. Thus, at present, the only study that actually examines the microscopic origin of the observed isotope effect by taking into account the entire protein, and more importantly the nature of the catalytic effect, is the study of Olsson et al. [50]. Here it was found that the QCP simulation reproduces the observed isotope effect and the corresponding reduction in activation free energy. Most significantly, the simulations show that the same NQM effect occurs in the enzyme and solution and that the reorganization energy in the protein and solution is extremely small ($\lambda_{out} \approx 2$ kcal mol^{-1}) since we are dealing with a hydrogen transfer (rather than PT) reaction. In view of the similarity between the NQM in the enzyme and solution, it seems that the catalysis does not involve optimization of NQM effects.

At this point one might wonder about our assertion that the reaction in solution and in the protein have similar quantum corrections since a reaction with a model compound that is assumed to be related to lipoxygenase show only an isotope effect of 6. Here, we can only bring out the following points: (i) we are quite convinced that the effect of the protein and the solution on the H atom transfer reaction is quite small, (ii), our studies have shown that the isotope effect depends strongly on the donor acceptor distance (see also above) and it is possible that the average distance is longer in the protein than in water, (iii), it is not clear that the model reaction in water reflects only the H-transfer step. Finally, our main point is that if the quantum corrections are smaller in water than in the protein (and the reaction is the same), we must have a larger donor–acceptor distance in the protein and thus a negative catalytic effect (the barrier is higher at a longer distance). In other words we do not see any simple way for the NQM effects to catalyze this reaction.

To conclude this section, it is useful to point out an interesting conclusion that emerged from the use of Eqs. (8.25) and (8.27). That is, using the above equations we found that the degree of tunneling (or at least the magnitude of the isotope effect) decreases, rather than increases, when the donor–acceptor distance is reduced. This reflects the reduction in the effective diabatic X–H stretching frequency. This means that the common idea that enzymes can catalyze H transfer or PT reactions by compressing the donor and acceptor complexes [101] is very problematic. Such a compression will in fact reduce the tunneling contributions, since the largest contribution to the KIH comes from $(S_{00}^2)_D/(S_{00}^2)_H$, whose value decreases drastically when Δ_H and ω_H decrease. Furthermore, as we have shown repeatedly, enzymes are flexible and unlikely to be able to change drastically the reaction surface [3, 105]. Thus it is quite likely that the main difference between reactions with large tunneling and small tunneling corrections is the intrinsic shape of the potential surface rather than the effect of the environment on this surface. It is thus possible that radical reactions involve steep potential surfaces with relatively small H_{12}, while other reactions involve large H_{12} and shallow surfaces with small effective ω_H.

8.8 Concluding Remarks

This work addressed the issue of proton transfer, focusing on clear microscopically based concepts and the power of computer simulation approaches. It was shown that when such concepts as reorganization energy and Marcus' parabolas are formulated in a consistent microscopic way, they could be used to explore the nature of PT and PTR in proteins. It was also clarified that phenomenological applications of the Marcus' formula or related expressions can lead to problematic conclusions.

The use of the EVB approach is shown to provide a powerful quantitative bridge between the classical concepts of physical organic chemistry and the actual energetics of enzymatic reactions. This approach provides quantitative LFERs for PT in enzymes and solution, and allows us to quantify catalytic effects and to define them in terms of the relevant reorganization energies, reaction free energies, and the preorganization of enzyme active sites.

Our studies have demonstrated that dynamical effects do not play an important role in enzyme catalysis. We also show that NQM effects do not contribute significantly since similar effects occur in the reference solution reactions. Nevertheless, one should be able to calculate NQM effects and to treat dynamical effects in order to reproduce the actual rate constant in enzymes and to analyze and discriminate between different catalytic proposals. The ability to address such problems has been provided by our approaches.

In summary, the approaches and concepts outlined in this work provide a powerful way to address different aspects of PT in proteins. Using the EVB and related approaches should allow one to resolve most of the open questions about PT and PTR in proteins. It also allows one to obtain a reliable structure function correlation in specific cases, and thus to convert qualitative concepts about proton transfer in biology to quantitative microscopic concepts.

Acknowledgements

This work was supported by NIH grants GM-24492 and GM-40283. One of us (JM) thanks the J. William Fulbright Scholarship Board for the award of a research scholarship. We also gratefully acknowledge the University of Southern California's High Performance Computing and Communications Center for computer time.

Abbreviations

α	Gas-phase shift
$\Delta G^{\ddagger}$	Reaction free energy barrier
Δg_1	Free energy function
λ	Reorganization energy
ω	Frequency
BD	Brownian dynamics
C_g	Ground state eigenvector
E_a	Activation energy
Eg	Ground state adiabatic energy surface
EVB	Empirical valence bond
F or κ	Transmission factor

FEP/US	Free energy perturbation/umbrella sampling
H_{EVB}	EVB Hamiltonian
K	Rate constant
k_B	Boltzmann's constant
LBHB	Low barrier hydrogen bond
LD	Langevin dynamics
LFER	Linear free energy relationships
MDQT	Molecular dynamics with quantum transition
MM	Molecular mechanics
NQM	Nuclear quantum mechanical
PT	Proton transfer
PTR	Proton translocations
QCP	Quantum classical path
QM	Quantum mechanics
R	Gas constant, same as k_B, but usually in different units
R	Solute coordinate
r	Solvent coordinate
S_{mn}	Franck–Condon factor for transition from state m to state n
S	Overlap matrix element
T	Temperature
TS	Transition state
TST	Transition state theory
VB	Valence bond
VET	Vibrationally enhanced tunneling
x	Reaction coordinate, energy gap
Z	Partitionfunction

References

1 Warshel, A., Dynamics of Reactions in Polar Solvents. Semiclassical Trajectory Studies of Electron-Transfer and Proton-Transfer Reactions, *J. Phys. Chem.* **1982**, 86, 2218–2224.

2 Warshel, A., Dynamics of Enzymatic Reactions, *Proc. Natl. Acad. Sci. USA* **1984**, 81, 444–448.

3 Warshel, A., *Computer Modeling of Chemical Reactions in Enzymes and Solutions*, John Wiley & Sons, New York, 1991.

4 Villa, J., Warshel, A., Energetics and Dynamics of Enzymatic Reactions, *J. Phys. Chem. B* **2001**, 105, 7887–7907.

5 Keck, J. C., Variational Theory of Reaction Rates, *Adv. Chem. Phys.* **1966**, 13, 85–121.

6 Bennett, C. H., Molecular dynamics and transition state theory: the simulation of infrequent events, in *Algorithms for Chemical Computations*, Christofferson, R. E. (Ed.), American Chemical Society, Washington, D. C., 1977, pp. 63–97.

7 Grimmelmann, E. K., Tully, J. C., Helfand, E., Molecular Dynamics of Infrequent Events: Thermal Desorption of Xenon from a Platinum Surface, *J. Chem. Phys.* **1981**, 74, 5300–5310.

8 Pople, J. A., Quantum chemical models (Nobel lecture), *Angew. Chem. Int. Ed. Engl.* **1999**, 38, 1894–1902.

9 Shurki, A., Warshel, A., Structure/Function Correlations of Protreins

using MM, QM/MM and Related Approaches; Methods, Concepts, Pitfalls and Current Progress, *Adv. Protein Chem.* **2003**, 66, 249–312.

10 Warshel, A., Levitt, M., Theoretical studies of enzymic reactions: dielectric, electrostatic and steric stabilization of the carbonium ion in the reaction of lysozyme, *J. Mol. Biol.* **1976**, 103, 227–249.

11 Théry, V., Rinaldi, D., Rivail, J.-L., Maigret, B., Ferenczy, G. G., Quantum Mechanical Computations on Very Large Molecular Systems: The Local Self-Consistent Field Method, *J. Comput. Chem.* **1994**, 15, 269–282.

12 Zhang, Y., Liu, H., Yang, W., Free energy calculation on enzyme reactions with an efficient iterative procedure to determine minimum energy paths on a combined *ab initio* QM/MM potential energy surface, *J. Chem. Phys.* **2000**, 112, (8), 3483–3492.

13 Gao, J., Hybrid quantum and molecular mechanical simulations: an alternative avenue to solvent effects in organic chemistry, *Acc. Chem. Res.* **1996**, 29, 298–305.

14 Bakowies, D., Thiel, W., Hybrid Models for Combined Quantum Mechanical and Molecular Approaches, *J. Phys. Chem* **1996**, 100, 10580–10594.

15 Field, M. J., Bash, P. A., Karplus, M., A Combined Quantum Mechanical and Molecular Mechanical Potential for Molecular Dynamics Simulations, *J. Comput. Chem.* **1990**, 11, 700–733.

16 Friesner, R., Beachy, M. D., Quantum Mechanical Calculations on Biological Systems, *Curr. Opin. Struct. Biol.* **1998**, 8, 257–262.

17 Monard, G., Merz, K. M., Combined Quantum Mechanical/Molecular Mechanical Methodologies Applied to Biomolecular Systems, *Acc. Chem. Res.* **1999**, 32, 904–911.

18 Garcia-Viloca, M., Gonzalez-Lafont, A., Lluch, J. M., A QM/MM study of the Racemization of Vinylglycolate Catalysis by Mandelate Racemase Enzyme, *J. Am. Chem. Soc* **2001**, 123, 709–721.

19 Martí, S., Andrés, J., Moliner, V., Silla, E., Tunon, I., Bertran, J., Transition structure selectivity in enzyme catalysis: a QM/MM study of chorismate mutase, *Theor. Chem. Acc.* **2001**, 105, 207–212.

20 Field, M., Stimulating Enzyme Reactions: Challenges and Perspectives. *J. Comp. Chem.* **2002**, 23, 48–58.

21 Mulholland, A. J., Lyne, P. D., Karplus, M., Ab Initio QM/MM Study of the Citrate Synthase Mechanism. A Low-Barrier Hydrogen Bond is not Involved, *J. Am. Chem. Soc.* **2000**, 122, 534–535.

22 Cui, Q., Elstner, M., Kaxiras, E., Frauenheim, T., Karplus, M., A QM/MM Implementation of the Self-Consistent Charge Density Functional Tight Binding (SCC-DFTB) Method, *J. Phys. Chem. B* **2001**, 105, 569–585.

23 Strajbl, M., Hong, G., Warshel, A., Ab-initio QM/MM Simulation with Proper Sampling: "First Principle" Calculations of the Free Energy of the Auto-dissociation of Water in Aqueous Solution, *J. Phys. Chem. B* **2002**, 106, 13333–13343.

24 Aqvist, J., Warshel, A., Simulation of Enzyme Reactions Using Valence Bond Force Fields and Other Hybrid Quantum/Classical Approaches, *Chem. Rev.* **1993**, 93, 2523–2544.

25 Hong, G., Rosta, E., Warshel, A., Using the Constrained DFT Approach in Generating Diabatic Surfaces and Off Diagonal Empirical Valence Bond Terms for Modeling Reactions in Condensed Phase, *J. Chem. Phys.* In press.

26 Wesolowski, T., Muller, R. P., Warshel, A., *Ab Initio* Frozen Density Functional Calculations of Proton Transfer Reactions in Solution, *J. Phys. Chem.* **1996**, 100, 15444–15449.

27 Hwang, J.-K., King, G., Creighton, S., Warshel, A., Simulation of Free Energy Relationships and Dynamics of S_N2 Reactions in Aqueous Solution, *J. Am. Chem. Soc.* **1988**, 110, (16), 5297–5311.

28 Warshel, A., Weiss, R. M., An Empirical Valence Bond Approach for Comparing Reactions in Solutions and

in Enzymes, *J. Am. Chem. Soc.* **1980**, 102, (20), 6218–6226.

29 Marcus, R. A., Chemical and electrochemical electron transfer theory, *Annu. Rev. Phys. Chem.* **1964**, 15, 155–196.

30 Marcus, R. A., Electron-transfer reactions in chemistry – theory and experiment (Nobel lecture), *Angew. Chem. Int. Ed. Engl.* **1993**, 32, 1111–1121.

31 Billeter, S. R., Webb, S. P., Agarwal, P. K., Iordanov, T., Hammes-Schiffer, S., Hydride Transfer in Liver Alcohol Dehydrogenase: Quantum Dynamics, Kinetic Isotope Effects, and Role of Enzyme Motion, *J. Am Chem. Soc.* **2001**, 123, 11262–11272.

32 Vuilleumier, R., Borgis, D., An Extended Empirical Valence Bond Model for Describing Proton Transfer in $H^+(H_2O)_n$ Clusters and Liquid Water, *Chem. Phys. Lett.* **1998**, 284, 71–77.

33 Smondyrev, A. M., Voth, G. A., Molecular dynamics simulation of proton transport near the surface of a phospholipid membrane, *Biophys. J.* **2002**, 82, (3), 1460–1468.

34 Billeter, S. R., Webb, S. P., Iordanov, T., Agarwal, P. K., Hammes-Schiffer, S., Hybrid approach for including electronic and nuclear quantum effects in molecular dynamics simulations of hydrogen transfer reactions in enzymes, *J. Chem. Phys.* **2001**, 114, 6925–6936.

35 Florian, J., Comment on Molecular Mechanics for Chemical Reactions, *J. Phys. Chem. A* **2002**, 106(19), 5046–5047.

36 Truhlar, D. G., Reply to Comment on Molecular Mechanics for Chemical Reactions, *J. Phys. Chem. A* **2002**, 106(19), 5048–5050.

37 Warshel, A., Bromberg, A., Oxidation of 4a,4b-Dihydrophenanthrenes. III. A Theoretical Study of the Large Kinetic Isotope Effects of Deuterium in the Initiation Step of the Thermal Reaction With Oxygen, *J. Chem. Phys.* **1970**, 52, 1262.

38 Raff, L. M., Theoretical investigations of the reaction dynamics of polyatomic systems: Chemistry of the hot atom (T^* + CH4) and (T^* + CD4), *J. Chem. Phys.* **1974**, 60, (6), 2220.

39 Schmitt, U. W., Voth, G. A., Multistate Empirical Valence Bond Model for Proton Transport in Water, *J. Phys. Chem. B* **1998**, 102, 5547–5551.

40 Warshel, A., Russell, S., Theoretical Correlation of Structure and Energetics in the Catalytic Reaction of Trypsin, *J. Am. Chem. Soc.* **1986**, 108, 6569–6579.

41 Cuma, M., Schmitt, U. W., Voth, G. A., A multi-state empirical valence bond model for weak acid dissociation in aqueous solution, *J. Phys. Chem. A* **2001**, 105, (12), 2814–2823.

42 Warshel, A., Chu, Z. T., Quantum Corrections for Rate Constants of Diabatic and Adiabatic Reactions in Solutions, *J. Chem. Phys.* **1990**, 93, 4003.

43 Geissler, P. L., Dellago, C., Chandler, D., Hutter, J., Parrinello, M., Autoionization in liquid water, *Science* **2001**, 291, (5511), 2121–2124.

44 Warshel, A., Parson, W. W., Dynamics of biochemical and biophysical reactions: insight from computer simluations, *Q. Rev. Biophys.* **2001**, 34, 563–670.

45 Tully, J. C., Preston, R. M., Trajectory surface hopping approach to nonadiabatic molecular collisions: the reaction of H^+ with D_2, *J. Chem. Phys.* **1971**, 55, 562–572.

46 Warshel, A., Hwang, J.-K., Simulation of the Dynamics of Electron Transfer Reactions in Polar Solvents: Semiclassical Trajectories and Dispersed Polaron Approaches, *J. Chem. Phys.* **1986**, 84, 4938–4957.

47 Kuznetsov, A. M., Ulstrup, J., Proton and hydrogen atom tunnelling in hydrolytic and redox enzyme catalysis, *Can. J. Chem.-Rev. Can. Chim.* **1999**, 77, (5–6), 1085–1096.

48 Borgis, D., Hynes, J. T., Molecular-dynamics simulation for a model nonadiabatic proton transfer reaction in solution, *J. Chem. Phys.* **1991**, 94, 3619–3628.

49 Antoniou, D., Schwartz, S. D., Large kinetic isotope effects in enzymatic proton transfer and the role of substrate oscillations, *Proc. Natl. Acad. Sci. USA* **1997**, 94, 12360–12365.

50 Olsson, M. H. M., Siegbahn, P. E. M., Warshel, A., Simulations of the large kinetic isotope effect and the temperature dependence of the hydrogen atom transfer in lipoxygenase, *J. Am. Chem. Soc.* **2004**, 126, (9), 2820–2828.

51 Smedarchina, Z., Siebrand, W., Fernández-Ramos, A., Cui, Q., Kinetic Isotope Effects for Concerted Multiple Proton Transfer: A Direct Dynamics Study of an Active-Site Model of Carbonic Anhydrase II, *J. Am. Chem. Soc* **2003**, 125, (1), 243–251.

52 Lehnert, N., Solomon, E. I., Density-functional investigation on the mechanism of H-atom abstraction by lipoxygenase, *J. Biol. Inorg. Chem.* **2003**, 8, (3), 294–305.

53 Knapp, M. J., Rickert, K., Klinman, J. P., Temperature-dependent isotope effects in soybean lipoxygenase-1: Correlating hydrogen tunneling with protein dynamics, *J. Am. Chem. Soc.* **2002**, 124, (15), 3865–3874.

54 Gillan, M. J., Quantum-Classical Crossover of the Transition Rate in the Damped Double Well, *J. Phys. C. Solid State Phys.* **1987**, 20, 3621–3641.

55 Voth, G. A., Path-integral centroid methods in quantum statistical mechanics and dynamics, *Adv. Chem. Phys.* **1996**, 93, 135–218.

56 Voth, G. A., Chandler, D., Miller, W. H., Rigorous Formulation of Quantum Transition State Theory and Its Dynamical Corrections, *J. Chem Phys.* **1989**, 91, 7749.

57 Hwang, J.-K., Warshel, A., A Quantized Classical Path Approach for Calculations of Quantum Mechanical Rate Constants, *J. Phys. Chem.* **1993**, 97, 10053–10058.

58 Hwang, J.-K., Warshel, A., How Important are Quantum Mechanical Nuclear Motions in Enzyme Catalysis? *J. Am. Chem. Soc.* **1996**, 118, 11745–11751.

59 Feynman, R. P., *Statistical Mechanics*, Benjamin, New York, 1972.

60 Hwang, J.-K., Chu, Z. T., Yadav, A., Warshel, A., Simulations of Quantum Mechanical Corrections for Rate Constants of Hydride-Transfer Reactions in Enzymes and Solutions, *J. Phys. Chem.* **1991**, 95, 8445–8448.

61 Tuckerman, M. E., Marx, D., Heavy-atom skeleton quantization and proton tunneling in "intermediate-barrier" hydrogen bonds, *Phys. Rev. Lett.* **2001**, 86, (21), 4946–4949.

62 King, G., Warshel, A., Investigation of the Free Energy Functions for Electron Transfer Reactions, *J. Chem. Phys.* **1990**, 93, 8682–8692.

63 Warshel, A., Computer Simulations of Enzyme Catalysis: Methods, Progress, and Insights, *Annu. Rev. Biophys. Biomol. Struct.* **2003**, 32, 425–443.

64 Warshel, A., Hwang, J. K., Åqvist, J., Computer Simulations of Enzymatic Reactions: Examination of Linear Free-Energy Relationships and Quantum-Mechanical Corrections in the Initial Proton-transfer Step of Carbonic Anhydrase, *Faraday Discuss.* **1992**, 93, 225.

65 Warshel, A., *Simulating the Energetics and Dynamics of Enzymatic Reactions*, Pontificiae Academiae Scientiarum Scripta Varia, 1984, Vol. 55.

66 Kong, Y., Warshel, A., Linear Free Energy Relationships with Quantum Mechanical Corrections: Classical and Quantum Mechanical Rate Constants for Hydride Transfer Between NAD^+ Analogues in Solutions, *J. Am. Chem. Soc.* **1995**, 117, 6234–6242.

67 Schweins, T., Warshel, A., Mechanistic Analysis of the Observed Linear Free Energy Relationships in p21 ras and Related Systems, *Biochemistry* **1996**, 35, 14232–14243.

68 Warshel, A., Schweins, T., Fothergill, M., Linear Free Energy Relationships in Enzymes. Theoretical Analysis of the Reaction of Tyrosyl-tRNA Synthetase, *J. Am. Chem. Soc.* **1994**, 116, 8437–8442.

69 Warshel, A., Chu, Z. T., Parson, W. W., Dispersed Polaron Simulations

of Electron Transfer in Photosynthetic Reaction Centers, *Science* **1989**, 246, 112–116.

70 Kiefer, P. M., Hynes, J. T., Nonlinear free energy relations for adiabatic proton transfer reactions in a polar environment. I. Fixed proton donor-acceptor separation, *J. Phys. Chem. A* **2002**, 106, (9), 1834–1849.

71 Silverman, D. N., Tu, C., Chen, X., Tanhauser, S. M., Kresge, A. J., Laipis, P. J., Rate-Equilibria Relationships in Intramolecular Proton Transfer in Human Carbonic Anhydrase III, *Biochemistry* **1993**, 34, 10757–10762.

72 Silverman, D. N., Marcus Rate Theory Applied to Enzymatic Proton Transfer, *Biochem. Biophys. Acta.* **2000**, 1458, (1), 88–103.

73 Åqvist, J., Fothergill, M., Warshel, A., Computer Simulation of the CO_2/$HCO^-{}_3$ Interconversion Step in Human Carbonic Anhydrase I, *J. Am. Chem. Soc.* **1993**, 115, 631–635.

74 Silverman, D. N., Lindskog, S., The Catalytic Mechanism of Carbonic Anhydrase: Implications of a Rate-Limiting Protolysis of Water, *Acc. Chem. Res.* **1988**, 21, 30–36.

75 Åqvist, J., Warshel, A., Computer Simulation of the Initial Proton Transfer Step in Human Carbonic Anhydrase I, *J. Mol. Biol.* **1992**, 224, 7–14.

76 Schweins, T., Geyer, M., Scheffzek, K., Warshel, A., Kalbitzer, H. R., Wittinghofer, A., Substrate-assisted Catalysis as a Mechanism for GTP Hydrolysis of p21 ras and Other GTP-binding Proteins, *Nature Struct. Biol.* **1995**, 2, (1), 36–44.

77 Warshel, A., Russell, S. T., Calculations of Electrostatic Interactions in Biological Systems and in Solutions, *Q. Rev. Biophys.* **1984**, 17, 283–421.

78 Schutz, C. N., Warshel, A., What are the dielectric "constants" of proteins and how to validate electrostatic models, *Proteins: Struc., Func., and Gen.* **2001**, 44, 400–417.

79 Schutz, C. N., Warshel, A., Analyzing Free Energy Relationships for Proton Translocations in Enzymes; Carbonic Anhydrase Revisted, *J. Phys. Chem. B* **2004**, 108, (6), 2066–2075.

80 Muegge, I., Qi, P. X., Wand, A. J., Chu, Z. T., Warshel, A., The Reorganization Energy of Cytochrome *c* Revisited, *J. Phys. Chem. B* **1997**, 101, 825–836.

81 Gerlt, J. A., Gassman, P. G., An Explanation for Rapid Enzyme-Catalyzed Proton Abstraction from Carbon Acids: Importance of Late Transition States in Concerted Mechanisms, *J. Am. Chem. Soc.* **1993**, 115, 11552–11568.

82 Yarnell, A., Blockade in the Cell's Waterway, *Chem. Eng. News* **2004**, 82, (4), 42–44.

83 Warshel, A., Conversion of Light Energy to Electrostatic Energy in the Proton Pump of Halobacterium halobium, *Photochem. Photobiol.* **1979**, 30, 285–290.

84 Burykin, A., Warshel, A., What Really Prevents Proton Transport Through Aquaporin? Charge Self-Energy versus Proton Wire Proposals, *Biophys. J.* **2003**, 85, 3696–3706.

85 Braun-Sand, S., Strajbl, M., Warshel, A., Studies of Proton Translocations in Biological Systems: Simulating Proton Transport in Carbonic Anhydrase by EVB Based Models, *Biophys. J.* **2004**, 87, 2221–2239.

86 Warshel, A., Calculations of Enzymic Reactions: Calculations of pK_a, Proton Transfer Reactions, and General Acid Catalysis Reactions in Enzymes, *Biochemistry* **1981**, 20, 3167–3177.

87 Warshel, A., Sussman, F., Hwang, J.-K., Evaluation of Catalytic Free Energies in Genetically Modified Proteins, *J. Mol. Biol.* **1988**, 201, 139–159.

88 Åqvist, J., Warshel, A., Free Energy Relationships in Metalloenzyme-Catalyzed Reactions. Calculations of the Effects of Metal Ion Substitutions in Staphylococcal Nuclease, *J. Am. Chem. Soc.* **1990**, 112, 2860–2868.

89 Warshel, A., Electrostatic Origin of the Catalytic Power of Enzymes and the Role of Preorganized Active Sites, *J. Biol. Chem.* **1998**, 273, 27035–27038.

90 Frey, P. A., Whitt, S. A., Tobin, J. B., A Low-Barrier Hydrogen Bond in the Catalytic Triad of Serine Proteases, *Science* **1994**, 264, 1927–1930.

91 Schutz, C. N., Warshel, A., The low barrier hydrogen bond (LBHB) proposal revisited: The case of the Asp.His pair in serine proteases, *Proteins* **2004**, 55, (3), 711–723.

92 Kohen, A., Klinman, J. P., Hydrogen tunneling in biology, *Chem. Biol.* **1999**, 6, R191–R198.

93 Sutcliffe, M. J., Scrutton, N. S., Enzyme catalysis: over-the-barrier or through-the-barrier? *Trends Biochem. Sci.* **2000**, 25, 405–408.

94 German, E. K., *J. Chem. Soc., Faraday Trans. 1* **1981**, 77, 397–412.

95 Feierberg, I., Luzhkov, V., Åqvist, J., Computer simulation of primary kinetic isotope effects in the proposed rate limiting step of the glyoxalase I catalyzed reaction, *J. Biol. Chem.* **2000**, 275, 22657–22662.

96 Alhambra, C., Corchado, J. C., Sanchez, M. L., Gao, J., Truhlar, D. G., Quantum Dynamics of Hydride Transfer in Enzyme Catalysis, *J. Am. Chem. Soc.* **2000**, 122, (34), 8197–8203.

97 Berendsen, H. J. C., Mavri, J., Quantum Simulation of Reaction Dynamics by Density-Matrix Evolution, *J. Phys. Chem.* **1993**, 97, (51), 13464–13468.

98 Mavri, J., Grdadolnik, J., Proton transfer dynamics in acetylacetone: A mixed quantum-classical simulation of vibrational spectra, *J. Phys. Chem. A* **2001**, 105, (10), 2045–2051.

99 Lensink, M. F., Mavri, J., Berendsen, H. J. C., Simulation of slow reaction with quantum character: Neutral hydrolysis of carboxylic ester, *J. Comput. Chem.* **1999**, 20, (8), 886–895.

100 Neria, E., Karplus, M., Molecular Dynamics of an Enzyme Reaction: Proton Transfer in TIM, *Chem. Phys. Lett.* **1997**, 267, 23–30.

101 Ball, P., Enzymes – By chance, or by design? *Nature* **2004**, 431, (7007), 396–397.

102 Kosloff, D., Kosloff, R., A Fourier Method Solution for the Time-Dependent Schrodinger-Equation as a Tool in Molecular-Dynamics, *J. Comput. Phys.* **1983**, 52, (1), 35–53.

103 Hatcher, E., Soudackov, A. V., Hammes-Schiffer, S., Proton-coupled electron transfer in soybean lipoxygenase, *J. Am. Chem. Soc.* **2004**, 126, (18), 5763–5775.

104 Olsson, M. H. M., Mavri, J., Warshel, A., Transition State Theory can be Used in Studies of Enzyme Catalysis: Lessons from Simulations of Tunneling and Dynamical Effects in Lipoxygenase and Other Systems, *Phil. Trans. R. Soc. B.* **2006**, 361, 1417–1432.

105 Shurki, A., Štrajbl, M., Villa, J., Warshel, A., How Much Do Enzymes Really Gain by Restraining Their Reacting Fragments? *J. Am. Chem. Soc.* **2002**, 124, 4097–4107.